Oxford Resources
for Cambridge

Core

Cambridge IGCSE®
Complete
Mathematics

David Rayner

Ian Bettison

Mathew Taylor

Editor: Deborah Barton

Neha Kawatra
Bhavana Kotwal
Domnic Odipo
Diano Pesidas
Preeti Verma

OXFORD
UNIVERSITY PRESS

OXFORD
UNIVERSITY PRESS

Great Clarendon Street, Oxford, OX2 6DP, United Kingdom

Oxford University Press is a department of the University of Oxford.

It furthers the University's objective of excellence in research, scholarship, and education by publishing worldwide. Oxford is a registered trade mark of Oxford University Press in the UK and in certain other countries.

© Oxford University Press 2023

First published in 2023

British Library Cataloguing in Publication Data

Data available

9781382042499

9781382042505 (ebook)

10 9 8 7 6 5 4 3 2

Paper used in the production of this book is a natural, recyclable product made from wood grown in sustainable forests.

The manufacturing process conforms to the environmental regulations of the country of origin.

Printed in China by Golden Cup

Acknowledgements

The publisher and authors would like to thank the following for permission to use photographs and other copyright material:

Cover: Dimitri Otis/Photodisc/Getty Images. Photos: p1: Tom Urbelionis / Shutterstock; p6: Hadrian/ Shutterstock; p9(t): Quang Ho/ Shutterstock; p9(b): FrameAngel/Shutterstock; p20: Mariontxa/Shutterstock; p41: Everett - Art / Shutterstock; p73: Offscreen / Shutterstock; p117(t): Science Photo Library; p117(b): Jaguar PS/Shutterstock; p128: Aphelleon/Shutterstock; p135: Everett Collection/Shutterstock; p181: Bettmann / Getty Images; p187(t): Aleksandar Todorov / Shutterstock; p187(b): Dado Photos / Shutterstock; p190: Supertrooper / Shutterstock; p194(t): John T Takai / Shutterstock; p194(b): baitong333 / Shutterstock; p199: Tarikdiz / Shutterstock; p203: acceptphoto / Shutterstock; p204: OlegSam / Shutterstock; p206: Eric Isselee / Shutterstock; p214: Roger Viollet Collection/Getty Images; p220: sizov / Shutterstock; p221: Clara / Shutterstock; p226: Gallinago_media / Shutterstock; p263(t): GL Archive / Alamy Stock Photo; p263(b): Tasos Katopodis / Belga News Agency / Alamy Stock Photo; p295: De Agostini Editorial/De Agostini/ DEA PICTURE LIBRARY/Getty Images; p297: A_Lesik / Shutterstock; p299: mahfud21 / Shutterstock; p301: Claudio Divizia/Shutterstock; p317(t): Science History Images / Alamy Stock Photo; p317(b): STEPHEN LAM / REUTERS / Alamy Stock Photo; p335(t): Science History Images / Alamy Stock Photo; p335(b): Young Ho/Sipa US/Alamy Stock Photo; p345: Thompson Digital.

Artwork by Aptara Inc., Thompson Digital, and Oxford University Press.

Every effort has been made to contact copyright holders of material reproduced in this book. Any omissions will be rectified in subsequent printings if notice is given to the publisher.

This Student Book refers to the Cambridge IGCSE® Mathematics (0580) Syllabus published by Cambridge Assessment International Education.

This work has been developed independently from and is not endorsed by or otherwise connected with Cambridge Assessment International Education.

Contents

Introduction

About this book

This book is designed specifically for the Cambridge IGCSE® Mathematics course. Experienced examiners have been involved in all aspects of the course, to ensure that the content adheres to the latest syllabus.

Using this book will ensure that you are well prepared for the exam at this level, and also studies beyond the IGCSE level in Mathematics. The features below are designed to make learning as interesting and effective as possible.

Finding your way around

To get the most out of this book when studying or revising, use the:

- **Contents list** to help you find the appropriate units
- **Index** to find key words so you can turn to any concept straight away.

Learning objectives

At the start of each chapter you will find a list of objectives. These will tell you what you should be able to do by the end of the chapter. They are based on what you need to cover for the Cambridge IGCSE syllabus.

Famous mathematicians

These are included at the start of each chapter to give you a brief insight into the life of a mathematician who played an important part in the development of the ideas contained in that chapter.

By finding out about the history of mathematics and considering a topic within the broader context of the subject, you can make connections between topics and develop a greater appreciation of how mathematics has developed over the centuries.

Worked examples

Worked examples are an important feature of the book and can be found in every sub-topic. These show you the important skills and techniques required in the exercises below and also provide a model for how to structure your solutions.

Exercises

There are thousands of questions in this book, providing ample opportunities to practise the skills and techniques required in the exam. The exercises contain questions of varying levels of difficulty, so that you can progress through a topic as your knowledge and confidence increases.

Each exercise has an icon to denote whether you can use a calculator or not. This 🖩 means you can use a calculator, while this 🚫 means you should not. The same icons also appear in the Revision Exercises.

Revision Exercise

At the end of each chapter, you will find revision questions to bring together all your knowledge and test your understanding of the contents of the chapter.

Examination-style questions

The revision exercises are followed by exam-style practice questions. These are very similar to the kind of questions you should expect to see in the real exam.

Tips

Yellow boxes throughout the exercises provide further information, hints on how to approach a question, or reminders of other concepts.

Answers

These can be found at the back of this book, so you can find out immediately whether or not you have answered a question correctly. Answers to all the numerical problems in the exercises, the review questions, and the exam-style questions are all included.

kerboodle

Additional support can be found on Kerboodle. There are resources for every sub-topic, including adaptive assessments, personalised Next Steps and data-rich reports. You can also access the Digital Student Book.

1 Number 1

Karl Friedrich Gauss (1777–1855) is thought by many to have been the greatest all-round mathematician of all time. He considered that his finest discovery was the method for constructing a regular seventeen-sided polygon. This was not of the slightest use outside the world of mathematics, but was a great achievement of the human mind. Gauss would not have understood the modern view held by many that mathematics must somehow be 'useful' to be worthy of study.

- Work with different types of numbers, e.g. natural numbers, integers, primes, squares, cubes, common factors, common multiples, rational numbers, irrational numbers, reciprocals, powers and roots.
- Calculate with and convert between the following, including in contexts: proper fractions, improper fractions, mixed numbers and decimals.
- Order quantities and understand the symbols $=$, $\neq$, $>$, $<$, $\geqslant$ and $\leqslant$.
- Calculate with integers, decimals and fractions including the correct order of operations and brackets.
- Understand and interpret positive, zero and negative indices including using the rules of indices.
- Understand and calculate with numbers in standard form $A \times 10^n$ where n is a positive or negative integer and $1 \leqslant A < 10$.

1.1 Place value

Whole numbers are made up from ones, tens, hundreds, thousands and so on.

thousands	hundreds	tens	ones
3	2	6	4

In the number 3264:

- the digit 3 means 3 thousands
- the digit 2 means 2 hundreds
- the digit 6 means 6 tens
- the digit 4 means 4 ones (units).

In words you write 'three thousand, two hundred and sixty-four'.

Exercise 1.1A

In Questions **1** to **8** state the value of the figure underlined.

1. 2̲7 **2.** 4̲16 **3.** 238̲2 **4.** 51̲6

5. 6̲008 **6.** 2̲6 104 **7.** 5̲ 250 000 **8.** 8̲26 111

In Questions **9** to **16** write down the number which goes in each box.

9. 293 = ☐ + 90 + 3

10. 574 = 500 + ☐ + 4

11. 816 = 800 + ☐ + 6

12. 899 = ☐ + 90 + 9

13. 6217 = ☐ + 200 + 10 + 7

14. 5065 = 5000 + ☐ + 5

15. 63 410 = 60 000 + 3000 + ☐ + 10

16. 75 678 = ☐ + 5000 + 600 + ☐ + 8

17. Write these numbers in figures:

a) Seven hundred and twenty

b) Five thousand, two hundred and six

c) Sixteen thousand, four hundred and thirty

d) Half a million

e) Three hundred thousand and ninety

f) Eight and a half thousand

g) Twelve billion.

18. Here are four number cards:

a) Use all the cards once only to make the largest possible number.

b) Use all the cards once only to make the smallest possible number.

19. Write these numbers in words:

    a) 4620   **b)** 607   **c)** 25 400

    d) 6 800 000   **e)** 21 425

20. Here are five number cards:

    a) Use all the cards to make the largest possible *odd* number.

    b) Use all the cards to make the smallest possible *even* number.

21. Write down the number that is ten more than:

    a) 247   **b)** 3211   **c)** 694

22. Write down the number that is one thousand more than:

    a) 392   **b)** 25 611   **c)** 256 900

23. a) Prini puts a 2-digit whole number into her calculator.

    She multiplies the number by 10.

    Fill in *one* other digit that you know must now be on the calculator.

    b) Prini starts again with the same 2-digit number and this time she multiplies it by 1000.

    Now, fill in all five digits on the calculator.

24. Write these numbers in order, from the smallest to the largest:

    a) 2142,   2290,   2058,   2136

    b) 5329,   5029,   5299,   5330

    c) 25 117,   25 200,   25 171,   25 000,   25 500

25. Find the number that goes in each box.

    a) $5 \times \boxed{} + 7 = 507$ **b)** $6 \times \boxed{} + 8 = 68$

26. Copy the statements and complete them by writing a number in each box.

    a) $8 \times \boxed{} + \boxed{} = 807$

    b) $7 \times \boxed{} + 5 \times \boxed{} = 7050$

1.2 Arithmetic without a calculator

Here are examples to remind you of non-calculator methods.

a)
$$\begin{array}{r} 4\ 2\ 7 \\ +\ 5\ 1\ 8\ 6 \\ \hline 5\ 6\ 1\ 3 \\ 1\ 1 \end{array}$$

b)
$$\begin{array}{r} 2\ 7\ \overset{7}{\cancel{8}}\ \overset{1}{4} \\ -\ 6\ 3\ 5 \\ \hline 2\ 1\ 4\ 9 \end{array}$$

c) $57 \times 100 = 5700$ [write two zeros on the end]

d)
$$\begin{array}{r} 3\ 7\ 4 \\ \times\ 6 \\ \hline 2\ 2\ 4\ 4 \\ 4\ 2 \end{array}$$

e) $7\overline{)3\ 7^2\ 9^1\ 4}$ gives $5\ 4\ 2$

f) $5\overline{)6^1 9^4 4}$ gives $1\ 3\ 8\ r\ 4$ or $138\frac{4}{5}$

> **Tip**
>
> In part **(f)**, the remainder written as a fraction is $4 \div 5 = \dfrac{4}{5}$

Exercise 1.2A

Work out, without a calculator:

1. $653 + 2844$
2. $64 + 214 + 507$
3. $65\,941 + 2580$
4. $387 - 175$
5. $927 - 68$
6. $1024 - 816$
7. 27×10
8. 5×1000
9. 316×8
10. $340 \div 4$
11. $1944 \div 6$
12. $364 \div 7$
13. $520 \div 10$
14. $9704 - 5135$
15. $6001 - 5994$
16. 54×20
17. $4716 \div 9$
18. $1504 \div 8$
19. $7 + 1609 + 25$
20. $7 + 295 - 48$

21. A coin collector has 2106 coins. He buys 329 more coins at a collectors' fair. How many coins does he now have in his collection?

22. A bag of flour contains 527 grams. Beatrice uses 486 grams to make a cake. How many grams are left in the bag?

23. There are 73 chocolates in a single box. How many chocolates are there in 5 boxes?

24. A computer hard drive costs $214. How much does it cost to buy 4 of these drives?

25. A length of wood needs to be cut to precisely 9224 mm. How much wood would be needed to cut 7 of these lengths? (You may ignore the wood lost through cutting.)

26. Tamara shares $3195 equally between her 5 grandchildren. How much does each grandchild get?

27. 2600 grams of sugar is shared equally between 8 recipes. How much sugar is used in each recipe?

28. Ravi earned $1714 this week for working Monday to Friday. He earned $289 for working on Sunday and he got a $15 tip.
    How much did Ravi earn in total this week?

29. There are 2906 supporters at a football match. 1414 of them support the away team.
    How many support the home team?

30. Jars of olive paste have a mass of 725 grams.
    What is the mass of 8 jars?

31. Ali has 289 comic books in his collection. At a comic book convention, he buys 154 more comics and sells 78.
    How many comic books does Ali have after the convention?

32. 53 chickens lay, on average, 400 eggs each per year.
    How many eggs are laid in total?

Exercise 1.2B

The table shows the price in dollars per person for two adults to share a twin/double room in different hotels in Hong Kong.

Hotels	E Hotel		Nina		Royal Plaza		Hari	
Holiday starting date	1 night	extra night	1 night	extra night	1 night	extra night	1 night	extra night
01 Apr–27 Oct	174	77	185	84	210	99	216	105
28 Oct–10 Nov	161	62	185	84	198	86	191	85
11 Nov–5 Dec	161	62	171	73	210	99	190	84
6 Dec–21 Dec	161	62	171	73	210	99	216	105
22 Dec–5 Jan	174	77	187	88	219	103	216	105
6 Jan–10 Jan	174	77	185	84	214	101	212	101
11 Jan–20 Jan	159	60	171	73	187	81	186	75
21 Jan–2 Mar	159	62	171	73	187	81	188	77
3 Mar–10 Mar	161	63	175	74	192	83	218	107
11 Mar–18 Mar	161	62	175	74	198	86	220	109
19 Mar–31 Mar	179	81	191	86	215	103	228	117

1. Two people want to spend 3 nights at the Hari hotel starting on 23 January.
How much will this cost?

2. Mr and Mrs Chen want to spend 2 nights at the Royal Plaza, starting on 10 May.
How much will their holiday cost?

3. Four people want to spend 5 nights at the E Hotel, starting on 3 March.
What is the total cost of this holiday?

4. A party of 10 people want to spend 2 nights at the Nina hotel starting on 27 August.
What is the total cost of this stay?

5. Chadresh and Mahima are celebrating their wedding anniversary by spending 3 nights at the Royal Plaza. They plan to begin their holiday on 9 December.
How much will this holiday cost?

6. Six people want to spend 7 nights at the Hari hotel, starting on 20 March.
What is the total cost of this holiday?

1.3 Inverse operations

The word **inverse** means 'opposite'.

- The inverse of adding is subtracting: $5 + 19 = 24$, $5 = 24 - 19$
- The inverse of subtracting is adding: $31 - 6 = 25$, $31 = 25 + 6$
- The inverse of multiplying is dividing: $7 \times 6 = 42$, $7 = 42 \div 6$
- The inverse of dividing is multiplying: $30 \div 3 = 10$, $30 = 10 \times 3$

Example

Find the missing digits.

a) $\boxed{}4 \div 6 = 14$

Work out 14×6 because multiplying is the inverse of dividing.
Since $14 \times 6 = 84$, the missing digit is 8.

b) $2\boxed{}8 \times 5 = 1340$

Work out $1340 \div 5$ because dividing is the inverse of multiplying.
Since $1340 \div 5 = 268$, the missing digit is 6.

c)

$$
\begin{array}{r}
3 \ \boxed{} \ 7 \\
+ \ 2 \ \ 5 \ \boxed{} \\
\hline
\boxed{} \ \ 3 \ \ 9
\end{array}
$$

Start from the right: $7 + 2 = 9$
Middle column: $8 + 5 = 13$
Check: 387
        $+ \ 252$
        $\overline{639} \checkmark$
        1

Exercise 1.3A

Find the missing digits.

1. a)
```
    2   8   5
+  □   1   4
───────────
    7  □   □
```

b)
```
    6   3  □
+  □   5   2
───────────
    8  □   9
```

c)
```
   □   3   5
+  3   4  □
───────────
    9  □   9
```

2. a)
```
    3   5   6
+  5  □   6
───────────
  □   8  □
```

b)
```
    2  □   4
+  5   3   7
───────────
  □   6   1
```

c)
```
    3   8   8
+  □   2  □
───────────
    8  □   3
```

3. a)
```
       4  □
  ×        3
───────────
    1   4   4
```

b)
```
       3  □
  ×        7
───────────
    2   3   1
```

c)
```
    □   □   1
  ×            5
───────────────
    1   6   0   5
```

4. a) $\boxed{}\boxed{}\boxed{} \div 3 = 50$ **b)** $\boxed{}\boxed{} \times 4 = 60$

   c) $9 \times \boxed{} = 81$ **d)** $\boxed{}\boxed{}\boxed{} \div 6 = 92$

5. a)
```
    4  □   5
+  2   8  □
───────────
  □   3   0
```

b)
```
    4  □   7
+  □   7  □
───────────
    6   0   4
```

c)
```
   □   3  □
+  2  □   4
───────────
    7   9   9
```

6. a) $\boxed{}\boxed{} \times 7 = 245$ **b)** $\boxed{}\boxed{} \times 10 = 580$

   c) $32 \div \boxed{} = 8$ **d)** $\boxed{}\boxed{}\boxed{} \div 5 = 190$

7. a) $\boxed{}\boxed{} + 29 = 101$ **b)** $\boxed{}\boxed{}\boxed{} - 17 = 91$

   c)
```
   □   8   9
−  3  □   6
───────────
    5   4  □
```

   d)
```
    3   3   5
−  2   1  □
───────────
  □  □   7
```

8. In each calculation the same number is missing from all three boxes. Find the missing number in each case.

a) $\square \times \square - \square = 12$

b) $\square \div \square + \square = 9$

c) $\square \times \square + \square = 72$

9. Copy and in the circle write $+$, $-$, $\times$ or $\div$ to make the calculation correct.

a) $7 \times 4 \bigcirc 3 = 25$

b) $8 \times 5 \bigcirc 2 = 20$

c) $7 \bigcirc 3 - 9 = 12$

d) $12 \bigcirc 2 + 4 = 10$

e) $75 \div 5 \bigcirc 5 = 20$

10. Write the following with the correct symbols.

a) $5 \times 4 \times 3 \bigcirc 3 = 63$

b) $5 + 4 \bigcirc 3 \bigcirc 2 = 4$

c) $5 \times 2 \times 3 \bigcirc 1 = 31$

1.4 Long multiplication and division

To work out 327×53 we will use the fact that $327 \times 53 = (327 \times 50) + (327 \times 3)$

Set out the working like this.

$$
\begin{array}{r}
327 \\
\times \ \ 53 \\
\hline
16350 \\
981 \\
\hline
17331 \\
\scriptstyle 1\ 1
\end{array}
$$

$16350 \rightarrow$ This is 327×50

$\ \ 981 \rightarrow$ This is 327×3

$17331 \rightarrow$ This is 327×53

Here is another example.

$$
\begin{array}{r}
541 \\
\times \ \ 84 \\
\hline
43280 \\
2164 \\
\hline
45444 \\
\scriptstyle 1
\end{array}
$$

$43280 \rightarrow$ This is 541×80

$\ \ 2164 \rightarrow$ This is 541×4

$45444 \rightarrow$ This is 541×84

Exercise 1.4A

Work out, without a calculator:

1. 35×23

2. 27×17

3. 26×25

4. 31×43

5. 45×61

6. 52×24

7. 323×14

8. 416×73

9. 504×56

10. 306×28

11. 624×75

12. 839×79

13. 694×83

14. 973×92

15. 415×235

With ordinary 'short' division, we divide and find remainders. The method for 'long' division is really the same but we set it out so that the remainders are easier to find.

Example

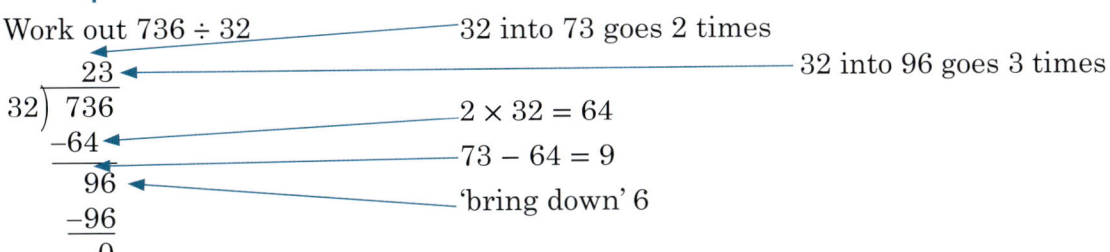

Work out $736 \div 32$

$$
\begin{array}{r}
23 \\
32\overline{)\ 736} \\
-64 \\
\hline
96 \\
-96 \\
\hline
0
\end{array}
$$

32 into 73 goes 2 times

32 into 96 goes 3 times

$2 \times 32 = 64$

$73 - 64 = 9$

'bring down' 6

Exercise 1.4B

Work out, without a calculator:

1. $672 \div 21$ **2.** $425 \div 17$ **3.** $576 \div 32$ **4.** $247 \div 19$ **5.** $875 \div 25$

6. $574 \div 26$ **7.** $806 \div 34$ **8.** $748 \div 41$ **9.** $666 \div 24$ **10.** $707 \div 52$

11. $951 \div 27$ **12.** $2917 \div 45$ **13.** $2735 \div 18$ **14.** $56\,274 \div 19$

Exercise 1.4C

Solve each problem without a calculator.

1. Eetu sells 56 bars of chocolate at 84c each. How much does he make altogether?

2. Eggs are packed eighteen to a box. How many boxes are needed for 828 eggs?

3. Simi makes 146 telephone calls a week. How many does he make in a year?

4. Soraga wants to buy as many 23c stamps as possible. She has $5 to buy them. How many can she buy and how much money does she have left?

5. How many 49-seater coaches will be needed to take 366 students on a school trip?

6. An office building has 24 windows on each of 8 floors. A window cleaner charges 42c for each window.
How much is he paid for the whole building?

7. $238 million is shared equally between 17 people. How much does each person receive?

8. It costs $7905 to hire a plane for a day. A trip is organised for 93 people. How much does each person pay?

9. A farmer finds an oil well on his land. The oil comes out of the well at a rate of $15-worth of oil every minute of the day and night. How much does he receive in a 24-hour day?

1.5 Negative numbers

Here are some examples of the use of negative numbers:

- If the weather is very cold and the temperature is 3 degrees below zero, it is written $-3\,°C$.

- If a golfer is 5 under par for his round, the scoreboard will show -5.

- On a bank statement if someone is $5 overdrawn [or 'in the red'] it would appear as $-\$5$.

An easy way to begin calculations with negative numbers is to think about changes in temperature:

a) Suppose the temperature is $-2\,°C$ and it rises by $7\,°C$.

The new temperature is $5\,°C$.

We can write $-2 + 7 = 5$.

b) Suppose the temperature is $-3\,°C$ and it falls by $6\,°C$.

The new temperature is $-9\,°C$.

We can write $-3 - 6 = -9$.

Exercise 1.5A

For Questions **1** to **12**, think about moving up or down a thermometer to find the new temperature.

1. The temperature is $+8\,°C$ and it falls by $3\,°C$.

2. The temperature is $+4\,°C$ and it falls by $5\,°C$.

3. The temperature is $+2\,°C$ and it falls by $6\,°C$.

4. The temperature is $-1\,°C$ and it falls by $6\,°C$.

5. The temperature is $-5\,°C$ and it rises by $1\,°C$.

6. The temperature is $-8\,°C$ and it rises by $4\,°C$.

7. The temperature is $-3\,°C$ and it rises by $7\,°C$.

8. The temperature is $+4\,°C$ and it rises by $8\,°C$.

9. The temperature is $+9\,°C$ and it falls by $14\,°C$.

10. The temperature is $-13\,°C$ and it rises by $13\,°C$.

11. The temperature is $-6\,°C$ and it falls by $5\,°C$.

12. The temperature is $-25\,°C$ and it rises by $10\,°C$.

In Questions **13** to **22** state whether the temperature has risen or fallen and by how many degrees.

13. It was $-5\,°C$ and it is now $-8\,°C$.
14. It was $5\,°C$ and it is now $-1\,°C$.

15. It was $9\,°C$ and it is now $-1\,°C$.
16. It was $-2\,°C$ and it is now $-7\,°C$.

17. It was $-11\,°C$ and it is now $-4\,°C$.
18. It was $-8\,°C$ and it is now $3\,°C$.

19. It was $-15\,°C$ and it is now $0\,°C$.
20. It was $-7\,°C$ and it is now $-2\,°C$.

21. It was $-3\,°C$ and it is now $-83\,°C$.
22. It was $4\,°C$ and it is now $-11\,°C$.

23. Copy each sequence and fill in the missing numbers.

    a) 9, 6, 3, ☐ ☐   **b)** ☐, -1, 3, 7, 11   **c)** ☐, ☐, -10, -5, 0

24. A diver is below the surface of the water at -15 m. He dives down 6 m, and then rises 4 m. How far below the surface is he now?

25. Some land in Bangladesh is below sea level.

    Here are the heights, above sea level, of five villages.

    A 1 m   **B** -4 m   **C** 21 m   **D** -2 m   **E** -1.5 m

    a) Which village is safest from flooding?

    b) Which village is most at risk from serious flooding?

26. Arjun is overdrawn at the bank by \$90 (this means he owes the bank \$90).

    If Arjun pays in \$150, how much money will he have in the bank?

For adding and subtracting, use a number line.

Example 1

Find: **a)** $-1 + 4$ **b)** $-2 - 3$ **c)** $4 - 6$

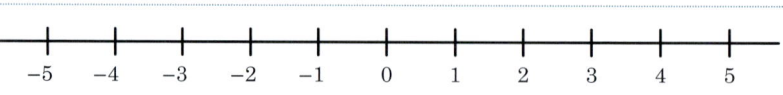

a) 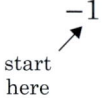 $-1 + 4 = 3$

b) 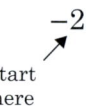 $-2 - 3 = -5$

c) 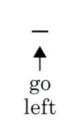 $4 - 6 = -2$

When you have two (+) or (−) signs together use this rule:

$$+ \; + \; = \; +$$ $$+ \; - \; = \; -$$
$$- \; - \; = \; +$$ $$- \; + \; = \; -$$

Example 2

a) $3 - (-6) = 3 + 6 = 9$
b) $-4 + (-5) = -4 - 5 = -9$
c) $-5 - (+7) = -5 - 7 = -12$

Exercise 1.5B

Work out:

1. $-6 + 2$ 2. $-7 - 5$ 3. $-3 - 8$ 4. $-5 + 2$

5. $-6 + 1$ 6. $8 - 4$ 7. $4 - 9$ 8. $11 - 19$

9. $4 + 15$ 10. $-7 - 10$ 11. $16 - 20$ 12. $-7 + 2$

13. $-6 - 5$ 14. $10 - 4$ 15. $-4 + 0$ 16. $-6 + 12$

17. $-7 + 7$ 18. $2 - 20$ 19. $8 - 11$ 20. $-6 - 5$

21. $-3 + (-5)$ 22. $-5 - (+2)$ 23. $4 - (+3)$ 24. $-3 - (-4)$

25. $6 - (-3)$ **26.** $16 + (-5)$ **27.** $-4 + (-4)$ **28.** $20 - (-22)$

29. $-6 - (-10)$ **30.** $95 + (-80)$ **31.** $-3 - (+4)$ **32.** $-5 - (+4)$

33. $6 + (-7)$ **34.** $-4 + (-3)$ **35.** $-7 - (-7)$ **36.** $3 - (-8)$

37. $-8 + (-6)$ **38.** $7 - (+7)$ **39.** $12 - (-5)$ **40.** $9 - (+6)$

When two numbers with the same sign are multiplied together, the answer is positive.

Example 1

a) $+7 \times (+3) = +21$

b) $-6 \times (-4) = +24$

When two numbers with different signs are multiplied together, the answer is negative.

Example 2

a) $-8 \times (+4) = -32$

b) $+7 \times (-5) = -35$

c) $-3 \times (+2) \times (+5) = -6 \times (+5) = -30$

When dividing numbers, the rules for signs are the same as for multiplication.

Example 3

a) $-70 \div (-2) = +35$

b) $+12 \div (-3) = -4$

c) $-20 \div (+4) = -5$

Exercise 1.5C

Work out:

1. -3×2 **2.** -4×1 **3.** $5 \times (-3)$ **4.** $-3 \times (-3)$

5. -4×2 **6.** -5×3 **7.** $6 \times (-4)$ **8.** 3×2

9. $-3 \times (-4)$ **10.** $6 \times (-3)$ **11.** -7×3 **12.** $-5 \times (-5)$

13. $6 \times (-10)$ **14.** $-3 \times (-7)$ **15.** 8×6 **16.** -8×2

17. -7×6 **18.** $-5 \times (-4)$ **19.** -6×7 **20.** $11 \times (-6)$

21. $8 \div (-2)$ **22.** $-9 \div 3$ **23.** $-6 \div (-2)$ **24.** $10 \div (-2)$

25. $-12 \div (-3)$ **26.** $-16 \div 4$ **27.** $4 \div (-1)$ **28.** $8 \div (-8)$

29. $16 \div (-8)$ **30.** $-20 \div (-5)$ **31.** $-16 \div 1$ **32.** $18 \div (-9)$

33. $36 \div (-9)$ **34.** $-45 \div (-9)$ **35.** $-70 \div 7$ **36.** $-11 \div (-1)$

37. $-16 \div (-1)$ **38.** $1 \div \left(-\dfrac{1}{2}\right)$ **39.** $-2 \div \dfrac{1}{2}$ **40.** $50 \div (-10)$

41. $-8 \times (-8)$ **42.** -9×3 **43.** $10 \times (-60)$ **44.** $-8 \times (-5)$

45. $-12 \div (-6)$ **46.** $-18 \times (-2)$ **47.** $-8 \div 4$ **48.** $-80 \div 10$

49. $-16 \times (-10)$ **50.** $32 \div (-16)$

Tip

It is helpful to write negative numbers in brackets, but brackets are not required around positive numbers.

Exercise 1.5D

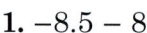

1. $-8.5 - 8$ **2.** $-2 \times (-7.5)$ **3.** -5×3.5 **4.** $-5.5 + 3$

5. $8 - (-7.5)$ **6.** $20.5 - 2.5$ **7.** $-7 \div (-3.5)$ **8.** $4.5 + (-10.5)$

9. $-2 + 13 - 8$ **10.** $8 \times (-6) \times 2$ **11.** $-9 + 2 - 7$ **12.** $-2 - (-11) + 6$

13. $-6 \times (-1) \div 2$ **14.** $2 - 20 - 8$ **15.** $-14 - (-4) + 6$ **16.** $-40 \div (-5) \div (-2)$

17. $5 - 11 - 20$ **18.** $-3 \times 10 \div 5$ **19.** $9 + (-5) - 6.5$ **20.** $7 \div (-7) \times (-8)$

1.6 Decimals

Look at these numbers.

tens	ones	•	$\dfrac{1}{10}$	$\dfrac{1}{100}$	$\dfrac{1}{1000}$
5	3	•	6	2	
	0	•	8	7	3

Notice that $53.62 = 50 + 3 + \dfrac{6}{10} + \dfrac{2}{100}$

$$0.873 = \dfrac{8}{10} + \dfrac{7}{100} + \dfrac{3}{1000}$$

When writing decimals in order of size, it is helpful to write them with the same number of figures after the decimal point.

Example

Write these three numbers in order from lowest to highest:
0.08, 0.107, 0.1

$$0.08 \rightarrow 0.080$$
$$0.107 \rightarrow 0.107$$
$$0.1 \rightarrow 0.100$$

Now you can see that the correct order is 0.08, 0.1, 0.107.

You could write $0.08 < 0.1 < 0.107$

Exercise 1.6A

In Questions **1** to **8**, write down each statement and decide whether it is true (T) or false (F). Rewrite the statements that are true using <, > or =.

1. 0.3 is less than 0.31.

2. 0.82 is more than 0.825.

3. 0.7 is equal to 0.70.

4. 0.17 is less than 0.71.

5. 0.02 is more than 0.002.

6. 0.6 is less than 0.06.

7. 0.1 is equal to $\frac{1}{10}$.

8. 5 is equal to 5.00.

> **Tip**
>
> < means 'less than'
>
> > means 'more than'
>
> = means 'equal to'

9. The number 43.6 can be written $40 + 3 + \frac{6}{10}$.

Write the number 57.2 in this way.

10. Write the decimal numbers for these additions.

a) $200 + 30 + 5 + \frac{1}{10}$

b) $60 + 7 + \frac{2}{10} + \frac{3}{100}$

c) $90 + 8 + \frac{3}{10} + \frac{2}{100}$

d) $3 + \frac{1}{10} + \frac{6}{100} + \frac{7}{1000}$

In Questions **11** to **18**, arrange the numbers in order of size, smallest first.

11. 0.41, 0.31, 0.2

12. 0.75, 0.58, 0.702

13. 0.43, 0.432, 0.41

14. 0.609, 0.61, 0.6

15. 0.04, 0.15, 0.2, 0.35

16. 1.8, 0.18, 0.81, 1.18

17. 0.7, 0.061, 0.07, 0.1

18. 0.2, 0.025, 0.03, 0.009

19. Here are some numbers with letters. Put the numbers in order and then write down the letters to make a word.

A 0.05   S 0.205   H 0.25   A 0.2   W 0.11   C 0.01   R 0.1

20. Increase each of these numbers by $\frac{1}{10}$:

    a) 32.41   **b)** 0.753   **c)** 1.06

21. Increase each of these numbers by $\frac{1}{100}$:

    a) 5.68   **b)** 0.542   **c)** 1.29

22. Write the following amounts in dollars:

    a) 350 cents   **b)** 15 cents   **c)** 3 cents

    d) 10 cents   **e)** 1260 cents   **f)** 8 cents

23. Copy each statement and say whether it is true or false.

    a) \$5.4 = \$5 + 40c   **b)** \$0.6 = 6c

    c) 5c = \$0.05   **d)** 50c is more than \$0.42

Scale readings

When reading a scale, divide the difference between the numbers by the number of gaps to find what one gap is worth.

For example, in Question **1** below, the difference between the numbers is 1 and there are 10 gaps, so each gap is equal to $1 \div 10 = 0.1$.

Exercise 1.6B

Work out the value indicated by the arrow.

1.

2.

3.

4.

5.

6.

7.

8.

9.

10.

11.

12.

13.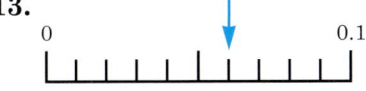

14.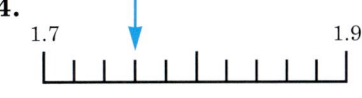

15.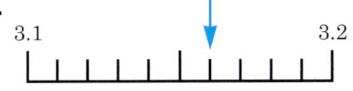

16.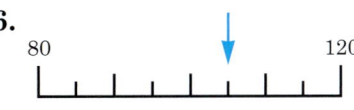

17.

18.

Multiplying and dividing decimals by 10, 100, 1000

To **multiply**, think of it as moving the decimal point to the *right*.

$3.24 \times 10 = 32.4$

$10.61 \times 10 = 106.1$

$4.134 \times 100 = 413.4$

$8.2 \times 100 = 820$

To **divide**, think of it as moving the decimal point to the *left*.

$15.2 \div 10 = 1.52$

$624.9 \div 100 = 6.249$

$509 \div 1000 = 0.509$

Exercise 1.6C

Work out:

1. 0.634×10

2. 0.838×10

3. 0.815×100

4. 0.074×100

5. 7.245×1000

6. 0.032×1000

7. 0.63×-10

8. -1.42×100

9. 0.041×-100

10. 0.3×100

11. 0.71×-1000

12. 3.95×10

13. $6.24 \div 10$

14. $8.97 \div 10$

15. $17.5 \div 100$

16. $23.6 \div -100$

17. $127 \div 1000$

18. $705 \div -1000$

19. $13 \div -10$

20. $0.8 \div 10$

21. $0.7 \div -100$

22. $-218 \div -10$

23. $35 \div 1000$

24. $-8.6 \div -1000$

25. -0.95×100

26. 11.11×10

27. $-3.2 \div 10$

28. 0.07×1000

29. $-57.6 \div 10$

30. $999 \div 100$

31. -66×-10

32. $100 \div 100$

33. $-42 \div -1000$

34. $0.62 \times -10\,000$

35. $-0.9 \div 100$

36. $-555 \div -10\,000$

37. Here are some number cards:

      and

a) Jason picks the cards [1] [3] and [4] to make the number 314.

Which extra card should he take to make a number ten times as big as 314?

b) Mel chooses three cards to make 5.2.

i) Which cards should she take to make a number ten times as big as 5.2?

ii) Which cards should she take to make a number 100 times as big as 5.2?

iii) Which cards should she take to make a number which is $\frac{1}{100}$ of 52?

Adding and subtracting decimals

Remember: line up the decimal points.

a) $4.2 + 1.76$

```
  4.20  put a zero
+ 1.76
------
  5.96
```

b) $26 - 1.7$

```
   5 1
  2 6.0
-   1.7
------
  24.3
```

c) $0.24 + 5 + 12.7$

```
    0.24
    5.00  extra
+  12.70  zeros
-------
   17.94
```

Exercise 1.6D

Work out, without a calculator:

1. $2.84 + 7.3$

2. $18.6 + 2.34$

3. $25.96 + 0.75$

4. $212.7 + 4.25$

5. $3.6 + 6$

6. $7 + 16.1$

7. $8 + 0.34 + 0.8$

8. $12 + 5.32$

9. $0.004 + 0.058$

10. $4.81 - 3.7$

11. $6.92 - 2.56$

12. $8.27 - 5.86$

13. $3.6 - 2.24$

14. $8.4 - 2.17$

15. $8.24 - 5.78$

16. $15.4 - 7$

17. $8 - 5.2$

18. $13 - 2.7$

19. $0.5 - 0.32$

20. $5 - 0.99$

21. $6 + 0.06 + 0.6$

22. $12.4 + 28.71$

23. $11 - 7.4$

24. $8.2 + 9.54 - 11.3$

Multiplying decimals

Count up the number of figures to the right of the decimal points in the question.
Put the same number of figures to the right of the decimal point in the answer.

Example

a) $0.\underline{2} \times 0.\underline{8}$

b) $0.\underline{4} \times 0.\underline{07}$

2 figures in total after the decimal points

$[2 \times 8 = 16]$

So $0.2 \times 0.8 = 0.\underline{16}$

3 figures in total after the decimal points

$[4 \times 7 = 28]$

So $0.4 \times 0.07 = 0.\underline{028}$

Exercise 1.6E

Work out, without a calculator:

1. 0.2×0.3	**2.** 0.5×0.3	**3.** 0.4×0.3	**4.** 0.2×0.03
5. 0.6×3	**6.** 0.7×5	**7.** 0.9×2	**8.** 8×0.1
9. 0.4×0.9	**10.** 0.02×0.7	**11.** 2.1×0.6	**12.** 4.7×0.5
13. 21.3×0.4	**14.** 5.2×0.6	**15.** 4.2×0.03	**16.** 212×0.6
17. 0.85×0.2	**18.** 3.27×0.1	**19.** 12.6×0.01	**20.** 0.02×17
21. 0.05×1.1	**22.** 52×0.01	**23.** 65×0.02	**24.** 0.5×0.002

Dividing by a decimal

Example

a) $9.36 \div 0.4$ Multiply both numbers by 10 so that you can divide by a whole number. [Think of it as moving the decimal points one place to the right.]

So work out $93.6 \div 4$.

$$\begin{array}{r} 2\,3.\,4 \\ 4{\overline{\smash{\big)}\,9^{1}3.^{1}6}} \end{array}$$

b) $0.0378 \div 0.07$ Multiply both numbers by 100 so that you can divide by a whole number. [Think of it as moving the decimal points two places to the right.]

So work out $3.78 \div 7$.

$$\begin{array}{r} 0.5\,4 \\ 7{\overline{\smash{\big)}\,3.7^{2}8}} \end{array}$$

Exercise 1.6F

Work out, without a calculator:

1. 0.84 ÷ 0.4

2. 0.93 ÷ 0.3

3. 0.872 ÷ 0.2

4. 0.8 ÷ 0.2

5. 2.8 ÷ 0.7

6. 1.25 ÷ 0.5

7. 8 ÷ 0.5

8. 40 ÷ 0.2

9. 7 ÷ 0.1

10. 0.368 ÷ 0.4

11. 0.915 ÷ 0.03

12. 0.248 ÷ 0.04

13. 0.625 ÷ 0.05

14. 8.54 ÷ 0.07

15. 1.272 ÷ 0.006

16. 4.48 ÷ 0.08

17. 0.12 ÷ 0.002

18. 7.5 ÷ 0.005

19. 0.09 ÷ 0.3

20. 0.77 ÷ 1.1

21. 0.055 ÷ 0.11

22. 21.28 ÷ 7

23. 22.48 ÷ 4

24. 3.12 ÷ 4

25. 0.7 ÷ 5

26. 3 ÷ 0.8

27. 0.3 ÷ 4

28. 1.2 ÷ 8

29. 0.732 ÷ 0.6

30. 0.1638 ÷ 0.001

31. 1.05 ÷ 0.6

32. 7.52 ÷ 0.4

33. A cake of mass 7.2 kg is cut into several pieces, each of mass 0.6 kg. How many pieces are there?

34. A phone call costs $0.04. How many calls can I make if I have $3.52?

35. A sheet of paper is 0.01 cm thick. How many sheets are there in a pile of paper 5.8 cm thick?

1.7 Properties of numbers

- **Natural numbers** (0, 1, 2, …) are used for counting and ordering, e.g. she is 4th tallest in the class.
- **Integers** (…, −2, −1, 0, 1, 2, …) are the natural numbers together with the negatives of the non-zero natural numbers.

Order of operations

Some people use the word 'BIDMAS' to help them to remember the correct order of operations.

Tip	
Brackets	**M**ultiply
Indices	**A**dd
Divide	**S**ubtract

Here are four examples:

- $8 + 6 \div 6 = 8 + 1 = 9$
- $20 - 8 \times 2 = 20 - 16 = 4$
- $(13 - 7) \div (6 - 4) = 6 \div 2 = 3$
- $20 - \dfrac{8}{5 + 3} = 20 - \dfrac{8}{8} = 20 - 1 = 19$

Tip

The dividing line acts like there are brackets and means that you find $20 - 8 \div (5 + 3)$

Exercise 1.7A

Work out, without a calculator:

1. $5 + 7 \times 2$

2. $9 - 12 \div 2$

3. $24 + 6 \times 2$

4. $(8 + 9) \times 2$

5. $17 - (2 + 5)$

6. $5 \times 8 + 6 \times 3$

7. $5 \times 6 - \dfrac{8}{2}$

8. $102 \div (17 - 15)$

9. $\dfrac{15}{15} + 15$

10. $11 - 3.2 \times 2.5$

11. $7 + \dfrac{9}{3}$

12. $20.7 - (11.6 + (-3.7))$

13. $9.7 \times 4 - 20$

14. $100 - 10 \div 10$

15. $\dfrac{3.6}{5 + 7}$

16. $2 \times (8 + 4.5 \times 3) - 7.2$

17. $(8 + 5) \times (20 - 16)$

18. $18 \div \left(9 - \dfrac{12}{4} \right)$

Indices

3^2 means '3 to the power 2' or 3×3

5^4 means '5 to the power 4' or $5 \times 5 \times 5 \times 5$

In these examples, the '2' and the '4' are indices.

In the order of operations, BIDMAS, we perform operations involving indices after operations in brackets.

For example, $3^2 \times 5 = 9 \times 5 = 45$

$$(5 + 4) - 2^3 = 9 - 2^3$$
$$= 9 - 8 = 1$$

Exercise 1.7B

Work out the following using BIDMAS.

1. $33 + 4^2$

2. $(8 - 5)^2$

3. $(4^2 + 1) \times 2$

4. $5 \times 4 - 2^3$

5. $3^2 - 5 + 1$

6. $(4^2 - 12)$

7. $(11 - 3^2) \times 5$

8. $4^2 \div (12 - 8 \div 2)$

9. $(10^2 - 9^2) + 11$

- **Factors**   Any number which divides exactly into 8 is a **factor** of 8. The factors of 8 are 1, 2, 4, 8.
- **Multiples**   Any number in the 8-times table is a **multiple** of 8. The first five multiples of 8 are 8, 16, 24, 32, 40.
- **Prime**   A **prime** number has exactly two different factors: 1 and itself. The number 1 is *not* prime. [It does not have two different factors.] The first five prime numbers are 2, 3, 5, 7, 11.
- **Prime factor**   The factors of 8 are 1, 2, 4, 8. The only **prime factor** of 8 is 2. It is the only prime number which is a factor of 8.

LCM and HCF

a) The first few multiples of 4 are 4, 8, 12, 16, ⃝20, 24, 28, 32, 36, 40, 44, 48, 52, 56, 60, …

The first few multiples of 5 are 5, 10, 15, ⃝20, 25, 30, 35, 40, 45, 50, 55, 60, …

Common multiples of 4 and 5 are 20, 40, 60, etc.

The **Lowest Common Multiple (LCM)** of 4 and 5 is 20.

It is the lowest number that is in both lists.

b) The factors of 12 are 1, 2, 3, ⃝4, 6, 12

The factors of 20 are 1, 2, ⃝4, 5, 10, 20

Common factors of 12 and 20 are 1, 2 and 4.

The **Highest Common Factor (HCF)** of 12 and 20 is 4.

It is the highest number that is in both lists.

c) Prime factors can be found using a 'factor tree'.

- Here is a factor tree for 140.

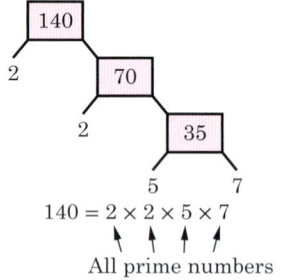

$140 = 2 \times 2 \times 5 \times 7$
All prime numbers

- Here is a factor tree for 40.

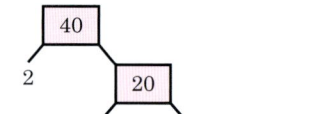

$40 = 2 \times 2 \times 2 \times 5$

You can use prime factors to find the HCF and the LCM of two numbers.

The HCF is the product of all the prime factors that the two numbers have in common.

In the example above, 140 and 40 both have two 2s and one 5 in their prime factors,

 $140 = ⃝2 \times ⃝2 \times ⃝5 \times 7$ and $40 = ⃝2 \times ⃝2 \times 2 \times ⃝5$

so the HCF is $2 \times 2 \times 5 = 20$

The LCM is found by multiplying the HCF by the prime factors not already used.

So the LCM of 140 and 40 is $20 \times 7 \times 2 = 280$

Exercise 1.7C

1. Write down all the factors of the following numbers:

   a) 6   **b)** 15   **c)** 18   **d)** 21   **e)** 40

2. Write down all the prime numbers less than 20.

3. Write down two prime numbers which add up to another prime number. Do this in two different ways.

4. Use a calculator to find which of the following are prime numbers:

   a) 91   **b)** 101   **c)** 143   **d)** 151   **e)** 293

5. Draw a factor tree for each of these numbers. Remember to start with the smallest prime number in each case.

   a) 36   **b)** 60   **c)** 216   **d)** 200   **e)** 1500

> **Tip**
>
> Divide by the prime numbers 2, 3, 5, 7, 11 and so on.

6. Here is the number 600 written as the product of its prime factors.

   $600 = 2 \times 2 \times 2 \times 3 \times 5 \times 5$

   Use this information to write 1200 as a product of its prime factors.

7. Write down the first four multiples of:

   a) 3   **b)** 4   **c)** 10   **d)** 11   **e)** 20

8. Find the 'one that does not belong' in each list.

   a) multiples of 6: 12, 18, 24, 32, 48

   b) multiples of 9: 18, 27, 45, 56, 72

9. Write down two numbers that are multiples of both 3 and 4.

10. Copy and choose the correct word to complete each sentence:

    a) An [odd/even] number is exactly divisible by 2.

    b) An [odd/even] number leaves a remainder of 1 when divided by 2.

    c) All [odd/even] numbers are multiples of 2.

11. Copy the table and then write the numbers 1 to 9, *one in each box*, so that all the numbers satisfy the conditions for both the row and the column.

	Prime number	Multiple of 3	Factor of 16
Number greater than 5			
Odd number			
Even number			

12. Find each of the mystery numbers below.

a) I am an odd number and a prime number. I am a factor of 14.

b) I am a two-digit multiple of 50.

c) I am one less than an even prime number.

d) I am odd, greater than one and a factor of both 20 and 30.

13. a) Write down the first six multiples of 2.

b) Write down the first six multiples of 5.

c) Write down the LCM of 2 and 5.

14. a) Write down the first four multiples of 4.

b) Write down the first four multiples of 12.

c) Write down the LCM of 4 and 12.

15. Find the LCM of:

a) 6 and 9   **b)** 8 and 12   **c)** 14 and 35

d) 2, 4 and 6   **e)** 3, 5 and 10   **f)** 4, 7 and 9

16. The table shows the factors and common factors of 24 and 36.

Number	Factors	Common factors
24	1, 2, 3, 4, 6, 8, 12, 24	1, 2, 3, 4, 6, 12
36	1, 2, 3, 4, 6, 9, 12, 18, 36	

Write down the HCF of 24 and 36.

17. The table shows the factors and common factors of 18 and 24.

Number	Factors	Common factors
18	1, 2, 3, 6, 9, 18	} 1, 2, 3, 6
24	1, 2, 3, 4, 6, 8, 12, 24	

Write down the HCF of 18 and 24.

18. Find the HCF of:

a) 12 and 18   **b)** 22 and 55   **c)** 45 and 72

d) 12, 18 and 30   **e)** 36, 60 and 72   **f)** 20, 40 and 50

19. Don't confuse LCMs with HCFs!

a) Find the HCF of 12 and 30.

b) Find the LCM of 8 and 20.

c) Write down two numbers whose HCF is 11.

d) Write down two numbers whose LCM is 10.

20. a) Write 30 and 165 as the product of prime factors.

b) Find the highest common factor of 30 and 165.

c) Find the lowest common multiple of 30 and 165.

21. a) Write 315 and 273 as the product of prime factors.

b) Find the highest common factor of 315 and 273.

c) Find the lowest common multiple of 315 and 273.

Rational and irrational numbers

- A rational number can always be written exactly in the form $\frac{a}{b}$ where a and b are whole numbers and b is not equal to zero.

$$\frac{3}{7} \qquad 1\frac{1}{2} = \frac{3}{2} \qquad 5.14 = \frac{257}{50} \qquad 0.\dot{6} = \frac{2}{3}$$

All these are rational numbers.

- An irrational number cannot be written in the form $\frac{a}{b}$. $\sqrt{2}, \sqrt{5}, \pi, \sqrt[3]{2}$ are all irrational numbers.

- In general, $\sqrt{n}$ is irrational unless n is a square number.

In this triangle, the length of the hypotenuse is exactly $\sqrt{5}$.

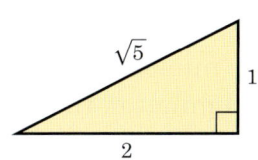

On a calculator, $\sqrt{5} = 2.236068$. This value of $\sqrt{5}$ is *not* exact and is correct only to 6 decimal places.

Tip

Exact here means giving your answer as a fraction or with the square root sign.

Exercise 1.7D

1. Which of the following numbers are rational?

$$\frac{\pi}{2} \qquad \sqrt{5} \qquad \left(\sqrt{17}\right)^2 \qquad \sqrt{3}$$

$$3.14 \qquad \sqrt{12} \qquad \pi^2 \qquad \frac{1}{3} + \frac{1}{9}$$

$$\sqrt{49} \qquad \frac{22}{7} \qquad \sqrt{100} \qquad \sqrt{82}$$

Tip

Remember: squaring is the inverse of finding a square root.

2. a) Write down any rational number between 4 and 6.

　　b) Write down any irrational number between 4 and 6.

3. a) For each shape state whether the **perimeter** is rational or irrational.

　　b) For each shape state whether the **area** is rational or irrational.

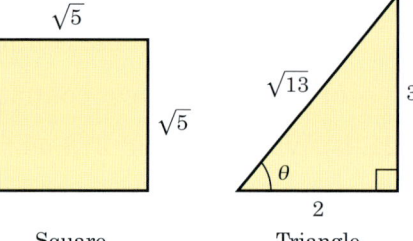

Square　　　　Triangle

1.8 Fractions

Fractions such as $\frac{3}{5}, \frac{2}{11}, \frac{1}{9}$ are called **proper** fractions.

Where the numerator (the top number) is greater than the denominator (the bottom number) the fraction is called an **improper** fraction. For example, $\frac{9}{4}$ or $\frac{11}{3}$.

An improper fraction can be written as a **mixed number**. For example $\frac{5}{3} = 1\frac{2}{3}$; $\frac{13}{3} = 4\frac{1}{3}$.

You can convert from a mixed number to an improper fraction.

The numerator of the improper fraction is equal to the whole number, multiplied by the denominator, then add the numerator. In the two cases above:

$$1\frac{2}{3} = \frac{1 \times 3 + 2}{3} = \frac{5}{3} \text{ and } 4\frac{1}{3} = \frac{4 \times 3 + 1}{3} = \frac{13}{3}$$

Equivalent fractions are fractions that have the same value. For example $\frac{1}{2}$ and $\frac{2}{4}$ are equivalent since both fractions are equal to 0.5.

A fraction is said to be in **simplest form** if the numerator and denominator do not share any common factors. For example $\frac{2}{4}$ is not in simplest form since 2 and 4 can both be divided by 2.

When adding or subtracting fractions, you should use a **common denominator**.

Example

a) $\dfrac{3}{4} + \dfrac{2}{5} = \dfrac{15}{20} + \dfrac{8}{20}$ Write both fractions with a common denominator.

$\qquad\qquad = \dfrac{23}{20}$ Add the numerators and write as a mixed number.

$\qquad\qquad = 1\dfrac{3}{20}$

b) $2\dfrac{3}{8} - 1\dfrac{5}{12} = \dfrac{19}{8} - \dfrac{17}{12}$ Write as improper fractions.

$\qquad\qquad\quad = \dfrac{57}{24} - \dfrac{34}{24}$ Use a common denominator and subtract the numerators.

$\qquad\qquad\quad = \dfrac{23}{24}$

c) $\dfrac{2}{5} \times \dfrac{6}{7} = \dfrac{12}{35}$ Multiply the numerators and the denominators.

d) $2\dfrac{2}{5} \div 6 = \dfrac{12}{5} \div \dfrac{6}{1}$ Write 6 as a fraction and then turn it upside down.

$\qquad\quad = \dfrac{\overset{2}{\cancel{12}}}{5} \times \dfrac{1}{\underset{1}{\cancel{6}}} = \dfrac{2}{5}$ Cancel a factor of 6 before multiplying.

Exercise 1.8A

1. Complete these pairs of equivalent fractions.

  a) $\dfrac{1}{3} = \dfrac{\square}{9}$ **b)** $\dfrac{3}{4} = \dfrac{\square}{16}$ **c)** $\dfrac{4}{5} = \dfrac{12}{\square}$ **d)** $\dfrac{3}{8} = \dfrac{30}{\square}$ **e)** $\dfrac{\square}{3} = \dfrac{16}{24}$ **f)** $\dfrac{7}{\square} = \dfrac{84}{108}$

2. Write each fraction in its simplest form.

  a) $\dfrac{6}{8}$ **b)** $\dfrac{8}{16}$ **c)** $\dfrac{50}{60}$ **d)** $\dfrac{24}{36}$ **e)** $\dfrac{121}{132}$ **f)** $\dfrac{1200}{1600}$

3. Work out:

a) $36 \times \dfrac{3}{4}$ **b)** $40 \times \dfrac{5}{8}$ **c)** $\dfrac{5}{6} \times 66$

d) $\dfrac{3}{7} \times 35$ **e)** $\dfrac{8}{9} \times 72$ **f)** $40 \times \dfrac{4}{5} \times \dfrac{3}{8}$

Work out and simplify where possible.

4. $\dfrac{1}{3} + \dfrac{1}{2}$ **5.** $\dfrac{1}{3} \times \dfrac{1}{2}$ **6.** $\dfrac{1}{3} \div \dfrac{1}{2}$ **7.** $\dfrac{3}{4} - \dfrac{1}{3}$ **8.** $\dfrac{3}{4} \times \dfrac{1}{3}$

9. $\dfrac{3}{4} \div \dfrac{1}{3}$ **10.** $\dfrac{2}{5} + \dfrac{1}{2}$ **11.** $\dfrac{2}{5} \times \dfrac{1}{2}$ **12.** $\dfrac{2}{5} \div \dfrac{1}{2}$ **13.** $\dfrac{3}{7} + \dfrac{1}{2}$

14. $\dfrac{3}{7} \times \dfrac{1}{2}$ **15.** $\dfrac{3}{7} \div \dfrac{1}{2}$ **16.** $\dfrac{5}{8} - \dfrac{1}{4}$ **17.** $\dfrac{5}{8} \times \dfrac{1}{4}$ **18.** $\dfrac{5}{8} \div \dfrac{1}{4}$

19. $\dfrac{1}{6} + \dfrac{4}{5}$ **20.** $\dfrac{1}{6} \times \dfrac{4}{5}$ **21.** $\dfrac{1}{6} \div \dfrac{4}{5}$ **22.** $\dfrac{3}{7} + \dfrac{1}{3}$ **23.** $\dfrac{3}{7} \times \dfrac{1}{3}$

24. $\dfrac{3}{7} \div \dfrac{1}{3}$ **25.** $\dfrac{4}{5} - \dfrac{1}{4}$ **26.** $\dfrac{4}{5} \times \dfrac{1}{4}$ **27.** $\dfrac{4}{5} \div \dfrac{1}{4}$ **28.** $\dfrac{2}{3} - \dfrac{1}{8}$

29. $\dfrac{2}{3} \times \dfrac{1}{8}$ **30.** $\dfrac{2}{3} \div \dfrac{1}{8}$ **31.** $\dfrac{5}{9} + \dfrac{1}{4}$ **32.** $\dfrac{5}{9} \times \dfrac{1}{4}$ **33.** $\dfrac{5}{9} \div \dfrac{1}{4}$

34. $2\dfrac{1}{2} - \dfrac{1}{4}$ **35.** $2\dfrac{1}{2} \times \dfrac{1}{4}$ **36.** $2\dfrac{1}{2} \div \dfrac{1}{4}$ **37.** $3\dfrac{3}{4} - \dfrac{2}{3}$ **38.** $3\dfrac{3}{4} \times \dfrac{2}{3}$

39. $3\dfrac{3}{4} \div \dfrac{2}{3}$ **40.** $\dfrac{\frac{1}{2} + \frac{1}{5}}{\frac{1}{2} - \frac{1}{5}}$ **41.** $\dfrac{\frac{3}{4} - \frac{1}{3}}{\frac{3}{4} + \frac{1}{3}}$ **42.** $\dfrac{2\frac{1}{4} \times \frac{4}{5}}{\frac{3}{5} - \frac{1}{2}}$ **43.** $\dfrac{3\frac{1}{2} \times 2\frac{2}{3}}{\frac{1}{2} + 1\frac{1}{18}}$

44. $\dfrac{3}{2} + \dfrac{6}{4}$ **45.** $\dfrac{5}{3} \times \dfrac{8}{5}$ **46.** $\dfrac{4}{3} + \dfrac{10}{6}$ **47.** $\dfrac{11}{2} - \dfrac{5}{4}$ **48.** $\dfrac{8}{3} \div \dfrac{9}{4}$

49. $\dfrac{20}{12} \times \dfrac{18}{5}$ **50.** $\dfrac{11}{4} \div \dfrac{7}{3}$ **51.** $\dfrac{16}{9} \times \dfrac{15}{10}$ **52.** $\dfrac{14}{6} - \dfrac{15}{8}$ **53.** $\dfrac{10}{4} + \dfrac{14}{6}$

Exercise 1.8B

1. Arrange the fractions in order of size, starting with the smallest:

a) $\dfrac{7}{12}, \dfrac{1}{2}, \dfrac{2}{3}$ **b)** $\dfrac{3}{4}, \dfrac{2}{3}, \dfrac{5}{6}$

c) $\dfrac{1}{3}, \dfrac{17}{24}, \dfrac{5}{8}, \dfrac{3}{4}$ **d)** $\dfrac{5}{6}, \dfrac{8}{9}, \dfrac{11}{12}$

> **Tip**
>
> Write all the fractions with the same denominator, so in part **(a)** write $\dfrac{7}{12}, \dfrac{6}{12}, \dfrac{8}{12}$.

2. Find the fraction that is mid-way between the two fractions given:

a) $\dfrac{2}{5}$, $\dfrac{3}{5}$

b) $\dfrac{5}{8}$, $\dfrac{7}{8}$

c) $\dfrac{2}{3}$, $\dfrac{3}{4}$

d) $\dfrac{1}{3}$, $\dfrac{4}{9}$

e) $\dfrac{4}{15}$, $\dfrac{1}{3}$

f) $\dfrac{3}{8}$, $\dfrac{11}{24}$

3. For each pair of fractions, copy and fill in the box using either >, < or =.

a) $\dfrac{4}{7}$ ☐ $\dfrac{5}{7}$

b) $\dfrac{6}{11}$ ☐ $\dfrac{6}{13}$

c) $\dfrac{2}{3}$ ☐ $\dfrac{14}{21}$

d) $\dfrac{5}{6}$ ☐ $\dfrac{6}{7}$

e) $\dfrac{11}{12}$ ☐ $\dfrac{13}{14}$

f) $\dfrac{9}{20}$ ☐ $\dfrac{45}{100}$

4. Vishala earns \$36 for her Saturday job. She got $\dfrac{1}{3}$ extra as a bonus.

How much did she receive including the bonus?

5. Copy and complete.

a) $2\dfrac{1}{4} = \dfrac{\square}{4}$

b) $\dfrac{15}{7} = \square\dfrac{\square}{7}$

c) $3\dfrac{1}{10} = \dfrac{\square}{10}$

6. Of the cars in a mechanic's workshop:

$\dfrac{1}{4}$ had broken brakes

$\dfrac{1}{3}$ had broken steering

$\dfrac{1}{6}$ had broken lights

the rest had worn tyres.

What fraction of the cars had worn tyres?

7. A rubber ball is dropped from a height of 300 cm. After each bounce, the ball rises to $\dfrac{4}{5}$ of its previous height.

a) How high, to the nearest cm, will it rise after the first bounce?

b) How high, to the nearest cm, will it rise after the fourth bounce?

8. Karol spends his income as follows:

$\dfrac{2}{15}$ of his income goes in tax

$\dfrac{2}{3}$ of what is left goes on food, rent and transport

he spends the rest on gym membership and computer games.

What fraction of his income is spent on gym membership and computer games?

9. Here is a list of the number of hours for which a computer was used over 5 days:
$6\dfrac{1}{2}$, $6\dfrac{1}{4}$, $5\dfrac{1}{2}$, $5\dfrac{3}{4}$, $6\dfrac{1}{2}$.

a) For how many hours was the computer used altogether?

b) The computer is rented at \$10 per hour. What is the total rent for the week?

10. In a set of drills, the smallest is $\frac{1}{8}$ cm and the sizes go up to $\frac{5}{8}$ cm in steps of $\frac{1}{16}$ cm.

    a) How many drills are there in the full set?

    b) Which size is half-way between $\frac{1}{4}$ cm and $\frac{3}{8}$ cm?

11. A fraction is equivalent to $\frac{2}{3}$ and its denominator is 8 more than its numerator. What is the fraction?

12. In this equation, all the asterisks stand for the same number. What is the number?

$$\left[\frac{*}{*} - \frac{*}{6} = \frac{*}{30} \right]$$

13. Work out one-half of one-third of three-quarters of $360.

14. Figures 1 and 2 show an equilateral triangle divided into thirds and quarters. They are combined in Figure 3. Calculate the fraction of Figure 3 that is shaded.

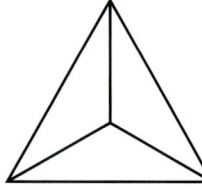
Figure 1

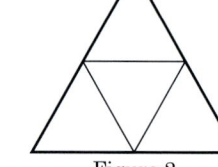
Figure 2

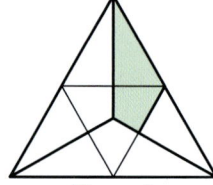
Figure 3

1.9 Indices 1

Indices

- Indices are used as a neat way of writing products.

    $2 \times 2 \times 2 \times 2 = 2^4$ [2 to the power 4]

    $5 \times 5 \times 5 = 5^3$ [5 to the power 3]

    $3 \times 3 \times 3 \times 3 \times 3 \times 10 \times 10 = 3^5 \times 10^2$

- Numbers such as 9, 25, 121 are **square numbers** (from 3^2, 5^2, 11^2).

    You need to know all the square numbers up to 225 (15^2).

- Numbers such as 8, 216, 1331 are **cube numbers** (from 2^3, 6^3, 11^3).

    You need to know all the cube numbers up to 125 (5^3) and you need to know that $10^3 = 1000$.

- To work out 3.2^2 on a calculator, press $\boxed{3}\ \boxed{\cdot}\ \boxed{2}\ \boxed{x^2}\ \boxed{=}$

    To work out 3^4 on a calculator, press $\boxed{3}\ \boxed{x^y}\ \boxed{4}\ \boxed{=}$ or $\boxed{3}\ \boxed{\wedge}\ \boxed{4}\ \boxed{=}$

Exercise 1.9A

Write in a form using indices.

1. $3 \times 3 \times 3 \times 3$
2. 5×5
3. $6 \times 6 \times 6$
4. $10 \times 10 \times 10 \times 10 \times 10$
5. $1 \times 1 \times 1 \times 1 \times 1 \times 1 \times 1$
6. $8 \times 8 \times 8 \times 8$
7. $7 \times 7 \times 7 \times 7 \times 7 \times 7$
8. $2 \times 2 \times 2 \times 5 \times 5$
9. $3 \times 3 \times 7 \times 7 \times 7 \times 7$
10. $3 \times 3 \times 10 \times 10 \times 10$
11. $5 \times 5 \times 5 \times 5 \times 11 \times 11$
12. $2 \times 3 \times 2 \times 3 \times 3$
13. $5 \times 3 \times 3 \times 5 \times 5$
14. $2 \times 2 \times 3 \times 3 \times 3 \times 11 \times 11$

15. Work out without a calculator:

    a) 4^2 **b)** 6^2 **c)** 10^2 **d)** 3^3 **e)** 10^3 **f)** 14^2 **g)** 5^3

16. Use the x^2 button to work out:

    a) 9^2 **b)** 21^2 **c)** 1.2^2 **d)** 0.2^2 **e)** 3.1^2

    f) 100^2 **g)** 25^2 **h)** 8.7^2 **i)** 0.9^2 **j)** 81.4^2

17. Find the areas of these squares.

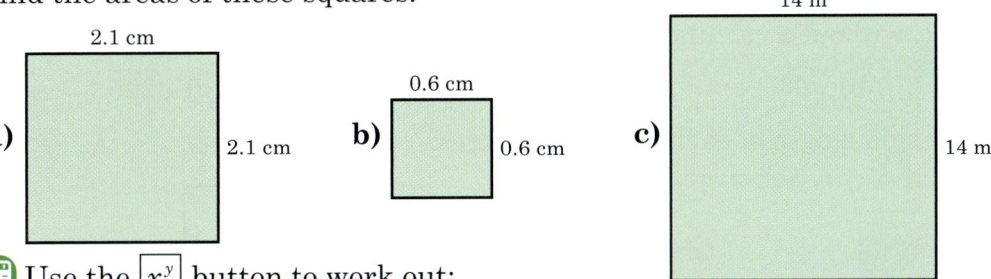

a) 2.1 cm, 2.1 cm   **b)** 0.6 cm, 0.6 cm   **c)** 14 m, 14 m

18. Use the x^y button to work out:

    a) 6^3 **b)** 2^8 **c)** 3^5 **d)** 10^5 **e)** 4^3

    f) 0.1^3 **g)** 1.7^4 **h)** $3^4 \times 7$ **i)** $5^3 \times 10$

19. A scientist has a dish containing 10^9 germs.
One day later there are 10 times as many germs.
How many germs are in the dish now?

20. A field has 2^8 daisies growing on the grass.
A cow eats half of the daisies.
How many daisies are left?

21. Abdul says, 'If you work out the product of any four consecutive numbers and then add one, the answer will be a square number'.

    For example: $1 \times 2 \times 3 \times 4 = 24$

            $24 + 1 = 25$, which is a square number.

    Is Abdul right? Test his theory on four (or more) sets of four consecutive numbers.

Square roots and cube roots

A square has an area of 529 cm².

How long is each side of the square?

In other words, what number *multiplied by itself* makes 529?

The answer is the **square root** of 529.

On a calculator press $\sqrt{}$ 5 2 9 =

Each side of the square is 23 cm.

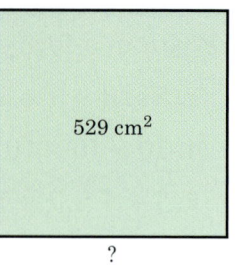

529 cm²

?

A cube has a volume of 512 cm³.

How long is each side of the cube?

The answer is the **cube root** of 512.

On a calculator press $\sqrt[3]{}$ 5 1 2 =

Each side of the cube is 8 cm. [Check: $8 \times 8 \times 8 = 512$]

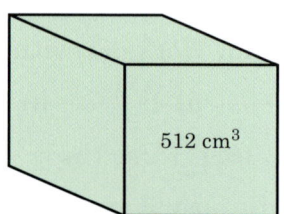

512 cm³

Exercise 1.9B

1. Work out, without a calculator:

   a) $\sqrt{16}$ **b)** $\sqrt{36}$ **c)** $\sqrt{1}$ **d)** $\sqrt{100}$ **e)** $\sqrt{225}$

2. Find the side length of each of these squares.

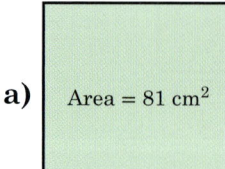

a) Area = 81 cm²

b) Area = 49 cm²

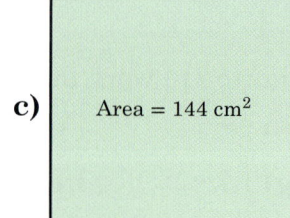

c) Area = 144 cm²

3. Use a calculator to find the following. Write your answer as a decimal and write down all of the numbers on the calculator display.

   a) $\sqrt{10}$ **b)** $\sqrt{29}$ **c)** $\sqrt{107}$ **d)** $\sqrt{19.7}$

   e) $\sqrt{2406}$ **f)** $\sqrt{58.6}$ **g)** $\sqrt{0.15}$ **h)** $\sqrt{0.727}$

4. A square photo has an area of 150 cm². Find the length of each side of the photo, correct to the nearest mm.

5. A square field has an area of 20 hectares. How long is each side of the field, correct to the nearest metre?

   [1 hectare = 10 000 m²]

6. The area of square A is equal to the sum of the areas of squares B and C.

Find the length x, correct to 1 decimal place.

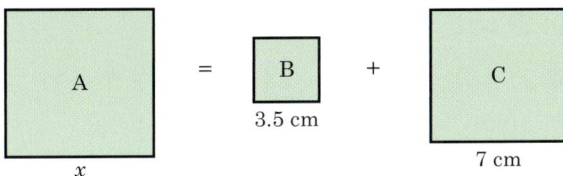

7. Find the following without using a calculator:

a) $\sqrt[3]{64}$ **b)** $\sqrt[3]{125}$ **c)** $\sqrt[3]{1000}$

8. A cube has a volume of 200 cm³.

Find the length of each side of the cube, correct to 1 decimal place.

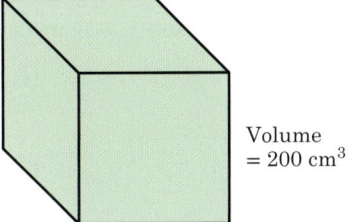

Volume
= 200 cm³

Negative indices and the zero index

- Look at this sequence.

From left to right the index goes down by one each time. Also we divide by two each time.

$2^4 \quad 2^3 \quad 2^2 \quad 2^1 \quad 2^0 \quad 2^{-1}$

$16 \rightarrow 8 \rightarrow 4 \rightarrow 2 \rightarrow \boxed{1} \rightarrow \boxed{\frac{1}{2}}$

We see that $2^0 = 1$ and that $2^{-1} = \frac{1}{2}$

- In general, $a^0 = 1$ for any value of a which is not zero.

In general, $a^{-1} = \frac{1}{a}$

Also $2^{-3} = \frac{1}{2^3}$ $3^{-2} = \frac{1}{3^2}$

> **Tip**
>
> Note also that $a^1 = a$ for all a.

- The **reciprocal** of 3 is $\frac{1}{3}$. The reciprocal of 10 is $\frac{1}{10}$.

- The reciprocal of n is $\frac{1}{n}$ (which can be written n^{-1}) and the reciprocal of $\frac{1}{n}$ is n.

- The reciprocal of a fraction $\frac{a}{b}$ is $\frac{b}{a}$ (think of it as flipping the fraction upside down).

Exercise 1.9C

For Questions **1** to **12**, work out the value of the number given.

1. 3^{-1} **2.** 4^{-1} **3.** 10^{-1} **4.** 1^{-4} **5.** 3^{-2} **6.** 4^{-2}

7. 10^{-2} **8.** 8^0 **9.** 7^{-2} **10.** $(-6)^0$ **11.** 9^{-2} **12.** 1^{-7}

In Questions **13** to **32** answer 'true' or 'false'.

13. $2^3 = 8$

14. $3^2 = 6$

15. $5^3 = 125$

16. $2^{-1} = \frac{1}{2}$

17. $10^{-2} = \frac{1}{20}$

18. $3^{-3} = \frac{1}{9}$

19. $2^2 > 2^3$

20. $2^3 < 3^2$

21. $2^{-2} > 2^{-3}$

22. $3^{-2} < 3^3$

23. $1^9 = 9$

24. $(-3)^2 = -9$

25. $5^{-2} = \frac{1}{10}$

26. $10^{-3} = \frac{1}{1000}$

27. $10^{-2} > 10^{-3}$

28. $5^{-1} = 0.2$

29. $10^{-1} = 0.1$

30. $2^{-2} = 0.25$

31. $5^0 = 1$

32. $16^0 = 0$

33. Write down the reciprocal of each of these numbers.

    a) 7

    b) 23

    c) $\frac{1}{5}$

    d) $\frac{1}{80}$

    e) $\frac{2}{3}$

    f) $\frac{5}{2}$

    g) $\frac{45}{71}$

    h) $\frac{1}{0.2}$

Multiplying and dividing

Example

To multiply powers of the same number **add** the indices.

$$3^2 \times 3^4 = (3 \times 3) \times (3 \times 3 \times 3 \times 3) = 3^6$$
$$2^3 \times 2^2 = (2 \times 2 \times 2) \times (2 \times 2) = 2^5$$
$$7^3 \times 7^5 = 7^8 \text{ [add the indices]}$$

To divide powers of the same number **subtract** the indices.

$$2^4 \div 2^2 = \frac{2 \times 2 \times 2 \times 2}{2 \times 2} = 2^2$$

$$\left.\begin{array}{l} 5^6 \div 5^2 = 5^4 \\ 7^8 \div 7^3 = 7^5 \end{array}\right\} \text{ [subtract the indices]}$$

Exercise 1.9D

Use the rules of indices to simplify these.

1. $5^2 \times 5^4$

2. $6^3 \times 6^2$

3. $10^4 \times 10^2 \times 10^3$

4. $7^5 \times 7 \times 7^2$

5. $3^3 \times 3^3 \times 3^4$

6. $8 \times 8^2 \times 8^2 \times 8$

7. $2^3 \times 2^{10}$

8. $3^6 \times 3^{-2}$

9. $5^4 \times 5^{-1}$

10. $7^7 \times 7^{-3}$

11. $5^{-3} \times 5^5$

12. $3^{-2} \times 3^{-2}$

13. $6^{-3} \times 6^5 \times 6^3$

14. $5^{-2} \times 5^{-5} \times 5^{-3}$

15. $7^{-3} \times 7 \times 7^8$

16. $7^4 \div 7^2$

17. $6^7 \div 6^2$

18. $8^5 \div 8^4$

19. $5^{10} \div 5^2$

20. $10^7 \div 10^5$

21. $9^6 \div 9^8$

22. $3^8 \div 3^{10}$

23. $2^6 \div 2^4 \times 2^2$

24. $3^3 \div 3^6 \times 3$

25. $7^2 \div 7^{10} \times 7^2$

26. $3^{-2} \div 3 \div 3$

27. $5^{-3} \div 5 \times 5^{-1}$

28. $8^{-1} \div 8^2 \div 8^2$

For Questions **29** to **36**, write down the number that would appear in the box to make these statements true.

29. $5^{-4} \div 5^{\square} = 5^{-5}$ **30.** $6^{\square} \div 6^{-2} = 6^4$ **31.** $3^4 \times 3^{\square} = 3^8$ **32.** $5^2 \times 5^{\square} = 5^7$

33. $\dfrac{3^{\square} \times 3^5}{3^2} = 3^3$ **34.** $\dfrac{2^8 \times 2^{\square}}{2^5} = 2^7$ **35.** $\dfrac{7^3 \times 7^3}{7^{\square}} = 7^2$ **36.** $\dfrac{5^9 \times 5^{\square}}{5^{20}} = 5^{-1}$

Further rules of indices

To raise a power of a number to a further power, **multiply** the indices.

> ### Example
> $(a^2)^3 = a^2 \times a^2 \times a^2 = a^6$ $(b^4)^2 = b^4 \times b^4 = b^8$

Exercise 1.9E

Use the rules of indices to simplify these.

1. $(3^3)^2$ **2.** $(5^4)^3$ **3.** $(7^2)^5$ **4.** $(8^2)^{10}$

5. $(2^{-1})^2$ **6.** $(3^{-2})^2$ **7.** $(7^{-1})^{-2}$ **8.** $(9^4)^{-12}$

9. $(7 \times 7^2)^3$ **10.** $(3^2 \times 3^4)^2$ **11.** $(5^2 \div 5^{-1})^4$ **12.** $(6^{-8} \times 6^5)^{-2}$

13. Write down the number that would appear in the box to make these statements true.

    a) $(7^2)^{\square} = 7^{10}$ **b)** $(8^{\square})^6 = 8^{18}$ **c)** $(3 \times 3^2)^{\square} = 3^{12}$ **d)** $(5^{\square} \div 5^3)^4 = 5^{-4}$

1.10 Standard form

When dealing with either very large or very small numbers, it is not convenient to write them out in full in the normal way. It is better to use **standard form**.

The number $A \times 10^n$ is in standard form when $1 \leqslant A < 10$ and n is a positive or negative integer.

> ### Example
> Write the following numbers in standard form:
>
> **a)** $2000 = 2 \times 1000 = 2 \times 10^3$
>
> **b)** $150 = 1.5 \times 100 = 1.5 \times 10^2$
>
> **c)** $0.0004 = 4 \times \dfrac{1}{10000} = 4 \times 10^{-4}$

Exercise 1.10A

For Questions **1** to **12** write the numbers in standard form.

1. 4000
2. 500
3. 70 000
4. 60
5. 2400
6. 380
7. 0.007
8. 0.0004
9. 0.0035
10. 0.421
11. 0.000 055
12. 0.01

13. These numbers have all been written incorrectly in standard form.
    Write them correctly.

    a) 56.4×10^4
    b) 0.19×10^9
    c) 50×10^{-3}

For Questions **14** to **19** write the numbers as normal numbers.

14. 4.6×10^4
15. 3.7×10^8
16. 9.21×10^{10}
17. 2.56×10^{-1}
18. 3.7×10^{-3}
19. 4.8×10^{-7}

20. The population of China is estimated at 1 400 000 000.
    Write this in standard form.

21. A hydrogen atom weighs 0.000 000 000 000 000 000 000 001 67 grams.
    Write this mass in standard form.

22. The area of the surface of the Earth is about 510 000 000 km².
    Express this in standard form.

23. A certain virus is 2.5×10^{-10} cm in diameter.
    Write this as a normal number.

24. Avogadro's number is 6.023×10^{23}.
    Write this as a normal number.

25. The speed of light is 300 000 km/s.
    Express this speed in cm/s in standard form.

Example 1

Work out $1500 \times 8 000 000$, giving your answer in standard form.

$$1500 \times 8 000 000 = (1.5 \times 10^3) \times (8 \times 10^6)$$
$$= 12 \times 10^9$$
$$= 1.2 \times 10^{10}$$

Notice that we multiply the numbers and the powers of 10 separately.

Tip

12×10^9 is not in standard form so needs to be adjusted.

Example 2

Work out $(8 \times 10^7) \div (2 \times 10^4)$, giving your answer as a normal number.

$$(8 \times 10^7) \div (2 \times 10^4) = (8 \div 2) \times (10^7 \div 10^4)$$
$$= 4 \times 10^3$$
$$= 4000$$

Example 3

Many calculators have a $\boxed{\times 10^x}$ button which is used for standard form.

a) To enter 1.6×10^7 into the calculator:

press $\boxed{1.6}$ $\boxed{\times 10^x}$ $\boxed{7}$

b) To enter 3.8×10^{-3} press $\boxed{3.8}$ $\boxed{\times 10^x}$ $\boxed{-3}$

c) To calculate $(4.9 \times 10^{11}) \div (3.5 \times 10^{-4})$:

$\boxed{4.9}$ $\boxed{\times 10^x}$ $\boxed{11}$ $\boxed{\div}$ $\boxed{3.5}$ $\boxed{\times 10^x}$ $\boxed{-4}$ $\boxed{=}$

The answer is 1.4×10^{15}.

Exercise 1.10B

In Questions **1** to **22**, give the answer in standard form.

Do not use a calculator for Questions **1** to **12**.

1. 5000×3000 **2.** $60\,000 \times 5000$ **3.** $0.000\,07 \times 400$ **4.** $0.0007 \times 0.000\,01$

5. $8000 \div 0.004$ **6.** $(0.002)^2$ **7.** 150×0.0006 **8.** $0.000\,03 \div 200$

9. $0.007 \div 20\,000$ **10.** $(0.0001)^4$ **11.** $(2000)^3$ **12.** $0.006 \div 3000$

13. $(1.4 \times 10^7) \times (3.5 \times 10^4)$ **14.** $(8.8 \times 10^{10}) \div (2 \times 10^{-2})$

15. $(1.2 \times 10^{11}) \div (8 \times 10^7)$ **16.** $(4 \times 10^5) \times (5 \times 10^{11})$

17. $(2.1 \times 10^{-3}) \times (8 \times 10^{15})$ **18.** $(8.5 \times 10^{14}) \div 2000$

19. $(3.3 \times 10^{12}) \times (3 \times 10^{-5})$ **20.** $(2.5 \times 10^{-8})^2$

21. $(1.2 \times 10^5)^2 \div (5 \times 10^{-3})$ **22.** $(6.2 \times 10^{-4}) \times (1.1 \times 10^{-3})$

23. If $a = 512 \times 10^2$ $b = 0.478 \times 10^6$ $c = 0.0049 \times 10^7$

arrange a, b and c in order of size (smallest first).

24. If the number 2.74×10^{15} is written out in full, how many zeros follow the 4?

25. If the number 7.31×10^{-17} is written out in full, how many zeros would there be between the decimal point and the digit 7?

26. If $x = 2 \times 10^5$ and $y = 3 \times 10^{-3}$, find the value of:

   a) xy

   b) $\dfrac{x}{y}$

27. Oil flows through a pipe at a rate of 40 m³/s. How long will it take to fill a tank of volume 1.2×10^5 m³?

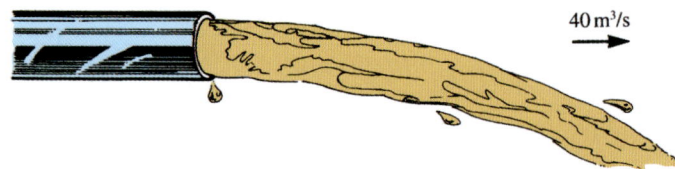

28. Given that $L = 2\sqrt{\dfrac{a}{k}}$, find the value of L in standard form when $a = 4.5 \times 10^{12}$ and $k = 5 \times 10^7$.

29. A light year is the distance travelled by a beam of light in a year. Light travels at a speed of approximately 3×10^5 km/s.

   a) Work out the length of a light year in km.

   b) Light takes about 8 minutes to reach Earth from the Sun. How far is Earth from the Sun in km?

30. A gardener counts 30 seeds from a packet of seeds and finds that their total mass is 6×10^{-2} g. The packet has a total mass of 50 grams. Use the sample to estimate the total number of seeds in the packet.

Revision exercise 1

1. Here are the first numbers of a pattern.

4, 13, 28, 49, 76, 109

   a) Which of these numbers are:

   i) odd   **ii)** square   **iii)** prime?

   b) The difference between the first and second numbers, $13 - 4$, is 9; between the second and the third it is 15, between the third and the fourth it is 21. Work out the difference between

   i) the fourth and the fifth

   ii) the fifth and the sixth.

   c) By considering your answers in part **(b)**, find the seventh and eighth numbers of the pattern.

   Explain how you reached this decision.

   d) Use the method you have described to write down the next two terms in the following pattern.

   1, 4, 12, 25, 43, 66, —, —.

2. A sports club decides to have a disco from 8 p.m. to midnight. The price of the tickets will be $5. The costs are as follows:

Disco and D.J., $250

Hire of hall, $50 an hour

200 cans of soft drinks at $0.50 each

200 packets of crisps at $0.60 each

Printing of tickets, $50

a) What is the total cost of putting on the disco?

b) How many tickets must be sold to cover the cost?

c) If 400 tickets are sold, all the drinks are sold at $0.75 each and all the packets of crisps at $1 each, calculate the profit or loss the sports club finally makes.

3. Filipe buys 500 marbles at $0.24 each. What change does he receive from $200?

4. Every day at school Omorede buys a roll for 28c, crisps for 22c and a drink for 42c. How much does he spend (in dollars) in the whole school year of 200 days?

5. A pile of 250 tiles is 2 m thick. What is the thickness of one tile in cm?

6. Copy the following bill and complete it by filling in the four blank spaces.

8 rolls of wallpaper at
$3.20 each   = $ ...

3 tins of paint at $ ... each   = $20.10

... brushes at $2.40 each   = $9.60

Total   = $ ...

7. Work out:

a) $-6 - 5$

b) $-7 + 30$

c) $-13 + 3$

d) -4×5

e) -3×-2

f) $-4 + -10$

8. Work out:

a) $\frac{3}{5} + \frac{1}{3}$

b) $\frac{3}{8} \times \frac{2}{3}$

c) $\frac{1}{5} - \frac{1}{10}$

d) $\frac{2}{3} \div \frac{1}{4}$

e) $1\frac{1}{2} - \frac{2}{5}$

f) $2\frac{1}{4} \times \frac{3}{4}$

9. Write in a form using indices:

a) $4 \times 4 \times 4 \times 4 \times 4$

b) $1 \times 1 \times 1 \times 1 \times 1 \times 1 \times 1$

c) $2 \times 2 \times 2 \times 5 \times 5$

10. Use the rules of indices to simplify these:

a) $6^2 \times 6^3$

b) $7^4 \times 7^4$

c) $3^{10} \div 3^3$

d) $10^4 \div 10^1$

e) $5^{-2} \times 5^6$

f) $2^4 \div 2^5$

11. Write in standard form:

a) $50\,000$

b) $610\,000$

c) 0.0003

d) 0.0015

e) 10 million

12. Work out each of these and give the answer in standard form:

a) $(2 \times 10^6) \times (1.5 \times 10^4)$

b) $(8 \times 10^9) \div (2 \times 10^5)$

c) $(4 \times 10^{-2}) \times (2 \times 10^8)$

d) $(1.5 \times 10^3) \times (3 \times 10^4)$

13. A mouse's heart beats at an average rate of 600 beats per minute. How many times does a mouse's heart beat in one day? Give your answer in standard form.

NON-CALCULATOR

1. The temperature at noon at an Arctic weather station was $-16\,°C$.
   At 11 p.m. it had fallen by $13\,°C$. What was the temperature at 11 p.m.? [1]

2. Work out the value of $\dfrac{16 + 4 \times 11}{5}$ [1]

3. 2, 4, 5, 8, 12, 16

   From the set of numbers above, write down

   a) a multiple of 6 [1]

   b) a prime factor of 35. [1]

4. Write down a multiple of both 4 and 10 that is less than 25. [1]

5. p is an integer between 20 and 40. Write down the value of p when it is:

   a) an even square number [1]

   b) an odd cube number [1]

   c) a prime factor of 155 [1]

   d) a multiple of 17. [1]

6. **a)** $4^a \times 4^5 = 4^{15}$. Find the value of a. [1]

   b) $2^7 \div 2^b = 2^4$. Find the value of b. [1]

   c) $5^c = \dfrac{1}{25}$. Find the value of c. [1]

7. There are 563 sheets of paper in a book.

   a) How many sheets of paper are there in 3000 of these books?
   Give your answer in standard form. [2]

   b) A pile of 563 sheets of paper is 24 millimetres high.

   Calculate the thickness of 1 sheet of paper.

   Give your answer in millimetres in standard form. [3]

8. Write 0.003 54 in standard form. [1]

9. **a)** Write the missing number: $\dfrac{3}{4} = \dfrac{\cdots}{16}$ [1]

   b) Without using your calculator, and writing down all your working,

   show that $1\dfrac{5}{8} - \dfrac{3}{4} = \dfrac{7}{8}$ [2]

10. A bottle of milk holds $1\dfrac{3}{4}$ litres. A glass holds $\dfrac{1}{8}$ of a litre.
    How many glasses can be filled from one bottle of milk? [2]

11. A piece of wood is 160 cm long. It has to be cut into equal lengths of $6\dfrac{2}{5}$ cm.

    ← 160 cm →

    How many of these lengths can be cut from the piece of wood? [2]

2 Algebra 1

Isaac Newton (1643–1727) is thought by many to have been one of the greatest intellects of all time. He went to Trinity College Cambridge in 1661 and by the age of 23 he had made three major discoveries: the nature of colours, calculus and the law of gravitation. He used his version of calculus to give the first satisfactory explanation of the motion of the Sun, the Moon and the stars. Because he was extremely sensitive to criticism, Newton was always very secretive, but he was eventually persuaded to publish his discoveries in 1687.

- Substitute into expressions and formulae.
- Simplify expressions, expand brackets and factorise.
- Construct and solve linear equations including those where x appears in the denominator as part of a linear expression.
- Solve simultaneous equations.
- Understand and use rules of indices including positive, zero and negative.
- Understand and use linear inequalities including representing on a number line.
- Understand sequences including continuing a sequence, recognising patterns, term-to-term rules, finding and using nth terms and relationships between sequences.

2.1 Introduction to algebra

Algebra is when you use letters to stand for generalised numbers.

3, 7, 8.3 and −4.61 are all numbers, whereas a letter can be used to stand for a general number.

The letter may take a single value, two possible values, or many more possible values, depending on the context in which it is used.

Algebra is made up of building blocks, called **terms**. A term is a single algebraic object.

These are examples of terms: $\quad 2a \quad x^2 \quad 3pq \quad \dfrac{x}{w}$

There are some conventions that you need to know when it comes to writing terms.

- If there is no number in front of the letter, for example x, this means $1 \times x$.

- If two letters are side by side, it means they are multiplied together, for example ab means $a \times b$.

- If a letter is squared, it means the letter multiplied by itself, for example p^2 means $p \times p$.

When terms are added or subtracted, they make **expressions**.

These are examples of expressions: $x + y \quad a^2 - 2 \quad pqr + 3s - 5t$

The number in front of a letter (or letters) is called the **coefficient**.

The coefficient of x in the expression $4x + y$ is 4.

If a number appears on its own in an expression, it is called a **constant term**.

The constant term in the expression $5y + 7$ is 7.

Exercise 2.1A

1. Consider this expression: $x + 3y$

   a) How many terms are there in the expression?

   b) What is the coefficient of

      i) y

      ii) x?

2. Consider this expression:
   $9x - 5y + 6z + 7$

   a) How many terms are there in the expression?

   b) What is the coefficient of

      i) x

      ii) y?

   c) What is the value of the constant term?

3. Consider this expression: $2a + 7b - 3ab + 4c - 6$

   a) How many terms are there in the expression?

   b) What is the coefficient of

      i) b

      ii) ab?

   c) What is the value of the constant term?

4. Bethia is considering the expression $x - 7y + 3$.

   She says, 'There are two terms in the expression since there are two letters. The coefficient of x is 0 since there is no number in front of it and the coefficient of y is 7 since there is a 7 in front of it'.

   Is Bethia correct? Explain your answer.

5. Maya has x counters. Jaya has 3 more counters than Maya. Keira has four times as many counters as Maya.

   a) Write down an expression for:

      i) the number of counters that Jaya has

      ii) the number of counters that Kiera has.

   b) Maya gives 8 counters to her brother.

      Write down an expression for the number of counters that Maya has now.

6. Amaar has a pears. Bharat has b pears. Chico has five times as many pears as Bharat.

   Write down an expression for:

   a) the number of pears that Amaar and Bharat have

   b) the number of pears that Amaar and Chico have

   c) the number of pears that Bharat and Chico have.

7. A square has side length p cm.

   Write down an expression for:

   a) the area of the square

   b) the perimeter of the square.

8. A rectangle has length x cm and width y cm.

Write down an expression for:

a) the area of the rectangle

b) the perimeter of the rectangle.

When you have lots of terms in an expression, some of them may be of the same type.

In the expression $4x + 3y + 2x$, there are two terms in x: the $4x$ term and the $2x$ term. These are called **like terms**.

You can collect like terms by adding (or subtracting) them to simplify an expression:

$$4x + 3y + 2x = 4x + 2x + 3y = 6x + 3y$$

Example

Simplify:

a) $4x + 13y + 5x - 2y - x$

b) $3x^2 + 5x + 6x^2 - 4x - 2x^2$

a) $4x + 5x - x + 13y - 2y$ Rewrite the terms so that like terms are together.

   $= 8x + 11y$ Then add or subtract the like terms.

b) $3x^2 + 6x^2 - 2x^2 + 5x - 4x$ Rewrite the order of the terms. Note that terms in x^2 and terms in x are *not* like terms.

   $= 7x^2 + x$ Add or subtract the like terms.

Exercise 2.1B

Simplify these expressions.

1. $3x + 2y + 4x$ **2.** $2x + 7y + 3x - 4y$ **3.** $8a + 4b - 6a - 7b$

4. $2p + 3q - 7p - q$ **5.** $4x^2 + 6x + 2x^2 - 11x$ **6.** $5a^2 - 3a - 2a^2 + 9a$

7. $xy + 3x + 2y + 4xy$ **8.** $2p^2 + 2q^2 + 4pq - p^2 - 5q^2$ **9.** $3c + 4d + 5e - 6c - 7d - e$

10. $14a - 11b + 3ab + 2a^2 - 7b - 3a + 4b^2 - 9ab$

You can also substitute numbers into an expression to work out the value.

Example

Find the value of $2x + 4y$ when:

a) $x = 4$, $y = 7$

b) $x = 3$, $y = -5$

---

a) $2 \times 4 + 4 \times 7$

$= 8 + 28$

$= 36$

b) $2 \times 3 + 4 \times (-5)$

$= 6 - 20$

$= -14$

Replace the letters with the numbers and evaluate. Remember to use BIDMAS.

> **Tip**
>
> Take care when substituting negative numbers into expressions.

Exercise 2.1C

If $a = -4$, $b = 5$, $c = -2$, work out:

1. $2a + 3$	**2.** $3b - 7$	**3.** $4a - 1$	**4.** $2b + c$
5. $5c - 2a$	**6.** $6a - 3$	**7.** $2c + b$	**8.** $3a - 2b$
9. $6c - 2b$	**10.** $3c + 4a$	**11.** $3c - 4$	**12.** $2a - 3c$
13. $7b + 3a$	**14.** $8a + 6c$	**15.** $2b - 4a$	**16.** $4b + 5$
17. $3a + 8$	**18.** $2c - a$	**19.** $5a - 2c$	**20.** $3b + 7$

If $n = 3$, $x = -1$, $y = 6$, work out:

21. $2x - 3$	**22.** $3y + 4n$	**23.** $5n + 2x$	**24.** $4y - x$
25. $7y - 2$	**26.** $3x + 2n$	**27.** $10x + 5$	**28.** $6x - y$
29. $4x - 5y$	**30.** $2y - 10$	**31.** $8n - 2y$	**32.** $7n + 3y$
33. $6y + 4$	**34.** $4n + 5x$	**35.** $2n + 3x$	**36.** $5y - 20$
37. $9y - n$	**38.** $8x + 2n$	**39.** $5x + 6$	**40.** $3n - 2x$

2.2 Sequences

A sequence is a list of numbers that follow a pattern.

Each number in the sequence is called a **term**.

How you get from one term in the sequence to the next is called the **term-to-term rule**.

Consider the sequence 4, 6, 8, 10, …

The term-to-term rule is 'add 2'.

Consider the sequence 5, 10, 20, 40, …

The term-to-term rule is '× 2'.

Exercise 2.2A

1. Find the next number in each sequence and write down the term-to-term rule.

a) 1, 5, 9, 13, …   **b)** 39, 36, 33, 30, …

c) 3, 6, 12, 24, …   **d)** 4, 9, 15, 22, …

e) 200, 100, 50, 25, …   **f)** 88, 99, 110, …

2. Write down each sequence, find the missing number and write down the term-to-term rule.

a) 1, 6, $\square$, 16   **b)** 1, 2, 4, 8, $\square$

c) $\square$, 2, 5, 8, 11   **d)** 2400, 240, 24, $\square$

e) 1, 2, 4, 7, $\square$, 16   **f)** 12, 8, 4, $\square$, −4

3. Here is the start of a sequence: 1, 3, 4, …

Each new term is found by adding the previous two terms. For example, $4 = 1 + 3$

The next term will be 7.

a) Write down the next six terms.

b) Use the same rule to write down the next four terms of the sequence which starts 2, 5, 7, …

c) The Fibonacci sequence is a special case of this type of sequence.

The Fibonacci sequence starts 1, 1, …

Write down the first ten terms of the Fibonacci sequence.

4. a) Write down the next two lines of the sequence:

$3 \times 4 = 3 + 3^2$

$4 \times 5 = 4 + 4^2$

$5 \times 6 = 5 + 5^2$

b) Complete the lines below:

$10 \times 11 =$

$30 \times 31 =$

5. Copy and complete the pattern and write down the next two lines.

$1 + 9 \times 0 \quad = \quad 1$

$2 + 9 \times 1 \quad = \quad 11$

$3 + 9 \times 12 \quad = \quad 111$

$4 + 9 \times 123 \quad = \quad 1111$

$5 + 9 \times 1234 \quad =$

6. For the sequence 2, 3, 8, ... each new term is found by squaring the previous term and then subtracting 1. Write down the next two terms.

7. The sequence 3, 3, 5, 4, 4 is obtained by counting the letters in 'one, two, three, four, five, ...'. Write down the next three terms.

8. The odd numbers 1, 3, 5, 7, 9, ... can be added to give an interesting sequence.

1	=	1	=	$1 \times 1 \times 1$
$3 + 5$	=	8	=	$2 \times 2 \times 2$
$7 + 9 + 11$	=	27	=	$3 \times 3 \times 3$
$13 + 15 + 17 + 19$	=	64	=	$4 \times 4 \times 4$

1, 8, 27, 64 are **cube** numbers.

You write $2^3 = 8$ ['two cubed equals eight']

$4^3 = 64$

Continue adding the odd numbers in the same way as shown above. Do you *always* get a cube number?

9. **a)** Write down the next three lines of this pattern.

$$1^3 = \qquad 1^2 \qquad = 1$$
$$1^3 + 2^3 = \quad (1 + 2)^2 \quad = 9$$
$$1^3 + 2^3 + 3^3 = (1 + 2 + 3)^2 = 36$$

b) Work out as simply as possible:

$$1^3 + 2^3 + 3^3 + 4^3 + 5^3 + 6^3 + 7^3 + 8^3 + 9^3 + 10^3$$

10. Here are the sequences of the first six odd and the first six even numbers.

	1st	2nd	3rd	4th	5th	6th
odd	1	3	5	7	9	11
even	2	4	6	8	10	12

Find: **a)** the 8th even number   **b)** the 8th odd number

     c) the 13th even number   **d)** the 13th odd number.

e) Explain in words the relationship between the nth odd number and the nth even number.

f) If the 57th even number is 114, what is the 57th odd number?

g) Write down:

     i) the 45th even number   **ii)** the 53rd odd number

     iii) the 100th odd number   **iv)** the 219th odd number.

11. Here the numbers 1 to 21 have been written in three columns.

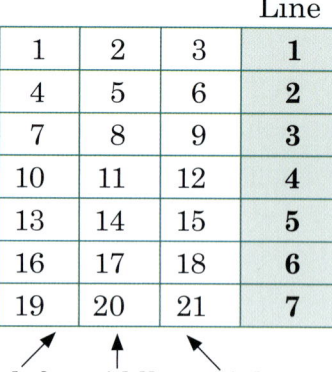

Line

1	2	3	1
4	5	6	2
7	8	9	3
10	11	12	4
13	14	15	5
16	17	18	6
19	20	21	7

left   middle   right

a) What number will you get on the right of:

i) line 8   **ii)** line 12   **iii)** line 25?

b) Write down the number in the middle of:

i) line 8   **ii)** line 12   **iii)** line 20.

c) What number will you get in:

i) line 10 on the left

ii) line 13 on the right

iii) line 17 in the middle

iv) line 30 on the left?

d) Find the missing number:

i) 120 is on the right of line _____.

ii) 61 is on the left of line _____.

iii) 92 is in the middle of line _____.

iv) 148 is on the left of line _____.

12. Here is a sequence:

1, 3, 6, 10, 15, …

a) Find the difference between each pair of successive terms of the sequence.

b) Describe the way in which the differences go up.

c) Use your answers to parts **(a)** and **(b)** to find:

i) the 8th term

ii) the 12th term

iii) the 15th term.

2.3 Finding a rule

For the sequence 3, 8, 13, 18, … the term-to-term rule is 'add 5'.
You can draw a mapping diagram with a column for 5 times the
term number (i.e. $5n$).

n	$5n$	term
1	5	→$^{-2}$ 3
2	10	→$^{-2}$ 8
3	15	→$^{-2}$ 13
4	20	→$^{-2}$ 18
n	$5n$	→$^{-2}$ $5n - 2$

You can see that each term is 2 less than $5n$.

So, the 10th term is $(5 \times 10) - 2 = 48$ and the 20th term is $(5 \times 20) - 2 = 98$

The nth term is $5 \times n - 2 = 5n - 2$

Sometimes the terms of the sequence decrease.

For the sequence 8, 6, 4, 2, ... the term-to-term rule is 'subtract 2'.

You can draw a mapping diagram with a column for -2 times the term number (i.e. $-2n$).

n	$-2n$	term
1	-2	→$^{+10}$ 8
2	-4	→$^{+10}$ 6
3	-6	→$^{+10}$ 4
4	-8	→$^{+10}$ 2
n	$-2n$	→$^{+10}$ $10 - 2n$

You can see that each term is 10 more than $-2n$.

So, the 5th term is $(-2 \times 5) + 10 = 0$ and the 9th term is $(-2 \times 9) + 10 = -8$

The nth term is $-2n + 10 = 10 - 2n$

In general, the number that n is multiplied by is the number in the term-to-term rule.

The nth term is called the **position-to-term** rule and means that you can find, for example, the 100th term in the sequence without having to work out all 100 terms.

For the first sequence above, with nth term $5n - 2$, the 100th term is $5 \times 100 - 2 = 498$.

For the second sequence above, with nth term $10 - 2n$, the 100th term is $10 - 2 \times 100 = -190$.

Exercise 2.3A

1. Look at the sequence 5, 8, 11, 14, ...

   The difference between terms is 3.

   Copy and complete: 'The nth term of the sequence is $3n + ...$'.

n	$3n$	term
1	3	→$^{+\square}$ 5
2	6	8
3	9	11
4	12	14

2. Look at the sequences and tables below. Find the nth term in each case.

a) Sequence 5, 9, 13, 17,...

n	$4n$	term
1	4	5
2	8	9
3	12	13
4	16	17

nth term =

b) Sequence 2, 8, 14, 20, ...

n	$6n$	term
1	6	2
2	12	8
3	18	14
4	24	20

nth term =

3. In the sequence 6, 11, 16, 21,...

the difference between terms is 5.
Copy and complete the table and write an
expression for the nth term of the sequence.

n		term
1		6
2		11
3		16
4		21
10		

4. Look at the sequence 6, 10, 14, 18,...

a) Write down the difference between terms. Make a table like the one in
Question **3** and use it to find an expression for the nth term.

b) Use your nth term to find the 20th term of the sequence.

5. Write down each sequence in a table. Find the nth term and the 50th term of
each sequence.

a) 5, 7, 9, 11,...

b) 3, 7, 11, 15,...

c) 2, 8, 14, 20,...

In Questions **6** to **17** you are given a sequence in a table. Copy the table and make
an extra column. Find an expression for the nth term of each sequence. [t stands
for 'term'.]

6.

n	t
1	3
2	8
3	13
4	18
5	23

7.

n	t
1	15
2	12
3	9
4	6
5	3

8.

n	t
1	7
2	13
3	19
4	25
5	31

9.

n	t
1	14
2	12
3	10
4	8
5	6

10.

n	t
1	6
2	7
3	8
4	9
5	10

11.

n	t
1	13
2	10
3	7
4	4
5	1

12.

n	t
1	5
2	13
3	21
4	29
5	37

13.

n	t
1	27
2	22
3	17
4	12
5	7

14.

n	t
6	26
7	23
8	20
9	17
10	14

15.

n	t
8	83
9	93
10	103
11	113
12	123

16.

n	t
4	15
5	18
6	21
7	24

17.

n	t
12	76
13	71
14	66
15	61

18. Make a table for each sequence. Find the nth term and the 20th term of each sequence.

a) 26, 18, 10, 2, …

b) 16, 13, 10, 7, …

c) 48, 39, 30, 21, …

> **Tip**
>
> A **linear sequence** is a sequence where the difference between one term and the next is constant.

19. The second term of a linear sequence is 16 and the fourth term is 30.

Find the 10th term of the sequence.

20. Hari says that 53 is a term in the sequence with nth term $6n + 2$.

Explain why Hari is incorrect.

21. a) Write down the nth term of the sequence 2, 4, 6, 8, …

b) Write down the nth term of the sequence 1, 3, 5, 7, …

c) Use your answers to **(a)** and **(b)** to explain why the sum of an odd number and an even number is always odd.

Here is a sequence of 'houses' made from matches.

   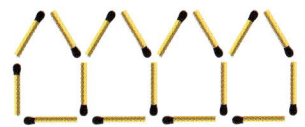

The table on the right records the number of houses h and the number of matches m.

h	m
1	5
2	9
3	13
4	17

Since the term-to-term rule is 'add 4' the sequence is based on 4h.

Now you can see that m is one more than 4h.

So the formula linking m and h is: $m = 4h + 1$

h	$4h$	m
1	4	5
2	8	9
3	12	13
4	16	27

Exercise 2.3B

1. Below is a sequence of diagrams showing blue tiles (b) and white tiles (w) with the related table.

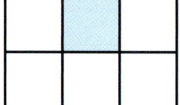 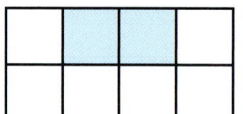 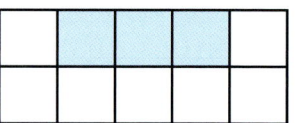

b	w
1	5
2	6
3	7
4	8

   What is the formula for w in terms of b? [i.e. write '$w = \ldots$']

2. This is a different sequence with blue tiles (b) and white tiles (w) and the related table.

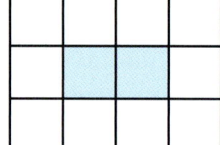 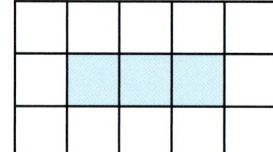 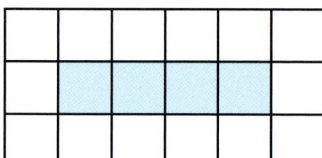

b	w
2	10
3	12
4	14
5	16

   What is the formula that links b and w? Write it as '$w = \ldots$'.

3. Here is a sequence of diagrams made from square tiles.

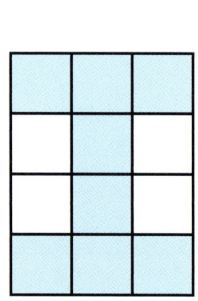  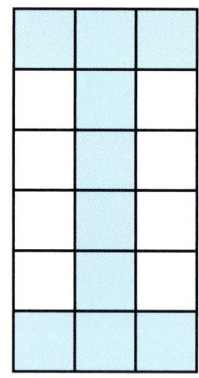

   Make your own table for blue tiles (b) and white tiles (w).
   What is the formula for w in terms of b?

4. In this sequence, the number of matches is m and the number of triangles is t.

t	m
1	3
2	5
.	.
.	.
.	.

Make a table for t and m. It starts like the table on the right:

Continue the table and find a formula for m in terms of t. Write '$m = \dots$'.

5. Here is a different sequence of matches and triangles.

  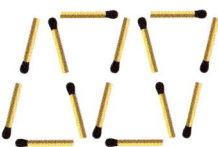

Make a table and find a formula connecting m and t.

6. In this sequence there are triangles (t) and squares (s) around the outside.

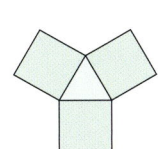 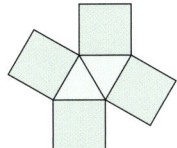 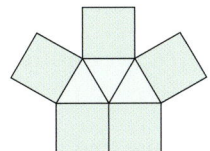 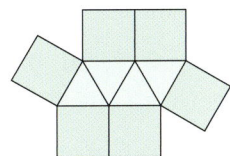

What is the formula connecting t and s?

7. Look at the tables below. In each case, find a formula connecting the two letters.

a)

n	p
1	3
2	8
3	13
4	18

write '$p = \dots$'

b)

n	k
2	17
3	24
4	31
5	38

write '$k = \dots$'

c)

n	w
3	17
4	19
5	21
6	23

write '$w = \dots$'

8. In these tables it is harder to find the formula because the numbers on the left do not go up by one each time. Try to find a formula in each case.

a)

n	y
1	4
3	10
7	22
8	25

b)

n	h
2	5
3	9
6	21
10	37

c)

n	w
3	14
7	26
9	32
12	41

9. This is one shape in a sequence of cubes made from matches.

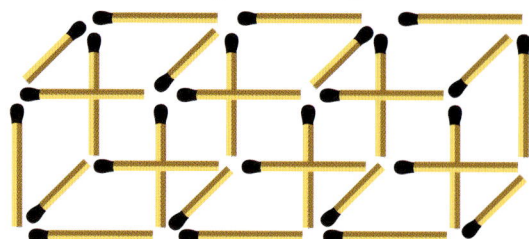

Find a formula connecting the number of matches (m) and the number of cubes (c).

Some sequences are not linear.

You already know the sequence of square numbers 1, 4, 9, 16,.... The nth term is n^2.

You already know the sequence of cube numbers 1, 8, 27, 64,.... The nth term is n^3.

You can use these sequences to find the nth term of other sequences.

The terms in the sequence 3, 6, 11, 18, ... are all two more than the terms in the sequence of square numbers, so the nth term is $n^2 + 2$.

Sequences that are based on the square numbers are called **quadratic sequences**.

The terms in the sequence 0, 7, 26, 63, ... are all one less than the terms in the sequence of cube numbers, so the nth term is $n^3 - 1$.

Sequences that are based on the cube numbers are called **cubic sequences**.

Exercise 2.3C

Find the first five terms of each of these sequences.

1. $n^2 + 4$ **2.** $n^2 + n$ **3.** $n^3 + 2$ **4.** $n^3 - n$

For each of these sequences, find an expression for the nth term and find the 20th term.

5. 4, 7, 12, 19,...

6. 2, 8, 18, 32,...

7. 0, 3, 8, 15,...

8. 0.5, 2, 4.5, 8,...

9. −6, −3, 2, 9,...

10. −1, −4, −9, −16,...

11. 0, −3, −8, −15,...

12. 2, 9, 28, 65,...

13. 2, 16, 54, 128,...

14. −1, 6, 25, 62,...

> **Tip**
>
> Write out the sequence of **cube** numbers and compare.

2.4 Brackets and factors

a) Expand $2(3x + 2)$

$$2(3x + 2) = \underline{2 \times 3x} + \underline{2 \times 2}$$
$$= \quad 6x \quad + \quad 4$$

Multiply each term in the bracket by the number in front.

b) Expand $2x(3 + 4x)$

$$2x(3 + 4x) = \underline{2x \times 3} + \underline{2x \times 4x}$$
$$= \quad 6x \quad + \quad 8x^2$$

Multiply each term in the bracket by the term in front.

c) Expand and simplify $3(2x + 5) + 4(x - 2)$

$$3(2x + 5) + 4(x - 2) = 6x + 15 + 4x - 8$$

Expand each bracket first.

$$= 10x + 7$$

Collect like terms.

Exercise 2.4A

Expand the brackets.

1. $3(x + 3)$ **2.** $4(x - 2)$ **3.** $5(2x + 1)$

4. $4(a + 7)$ **5.** $6(2x + 1)$ **6.** $10(5 - x)$

7. $3(4x + 5)$ **8.** $9(3 + x)$ **9.** $5(y - 2)$

10. $7(a - 2)$ **11.** $11(2x - y)$ **12.** $8(3x + 2y)$

Expand the brackets.

13. $x(4 + 2x)$ **14.** $x(3x - 2)$ **15.** $y(4y + 7)$

16. $a(2a - 6)$ **17.** $3x(4 - 6x)$ **18.** $4y(3y + 5)$

19. $5a(3a - 4)$ **20.** $7p(6 - 7p)$ **21.** $3x(4x + x^2)$

22. $4x(x^2 - 3x)$

Expand and simplify.

23. $3(4x + 1) + 4(2x - 1)$ **24.** $2(6x - 1) + 5(x - 3)$

25. $x(2x + 3) + x(4x + 1)$ **26.** $5(2x + 7) - 2(x + 4)$

27. $6(3x + 2) - 4(2x - 1)$ **28.** $2x(x + 3) - 3x(x - 4)$

> **Tip**
>
> Take care when expanding with negative numbers. Remember that multiplying a negative by a positive gives a negative answer and multiplying two negatives gives a positive answer.

Example

a) Expand $(x + 3)(x + 2)$

Method 1: Using a grid

	x	$+3$
x	x^2	$+3x$
$+2$	$+2x$	$+6$

So $(x + 3)(x + 2) = x^2 + 3x + 2x + 6$

$\qquad\qquad\qquad\quad = x^2 + 5x + 6$

Method 2: Using **FOIL**

$(x + 3)(x + 2) = x \times x + 2 \times x + 3 \times x + 3 \times 2$

$\qquad\qquad\quad = x^2 + 2x + 3x + 6 \qquad$ Add like terms.

$\qquad\qquad\quad = x^2 + 5x + 6$

b) Expand $(x - 4)(x + 7)$

Method 1: Using a grid

	x	-4
x	x^2	$-4x$
$+7$	$+7x$	-28

So $(x - 4)(x + 7) = x^2 - 4x + 7x - 28 \qquad$ Add like terms.

$\qquad\qquad\qquad\quad = x^2 + 3x - 28$

Method 2: Using **FOIL**

$(x - 4)(x + 7) = x \times x + 7 \times x - 4 \times x - 4 \times 7$

$\qquad\qquad\quad = x^2 + 7x - 4x - 28 \qquad$ Add like terms.

$\qquad\qquad\quad = x^2 + 3x - 28$

Tip

FOIL stands for **F**irst, **O**utside, **I**nside, **L**ast.

Multiply the **F**irst terms in each bracket.

Multiply the **O**utside terms in each bracket.

Multiply the **I**nside terms in each bracket.

Multiply the **L**ast terms in each bracket.

Exercise 2.4B

Expand and simplify.

1. $(x + 2)(x + 1)$ **2.** $(x + 4)(x + 6)$ **3.** $(x + 3)(x - 2)$

4. $(x - 2)(x + 6)$ **5.** $(x + 5)(x - 7)$ **6.** $(x - 8)(x + 3)$

7. $(x - 2)(x - 1)$ **8.** $(x - 5)(x - 3)$ **9.** $(x - 8)(x - 9)$

10. $(2x + 1)(x + 2)$ **11.** $(3x + 2)(x - 1)$ **12.** $(2x - 3)(x + 4)$

13. A rectangle has width $(2x + 1)$ cm and length $(2x + 3)$ cm.

Find an expression for the area of the rectangle.

14. A square has side length $(x + 5)$ cm.

Find an expression for the area of the square.

Factors

Factorising is the reverse process to expanding.

Find the highest common factor of the terms in the expression and then write this in front of the brackets.

The terms inside the brackets are found by dividing each term by the highest common factor.

Example

Factorise the following: **a)** $12a - 15b$ **b)** $3x^2 - 2x$ **c)** $2xy + 6y^2$

                               d) $12x + 16y - 8z$

a) $12a - 15b = 3(4a - 5b)$ The highest common factor of $12a$ and $15b$ is 3

You can check your factorisation by multiplying out your answer:

$3(4a - 5b) = 3 \times 4a - 3 \times 5b = 12a - 15b$ ✓

b) $3x^2 - 2x = x(3x - 2)$ The highest common factor of $3x^2$ and $2x$ is x.

c) $2xy + 6y^2 = 2y(x + 3y)$ The highest common factor of $2xy$ and $6y^2$ is $2y$.

d) $12x + 16y - 8z = 4(3x + 4y - 2z)$ You need to find the highest common factor of all three terms.

Note that the term factorise means *fully* factorise.

For example, $9a^2 + 15ab = 3(3a^2 + 5ab)$ is only partially factorised because $3a^2$ and $5ab$ still have the common factor a.

The highest common factor of $9a^2$ and $15ab$ is $3a$.

So $9a^2 + 15ab = 3a(3a + 5b)$ is fully factorised.

Exercise 2.4C

In Questions **1** to **10**, copy and complete the statement.

1. $6x + 4y = 2(3x + \square)$

2. $9x + 12y = 3(\square + 4y)$

3. $10a + 4b = 2(5a + \square)$

4. $4x + 12y = 4(\square + \square)$

5. $10a + 15b = 5(\square + \square)$

6. $18x - 24y = 6(3x - \square)$

7. $8u - 28v = \square(\square - 7v)$

8. $15s + 25t = \square(3s + \square)$

9. $24m + 40n = \square(3m + \square)$

10. $27c - 72d = \square(\square - 8d)$

In Questions **11** to **30** factorise the expression.

11. $20a + 8b$

12. $30x - 24y$

13. $27c - 33d$

14. $35u + 49v$

15. $12s - 32t$

16. $40x - 16t$

17. $24x + 84y$

18. $12x + 8y + 16z$

19. $12a - 6b + 9c$

20. $10x - 20y + 25z$

21. $20a - 12b - 28c$

22. $48m + 8n - 24x$

23. $42x + 49y - 21z$

24. $6x^2 + 15y^2$

25. $20x^2 - 15y^2$

26. $7a^2 + 28b^2$

27. $27a + 63b - 36c$

28. $12x^2 + 24xy + 18y^2$

29. $64p - 72q - 40r$

30. $36x - 60y + 96z$

In Questions **31** to **40** factorise the expression.

31. $a^2 + 4a$

32. $3x + 4x^2$

33. $4x^2 - x$

34. $7x - 3x^2$

35. $2x^2 + 4x$

36. $6x - 3x^2$

37. $12x + 16x^3$

38. $25x^2 - 15x^3$

39. $30x^3 + 10x^2$

40. $80y^3 - 30y^2$

2.5 Solving equations 1

In Section 2.1, you looked at expressions.

An expression is a mathematical statement that does not contain an 'equals' sign.

An **equation** is a mathematical statement that does contain an 'equals' sign.

$3x - 2$, $4x + 1$ and $6x^2 + 5x$ are all expressions.

$3x - 2 = 6$, $4x + 1 = 9$ and $6x^2 + 5x = 1$ are all equations.

You can solve equations by doing the same thing to both sides.

You use inverse operations (see Chapter 1). In the examples below, the inverse operations used are shown in the square brackets after each step of working.

Example

Solve the equations.

a) $x + 6 = 11$

$x + 6 - 6 = 11 - 6$ [subtract 6]

$x = 5$

b) $3x + 14 = 16$

$3x + 14 - 14 = 16 - 14$ [subtract 14]

$3x = 2$

$\dfrac{3x}{3} = \dfrac{2}{3}$ [divide both sides by 3]

$x = \dfrac{2}{3}$ You can leave the answer as a fraction.

c) $4x - 5 = -2$

$4x - 5 + 5 = -2 + 5$ [add 5]

$4x = 3$

$\dfrac{4x}{4} = \dfrac{3}{4}$ [divide by 4]

$x = \dfrac{3}{4} = 0.75$

d) $7 = 2x + 15$

$7 - 15 = 2x + 15 - 15$ [subtract 15]

$-8 = 2x$

$-\dfrac{8}{2} = \dfrac{2x}{2}$ [divide by 2]

$-4 = x$

or $x = -4$ The final answer is usually written with the letter on the left.

Exercise 2.5A

Solve the equations.

1. $x - 7 = 5$

2. $y + 11 = 20$

3. $x + 12 = 30$

4. $p - 6 = -2$

5. $x - 8 = 9$

6. $m + 5 = 0$

7. $x - 13 = -7$

8. $z + 10 = 3$

9. $5 + x = 9$

10. $9 + q = 17$

11. $y - 6 = 11$

12. $y + 8 = 3$

13. $3a + 1 = 16$

14. $4x + 3 = 27$

15. $2b - 3 = 1$

16. $5x - 3 = 1$

17. $3x - 7 = 0$

18. $2y + 5 = 20$

19. $6x - 9 = 2$

20. $7m + 6 = 6$

21. $9n - 4 = 1$

22. $11x - 10 = 1$

23. $15y + 2 = 5$

24. $7y + 8 = 10$

25. $4y - 11 = -8$ **26.** $3z - 8 = -6$ **27.** $4p + 25 = 30$ **28.** $5t - 6 = 0$

29. $9m - 13 = 1$ **30.** $4 + 3r = 5$ **31.** $7 + 2x = 8$ **32.** $5 + 20x = 7$

33. $3 + 8x = 0$ **34.** $50y - 7 = 2$ **35.** $200y - 51 = 49$ **36.** $5u - 13 = -10$

37. $9x - 7 = -11$ **38.** $11t + 1 = 1$ **39.** $3 + 8y = 40$ **40.** $12 + 7x = 2$

41. $6 = 3x - 1$ **42.** $8 = 4z + 5$ **43.** $9 = 2x + 7$ **44.** $11 = 5h - 7$

45. $0 = 3x - 1$ **46.** $40 = 11 + 14k$ **47.** $-4 = 5x + 1$ **48.** $-8 = 6x - 3$

49. $13 = 4x - 20$ **50.** $-103 = 2p + 7$

Equations with letters on both sides

Example

Solve the equations.

a) $8x - 3 = 3x + 1$ Add 3 to both sides and subtract $3x$
     $8x - 3x = 1 + 3$ from both sides.

          $5x = 4$ Simplify both sides of the equation.
           $x = \dfrac{4}{5} = 0.8$ Divide both sides by 5.

b) $3x + 9 = 18 - 7x$ Subtract 9 from both sides and add $7x$
     $3x + 7x = 18 - 9$ to both sides.

          $10x = 9$ Simplify both sides of the equation.
           $x = \dfrac{9}{10} = 0.9$ Divide both sides by 10.

Exercise 2.5B

Solve the equations.

1. $7x - 3 = 3x + 8$ **2.** $5y + 4 = 2y + 9$ **3.** $6x - 2 = x + 8$

4. $8a + 1 = 3a + 2$ **5.** $7b - 10 = 3b - 8$ **6.** $5x - 12 = 2x - 6$

7. $4m - 23 = m - 7$ **8.** $8z - 8 = 3z - 2$ **9.** $11x + 7 = 6x + 7$

10. $9n + 8 = 10$ **11.** $5 + 3x = x + 8$ **12.** $4 + 7g = g + 5$

13. $6x - 8 = 4 - 3x$ **14.** $5p + 1 = 7 - 2p$ **15.** $6x - 3 = 1 - x$

16. $3m - 10 = 2m - 3$ **17.** $5x + 1 = 6 - 3x$ **18.** $11r - 20 = 10r - 15$

19. $6 + 2x = 8 - 3x$ **20.** $7 + h = 9 - 5h$ **21.** $3y - 7 = y + 1$

22. $8y + 9 = 7y + 8$

23. $7y - 5 = 2y$

24. $3z - 1 = 5 - 4z$

25. $8 = 13 - 4t$

26. $10 = 12 - 2x$

27. $13 = 20 - 9u$

28. $8 = 5 - 2x$

29. $5 + v = 7 - 8v$

30. $3x + 11 = 2 - 3x$

2.6 Solving equations 2

Example

Solve the equations.

a) $3(x - 1) = 2(x + 7)$ Expand the brackets first.

$\quad 3x - 3 = 2x + 14$ Add 3 to both sides and subtract

$\quad 3x - 2x = 14 + 3$ $2x$ from both sides.

$\quad\quad x = 17$ Simplify.

b) $5(2x + 1) = 3(x - 2) + 20$ Expand the brackets first.

$\quad 10x + 5 = 3x - 6 + 20$ Collect x terms on the left and

$\quad 10x - 3x = -6 + 20 - 5$ numbers on the right.

$\quad\quad 7x = 9$ Simplify.

$\quad\quad x = \dfrac{9}{7} \text{ or } 1\dfrac{2}{7}$ Divide both sides by 7.

Exercise 2.6A

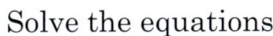

Solve the equations.

1. $2(x + 1) = x + 5$

2. $4(m - 2) = 2(m + 1)$

3. $5(x - 3) = 3(x + 2)$

4. $3(a + 2) = 2(a - 1)$

5. $5(x - 3) = 2(x - 7)$

6. $6(b + 2) = 2(b - 3)$

7. $10(x - 3) = x$

8. $3(2p - 1) = 4(p + 1)$

9. $4(2x + 1) = 5(x + 3)$

10. $3(n - 1) + 7 = 2(n + 1)$

11. $5(x + 1) + 3 = 3(x - 1)$

12. $7(y - 2) - 3 = 2(y + 2)$

13. $5(2x + 1) - 5 = 3(x + 1)$

14. $3(4z - 1) - 3 = z + 1$

15. $2(x - 10) = 4 - 3x$

16. $3r + 2(r + 1) = 3r + 12$

17. $4x - 2(x + 4) = x + 1$

18. $2s - 3(s + 2) = 2s + 1$

19. $5x - 2(x - 2) = 6 - 2x$

20. $3(x + 1) + 2(x + 2) = 10$

Equations with fractions

Example

Solve the equations.

a) $\dfrac{x}{7} = 8$

$x = 8 \times 7$ Multiply both sides by 7.

$x = 56$

b) $3 = \dfrac{8}{x}$

$3x = 8$ Multiply both sides by x.

$x = \dfrac{8}{3} = 2\dfrac{2}{3}$ Divide both sides by 3.

Exercise 2.6B

Solve the equations.

1. $\dfrac{x}{4} = 6$

2. $\dfrac{x}{5} = 3$

3. $\dfrac{y}{5} = -2$

4. $\dfrac{a}{7} = 3$

5. $\dfrac{t}{3} = 7$

6. $\dfrac{m}{4} = \dfrac{2}{3}$

7. $\dfrac{x}{7} = \dfrac{5}{8}$

8. $\dfrac{2x}{3} = 1$

9. $\dfrac{4x}{5} = 3$

10. $\dfrac{3y}{2} = 2$

11. $\dfrac{5t}{6} = 3$

12. $\dfrac{m}{8} = \dfrac{1}{4}$

13. $-6 = \dfrac{k}{4}$

14. $\dfrac{n}{7} = -10$

15. $\dfrac{x}{2} = 100$

16. $\dfrac{3}{x} = 5$

17. $\dfrac{4}{x} = 7$

18. $\dfrac{11}{x} = 12$

19. $\dfrac{6}{x} = 11$

20. $\dfrac{2}{x} = 3$

21. $\dfrac{5}{y} = 9$

22. $\dfrac{7}{y} = 9$

23. $\dfrac{4}{t} = 3$

24. $\dfrac{3}{a} = 6$

25. $\dfrac{8}{x} = 12$

26. $\dfrac{3}{p} = 1$

27. $\dfrac{15}{q} = 10$

28. $8 = \dfrac{5}{x}$

29. $19 = \dfrac{7}{y}$

30. $-5 = \dfrac{3}{a}$

31. $4 = \dfrac{33}{q}$

32. $\dfrac{500}{y} = -1$

33. $-99 = \dfrac{98}{f}$

34. $\dfrac{x}{3} + 5 = 7$

35. $\dfrac{x}{5} - 2 = 4$

Exercise 2.6C

In this exercise □, △, ○ and ＊ represent weights which are always balanced.

1.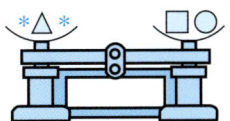

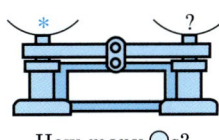

How many ○s?

2.

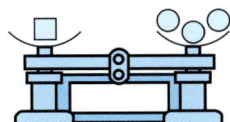

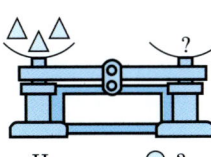

How many ○s?

3. ○ ○ □ = ＊ ＊

□ □ ○ = ＊ ＊ ○

□ = how many ○s?

4. □ ○ ○ = △ □ □ □

□ □ □ ○ = △ △ □

□ ○ = △ □

○ = how many □s?

5. □ □ = ○ △

○ ○ ○ □ = □ △

○ □ □ □ = △ △ ○

□ = how many ○s?

6. ○ ○ □ = ＊ ○

＊ ＊ = ○ ○ ○

□ ＊ = ○ ○

＊ = how many □s?

7. ○ □ □ = △ ＊

＊ ＊ ＊ = △ △

○ □ = △

△ △ △ △ = how many □s?

2.7 Solving problems with equations

Example

If I multiply a 'mystery' number by 2 and then add 3
the answer is 14. Find the 'mystery' number.

Let the mystery number be x.
Then $2x + 3 = 14$

$\qquad 2x = 14 - 3 \qquad$ Subtract 3 from both sides.

$\qquad 2x = 11$

$\qquad x = \dfrac{11}{2} = 5\dfrac{1}{2} \qquad$ Divide both sides by 2.

The 'mystery' number is $5\dfrac{1}{2}$ (or 5.5).

Exercise 2.7A

In each question, form an equation and then solve it.

1. If you multiply a number n by 3 and then add 4, the answer is 13.

   What is n?

2. Amy has n sweets, Ben has 2 more sweets than Amy and Carlos has twice as many sweets as Ben. Altogether they have 74 sweets.

   How many sweets does Amy have?

3. Fluffy the cat eats x biscuits. Otis the cat eats 3 fewer biscuits than Fluffy. Etsy the cat eats 4 times as many biscuits as Fluffy. Altogether they eat 99 biscuits.

   How many biscuits does Otis eat?

4. Bharat is playing a game. To get his overall score, the number of points he gets is doubled and then 15 is added. His total score is 179.

   How many points did Bharat get?

5. If you add 3 to a number n and then multiply the result by 4, the answer is 10.

   What is n?

6. If you subtract 11 from a number x and then treble the result, the answer is 20.

   What is x?

7. If you double a number p, add 4 and then multiply the result by 3, the answer is 13.

   What is p?

8. Eman has 800 grams of flour. She uses 55 grams to make cupcakes and then divides the rest evenly to make n cakes. Each cake uses 149 grams of flour.

   How many cakes did Eman make?

9. If you double a number m and subtract 7, you get the same answer as when you add 5 to the number.

   What is m?

10. If you multiply a number x by 5 and subtract 4, you get the same answer as when you add 3 to the number and then double the result.

    What is x?

11. If you multiply a number n by 6 and add 1, you get the same answer as when you add 5 to the number and then treble the result.

    What is n?

12. Carly bought 5 multipacks of bottles of juice to add to the 4 bottles she already had. Ambika bought 8 identical multipacks and drank 14 of the bottles. They then found that they had the same number of bottles.

How many bottles are in a multipack?

2.8 Indices 2

You have already used the rules of indices with numbers in Chapter 1.

Here is a recap of the rules.

- When multiplying powers of the same number, add the indices.

   $x^5 \times x^6 = x^{5+6} = x^{11}$

- When dividing powers of the same number, subtract the indices.

   $y^7 \div y^3 = y^{7-3} = y^4$

- Anything to the power of zero is equal to 1.

   $z^0 = 1$

- When raising a power of a number to a further power, multiply the indices.

   $(x^5)^3 = x^{15}$

- Negative indices mean '1 over'.

   $y^{-3} = \dfrac{1}{y^3}$

Exercise 2.8A

Simplify each of these using the rules of indices.

1. a) $a \times a \times a$ **b)** $n \times n \times n \times n$ **c)** $s \times s \times s \times s \times s$

    d) $p \times p \times q \times q \times q$ **e)** $b \times b \times b \times b \times b \times b \times b$

2. a) $x^4 \times x^3$ **b)** $y^6 \times y^2$ **c)** $a^9 \times a^{-4}$

    d) $x^{-2} \times x^6$ **e)** $\dfrac{x^{12}}{x^7}$ **f)** $\dfrac{p^{11}}{p^5}$

    g) $\dfrac{y^2}{y^5}$ **h)** $\dfrac{x^3}{x^{-1}}$ **i)** $\dfrac{x^{-2}}{x^{-5}}$

    j) $(x^2)^3$ **k)** $(a^5)^3$ **l)** $(n^7)^2$

    m) $(y^3)^3$ **n)** $(x^3)^{-1}$

Example 1

$3x^2 \times 4x^5 = 12x^7$ Multiply the numbers and add the indices.
(3×4) $(2 + 5)$

$4a^7 \times 5a^2 = 20a^9$ Multiply the numbers and add the indices.
(4×5) $(7 + 2)$

$12x^5 \div 3x^2 = 4x^3$ Divide the numbers and subtract the
$(12 \div 3)$ $(5 - 2)$ indices.

$(3a^2)^3 = 3^3 \times a^6 = 27a^6$ Cube the 3 and multiply the indices.
(2×3)

Example 2

Find the value of x in the equation $2^x = 4$

    4 is 2 squared, so $x = 2$

Find the value of x in the equation $2x^3 = 54$

    Dividing both sides by 2 gives $x^3 = 27$

    So $x = \sqrt[3]{27} = 3$

Exercise 2.8B

1. Simplify:

   a) $2a^2 \times 3a^3$ **b)** $4n^3 \times 5n^1$ **c)** $7x^4 \times 2x$ **d)** $8y^5 \times 3y^2$ **e)** $5n^3 \times n^4$ **f)** $6y^2 \times 2$

   g) $3p^3 \times 3p^2$ **h)** $2p \times 5p^5$ **i)** $(2x^2)^3$ **j)** $(3a^2)^3$ **k)** $(4y^3)^2$ **l)** $(5x^4)^2$

Find the value of x.

2. $x^2 = 9$ **3.** $x^5 = 1$ **4.** $x^3 = 27$ **5.** $x^5 = 0$

6. $2^x = 8$ **7.** $3^x = 3$ **8.** $5^x = 25$ **9.** $10^x = 1000$

10. $2^x = \dfrac{1}{2}$ **11.** $4^x = \dfrac{1}{4}$ **12.** $7^x = 1$ **13.** $3x^3 = 24$

14. $10x^3 = 640$ **15.** $2x^3 = 0$ **16.** $10^x = 0.1$ **17.** $5^x = 1$

2.9 Inequalities

Symbols

- There are four inequality symbols.

    $x < 4$ means 'x is **less than** 4'

    $y > 7$ means 'y is **greater than** 7'

    $z \leqslant 10$ means 'z is **less than or equal to** 10'

    $t \geqslant -3$ means 't is **greater than or equal to** -3'

    Also, $h \neq 4$ means 'h is **not equal to** 4'

- If there are two symbols in one statement, look at each part separately.

    For example, if n is an **integer** and $3 < n \leqslant 7$, n has to be greater than 3 but at the same time it has to be less than or equal to 7.

    So, n could be 4, 5, 6 or 7 only.

> **Tip**
>
> An integer is a whole number.

Example

Show on a number line the range of values of x stated.

a) $x > 1$ **b)** $x \leqslant -2$ **c)** $1 \leqslant x < 4$

a) $x > 1$ 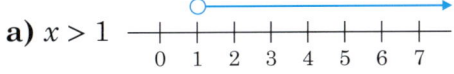   Use an open circle to show that 1 is not included.

b) $x \leqslant -2$ 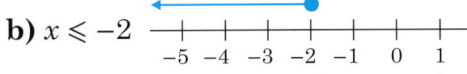   Use a filled-in circle to show that -2 is included.

c) $1 \leqslant x < 4$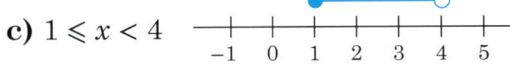

Exercise 2.9A

1. Write each statement with either $<$ or $>$ in the box.

    a) 3 ☐ 7   **b)** 0 ☐ -2 **c)** 3.1 ☐ 3.01

    d) -3 ☐ -5 **e)** 100 m ☐ 1 m   **f)** 1 kg ☐ 1 lb

2. Write the inequality shown. Use x for the variable.

a)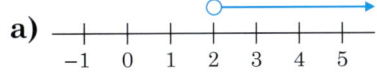

b)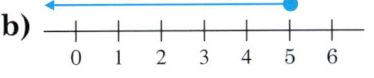

c)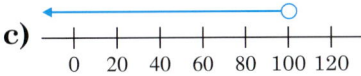

d)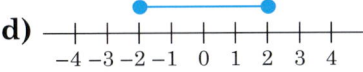

e)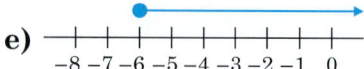

f)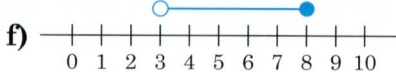

3. Draw number lines to show these inequalities.

a) $x \geqslant 7$ **b)** $x < 2.5$ **c)** $1 < x < 7$ **d)** $0 \leqslant x \leqslant 4$ **e)** $-1 < x \leqslant 5$

4. Write an inequality for each statement.

a) You must be at least 16 to get married.

[Use A for age.]

b) Vitamin J1 is not recommended for people over 70 or for children 3 years or under.

[Use A for age.]

c) To cook a jacket potato the oven temperature should be between 150 °C and 175 °C.

[Use T for temperature.]

d) To ride a rollercoaster, you must be at least 1.5 m tall.

[Use h for height.]

5. Answer 'true' or 'false'.

a) n is an integer and $1 < n \leqslant 4$, so n can be 2, 3 or 4 only.

b) x is an integer and $2 \leqslant x < 5$, so x can be 2, 3 or 4 only.

c) p is an integer and $p \geqslant 10$, so p can be 10, 11, 12, 13, …

6. Write each statement with either $=$ or $\neq$ in the box.

a) 15 cm ☐ 150 mm   **b)** 200 g ☐ 2 kg   **c)** 11 mm ☐ 1.1 cm

d) 350 ml ☐ 3.5 litres   **e)** $\dfrac{2}{100}$ ☐ 0.02   **f)** 0.1×0.1 ☐ 0.1

7. x is an integer. List all of the values of x that satisfy the inequality $-3 \leqslant x < 2$.

8. n is an integer. List all of the values of n that satisfy the inequality $7 < n \leqslant 11$.

9. p is an integer. List all of the values of p that satisfy the inequality $5 \leqslant 2p < 18$.

10. Given that $-4 \leqslant a \leqslant 3$ and $-5 \leqslant b \leqslant 4$, find

a) the largest possible value of a^2

b) the smallest possible value of ab

c) the largest possible value of ab

d) the value of b if $b^2 = 25$.

Revision exercise 2

1. Write down each sequence, find the next two numbers and write down the term-to-term rule.

   a) 2, 9, 16, 23, ...   **b)** 20, 18, 16, 14, ...

   c) −5, −2, 1, 4, ...   **d)** 128, 64, 32, 16, ...

   e) 8, 11, 15, 20, ...

2. Look at the number pattern below.

   $(2 \times 1) - 1 = 2 - 1$

   $(3 \times 3) - 2 = 8 - 1$

   $(4 \times 5) - 3 = 18 - 1$

   $(5 \times 7) - 4 = 32 - 1$

   $(6 \times a) - 5 = b - 1$

   a) What number does the letter a stand for?

   b) What number does the letter b stand for?

   c) Write down the next line in the pattern.

3. $1 + 3 = 2^2$

   $1 + 3 + 5 = 3^2$

   a) $1 + 3 + 5 + 7 = x^2$

   Calculate x.

   b) $1 + 3 + 5 + ... + n = 100$

   Calculate n.

4. Here is a sequence:

   $$1 + 2 + 1 = 2^2$$

   $$1 + 2 + 3 + 2 + 1 = 3^2$$

   $$1 + 2 + 3 + 4 + 3 + 2 + 1 = 4^2$$

   a) Write down the next two lines of the sequence.

   b) Complete the line below:

   $$1 + 2 + 3 + ... + 1 = 9^2$$

5. Here are three diagrams with lines and dots.

   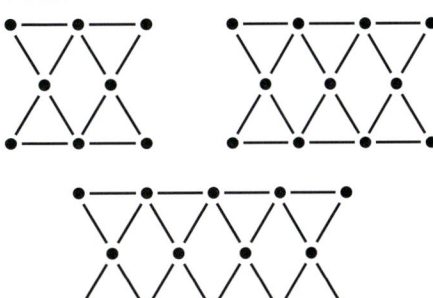

   a) Find a formula connecting the number of lines l and the number of dots d.

   b) How many dots are there in a diagram with 294 lines?

   c) Josh says that a pattern in this sequence will have 401 lines.

   Is Josh correct? Explain your answer.

6.

   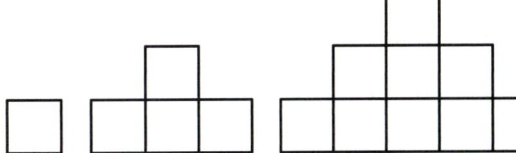

   a) Draw the next diagram in this sequence.

   b) Write down the number of squares in each diagram.

   c) Describe in words the sequence you obtained in part **(b)**.

   d) How many squares will there be in the diagram that has 13 squares on the base?

7. Each diagram in the sequence below consists of a number of dots.

Diagram number	1	2	3
Diagram			

a) Draw diagram number 4, diagram number 5 and diagram number 6.

b) Copy and complete the table below:

Diagram number	Number of dots
1	6
2	10
3	
4	
5	
6	

c) Without drawing the diagrams, state the number of dots in:

i) diagram number 10

ii) diagram number 15.

d) If you write x for the diagram number and n for the number of dots, write down a formula involving x and n.

8. Solve the equations.

a) $x - 6 = 3$ **b)** $x + 9 = 20$

c) $x - 5 = -2$ **d)** $3x + 1 = 22$

9. Solve the equations.

a) $3x - 1 = 20$ **b)** $4x + 3 = 4$

c) $5x - 7 = -3$

10. Nadia said: 'I thought of a number, multiplied it by 6, then added 15. My answer was less than 200'.

a) Write down Nadia's statement in symbols, using x as the starting number.

b) Nadia actually thought of a prime number. What was the largest prime number she could have thought of?

11. Solve the equations for x.

a) $x^3 = 8$ **b)** $3^x = 9$ **c)** $2^x = 16$

12. Simplify:

a) $(x^2)^4$ **b)** $(n^3)^3$ **c)** $4a^2 \times 3a$

Examination-style exercise 2

NON-CALCULATOR

1. Look at the sequence of numbers 9, 13, 17, 21, ...

   a) Write down the next number in the sequence. [1]

   b) Find the 10th number in the sequence. [1]

   c) Write an expression, in terms of n, for the nth number in the sequence. [1]

   d) Babar says that 84 is a term in the sequence.
   Is he correct? Explain your answer. [1]

2. **a)** The first four terms of a sequence are 11, 6, 1, −4.

   i) Write down the next two terms of the sequence. [2]

   ii) State the term-to-term rule for the sequence. [1]

   iii) Write down an expression for the nth term of this sequence. [2]

   b) The first four terms of another sequence are −4, 1, 6, 11.
   Write down an expression for the nth term of this sequence. [2]

   c) Add together the expressions for the nth terms of both sequences.
   Write your answer as simply as possible. [1]

3. Look at this sequence: 3, 8, 15, 24, ...

   a) Write down the next term. [1]

   b) Write down the 10th term. [1]

   c) Write down an expression for the nth term of the sequence. [2]

4. Factorise fully $4xy - 2x$. [1]

5. Expand the brackets and simplify $6x(x - y) + 3x^2$. [2]

6. **a)** Expand and simplify $4(3c - 4d) - 6c$. [2]

   b) Factorise $ab - a^2$. [1]

7. Solve the equations.

   a) $3x - 2 = 16$ [2]

   b) $\dfrac{y + 1}{4} = 3$ [2]

   c) $2(3z - 7) - 3(z - 4) = -7$ [3]

8. Gracie and Edith are each given x dollars.

   a) Edith spends 5 dollars. Write down an expression in terms of x for the
   number of dollars she has now. [1]

   b) Gracie doubles her money by working and then is given another 7 dollars.
   Write down an expression in terms of x for the number of dollars she has now. [1]

c) Gracie now has four times as much money as Edith. Write down an equation in x to show this. [1]

d) Solve the equation to find the value of x. [3]

9. Solve the equation $\dfrac{x}{4} + 7 = 12$. [2]

10. Solve the equation $4 - 5x = 2x + 7$. [2]

11. Write down the value of x when:

a) $3^x = 9$ [1]

b) $2^x = \dfrac{1}{16}$ [1]

12. Simplify:

a) p^0 [1]

b) $(x^3)^4$ [1]

13. Simplify:

a) $\left(\dfrac{1}{r}\right)^0$ [1]

b) $p^3 \times p^6$ [1]

c) $(x^3)^{-4}$ [1]

14. Three of the following five statements are correct.

A $0.06066 \leqslant 0.06606$

B $0.06066 \neq 0.06606$

C $0.06066 = 0.06606$

D $0.06066 < 0.06606$

E $0.06066 > 0.06606$

Write down the letters that correspond to the three correct statements. [2]

15. x is an integer.

Write down all the values of x that satisfy the inequality $-7 < 2x \leqslant 8$. [2]

3 Shape and space 1

Pythagoras (c. 570–490 BCE) was one of the first of the great mathematical names in Greek antiquity. He settled in southern Italy and formed a mysterious brotherhood with his students, who were bound by an oath not to reveal the secrets of numbers and who exercised great influence. They laid the foundations of arithmetic through geometry but failed to resolve the concept of irrational numbers. The work of these and others was brought together by Euclid at Alexandria in a book called 'The Elements', which was still studied in some schools as recently as 1900.

- Understand and use Cartesian coordinates.
- Use and interpret geometric terms and vocabulary.
- Understand and use geometrical drawings and constructions including triangles and nets.
- Understand and draw scale drawings including interpreting three-figure bearings.
- Recognise symmetry in 2D including line symmetry and order of rotational symmetry.
- Use and understand geometrical properties including:
  - angles at a point
  - angles on a straight line
  - vertically opposite angles
  - sum of angles in a triangle
  - sum of angles in a quadrilateral
  - angles in parallel lines (corresponding, alternate, co-interior)
  - angle properties of regular polygons.
- Understand and use geometrical properties of circles including:
  - angle between tangent and radius
  - angle in a semi-circle.

3.1 Accurate drawing

Some questions involving bearings or irregular shapes are made easier to solve by drawing an accurate diagram.

Navigators on ships use scale drawings to work out their position or their course.

To improve the accuracy of your work, follow these guidelines.

- Use a *sharp* HB pencil.
- Don't press too hard.
- If drawing an **acute** angle make sure your angle is less than $90°$.
- If you use a pair of compasses make sure they are fairly stiff so the radius does not change accidently.

Example

Draw the triangle ABC full size and measure the length x.

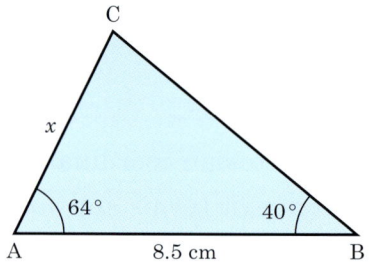

a) Draw a base line *longer than* 8.5 cm.

b) Put the centre of the protractor on A and measure an angle $64°$. Draw line AP.

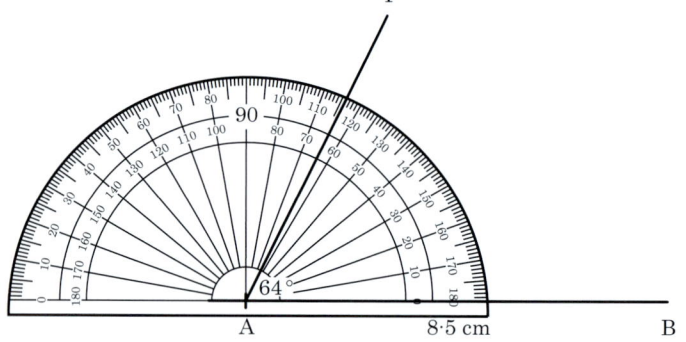

c) Use the same method to draw line BQ at an angle $40°$ to AB.

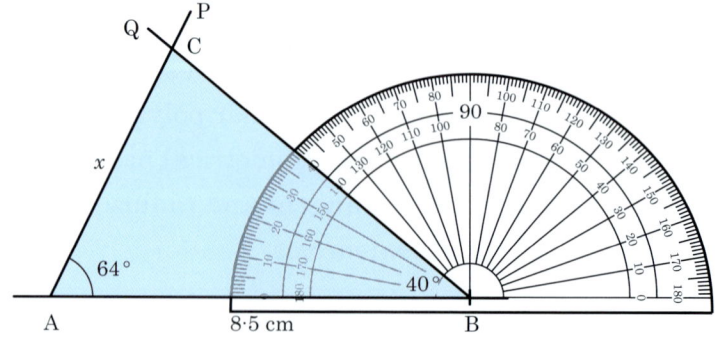

d) The triangle is formed.
Measure $x = 5.6$ cm.

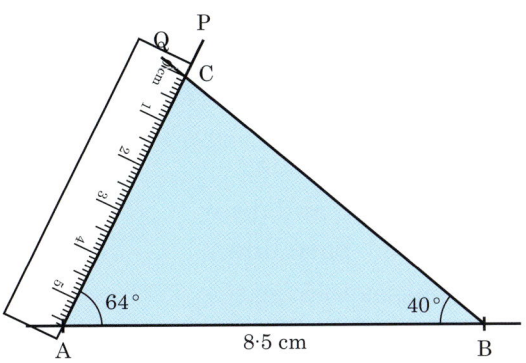

Exercise 3.1A

Use a protractor and ruler to draw full size diagrams and measure
the sides marked with letters.

1.

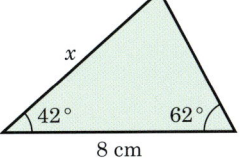

2.

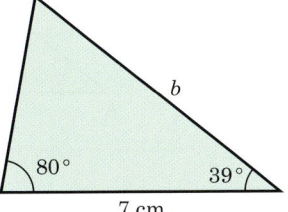

3.

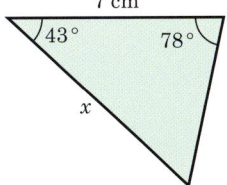

4.

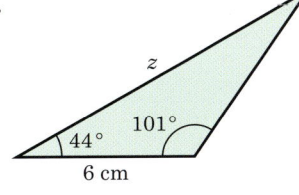

5.

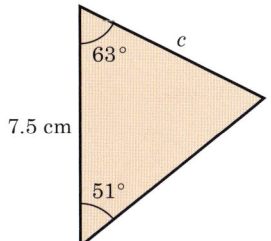

6.

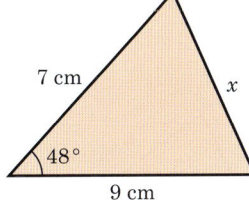

7.

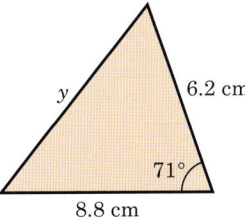

8.

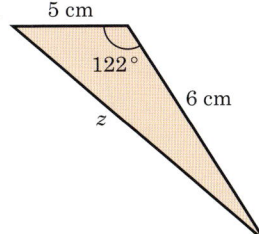

9.

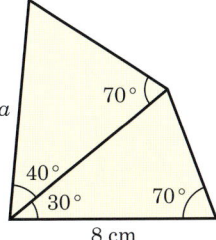

10.

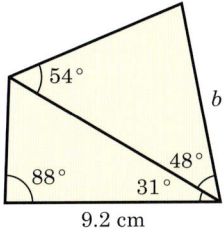

11.

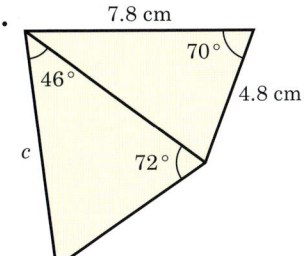

12.

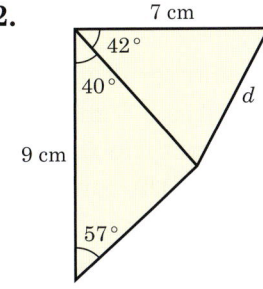

Example

Draw the triangle ABC full size and measure the angle x.

a) Draw a base line AB exactly 6 cm long.

b) Set a pair of compasses to 4 cm and draw an arc centred on A above the base line.

c) Similarly, set your compasses to 5 cm and draw another arc centred on B intersecting the first.

d) Join this crossing point to A and B to form the triangle.

e) Use a protractor to measure the angle marked x.

$x = 56°$.

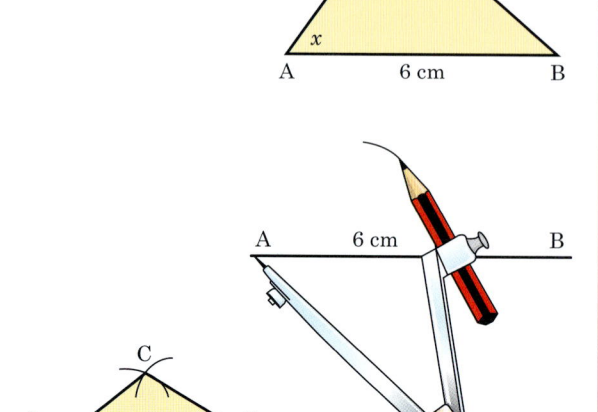

Exercise 3.1B

In Questions **1** to **6**, use a ruler and pair of compasses to make accurate drawings of these triangles. Measure each angle marked with x.

1.
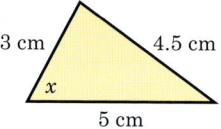
3 cm   4.5 cm
x
5 cm

2.
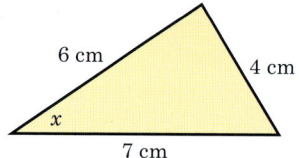
6 cm   4 cm
x
7 cm

3.
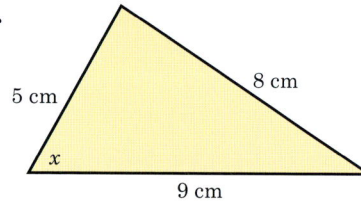
5 cm   8 cm
x
9 cm

4.
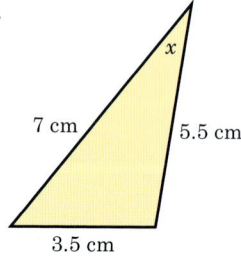
7 cm x 5.5 cm
3.5 cm

5.
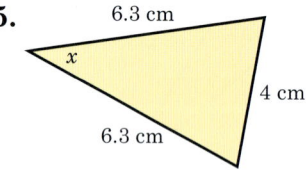
6.3 cm
x
6.3 cm   4 cm

6.
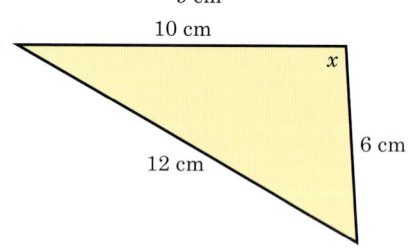
10 cm
x
12 cm   6 cm

7. The shorter diagonal of this rhombus measures 3.5 cm. Each side length is 3 cm. Construct the rhombus and measure the length of the longer diagonal.

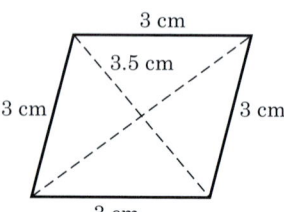
3 cm
3.5 cm
3 cm   3 cm
3 cm

Nets

If the cube here was made of cardboard, and you cut along some of the edges and laid it out flat, you would have the **net** of the cube.

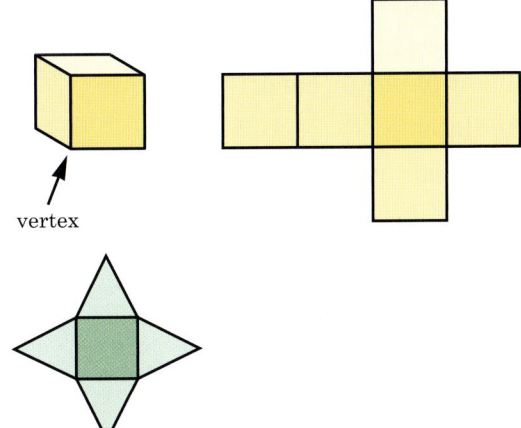

vertex

Here is the net for a square-based pyramid.

Exercise 3.1C

1. Which of the nets below can be used to make a cube?

a)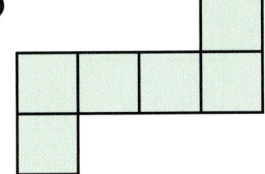

b)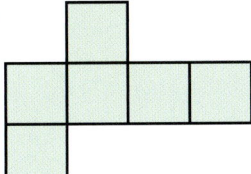

c)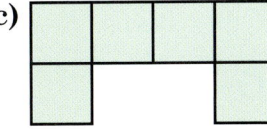

d)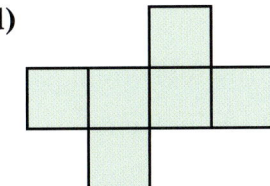

2. The numbers on opposite faces of a dice add up to 7. Draw one of the possible nets for a cube from Question **1** and show the number of dots on each face.

3. Here we have started to draw the net of a cuboid (a closed rectangular box) measuring 4 cm × 3 cm × 1 cm. Copy and then complete the net.

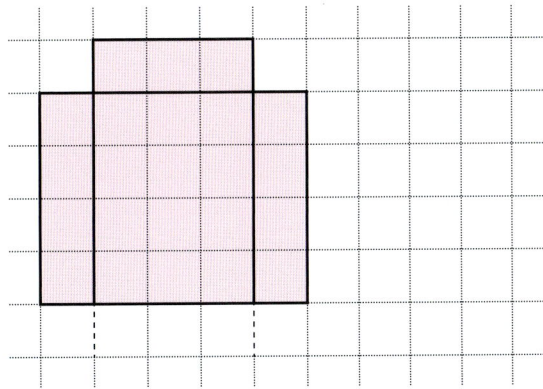

4. This diagram shows the net of a solid.

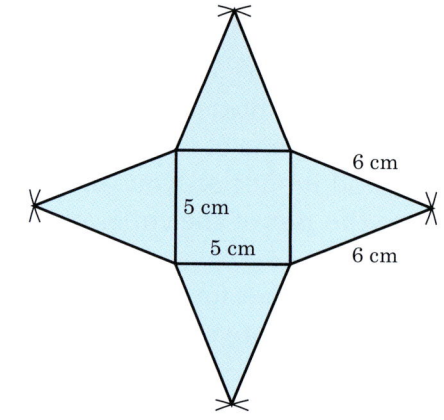

   a) Use a ruler and pair of compasses to draw the net accurately on paper or card.

   b) Draw on some flaps.

   c) Cut out the net, fold and glue it to make the solid.

   d) What is the name of the solid?

5. A cube can be made from three identical pyramids.

Make three 3D-shapes from the net shown and fit them together to make a cube. All lengths are in cm.

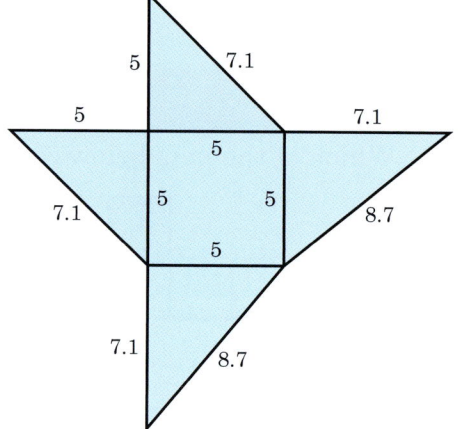

6. Sketch a possible net for each of the following:

   a) a cuboid measuring 5 cm by 2 cm by 8 cm

   b) a cuboid with sides 3 cm, 4 cm and 5 cm.

7. Construct accurately the net of this triangular prism.

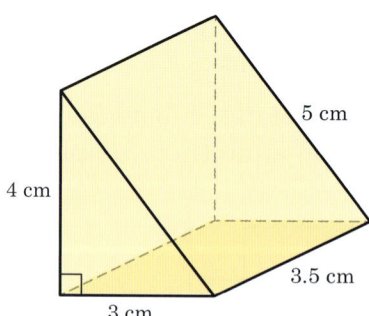

3.2 Angle facts

- The angles at a point add up to $360°$.

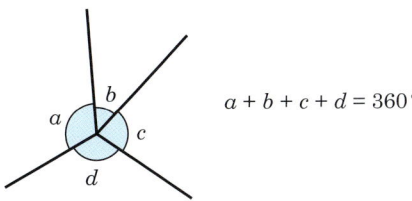

$$a + b + c + d = 360°$$

- The angles on a straight line add up to $180°$.

$$a + b = 180°$$

- An angle of $90°$ is called a **right angle**.

$90°$

- An **acute** angle is less than $90°$.

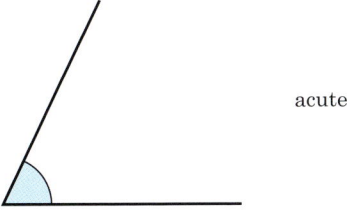

acute

- An **obtuse** angle is between $90°$ and $180°$.

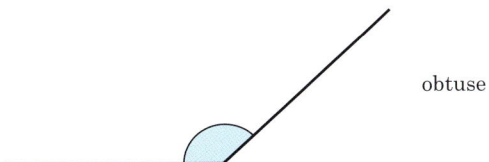

obtuse

- A **reflex** angle is greater than $180°$.

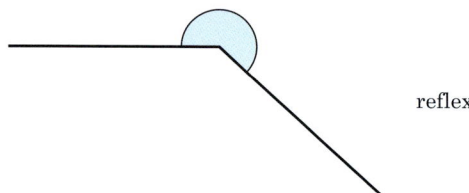

reflex

- **Vertically opposite** angles are equal.

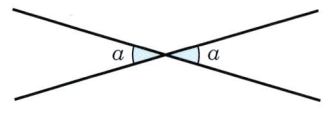

Example

Find the unknown angles.

a)

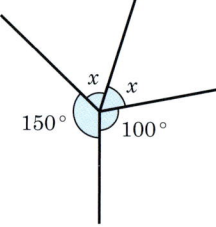

$x + x + 150° + 100° = 360°$

Angles at a point add up to $360°$

$2x = 360° − 250° = 110°$

$x = 55°$

b)

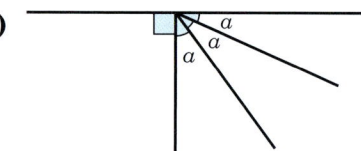

$3a + 90° = 180°$

Angles on a straight line add up to $180°$

$3a = 90°$

$a = 30°$

c)

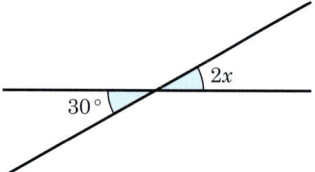

Vertically opposite angles are equal

$2x = 30°$

$x = 15°$

Exercise 3.2A

Find the angles marked with letters. The lines AB and CD are straight.

1.
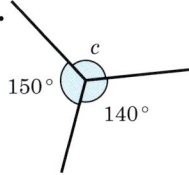

2.
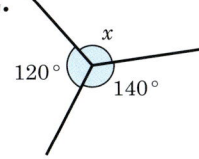

3.
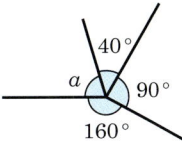

4.
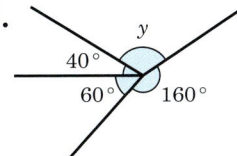

5.
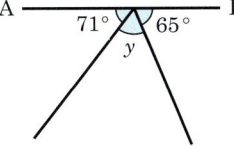

6.
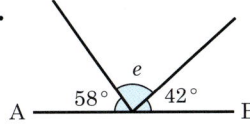

7.
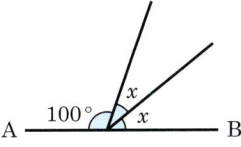

8.
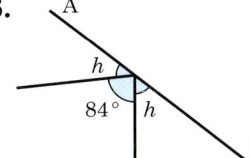

9.
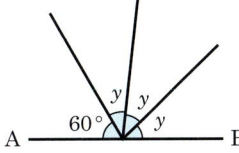

10.
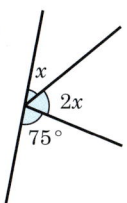

11.
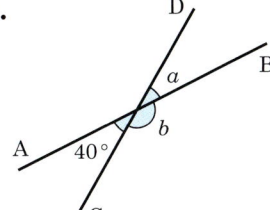

12.
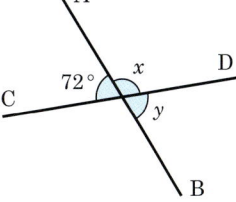

Triangles

The angles in a triangle add up to $180°$.

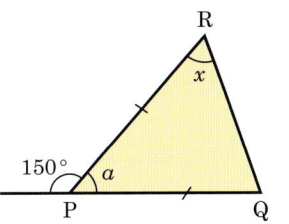

Example

Find the unknown angles.

$a = 180° - 150° = 30°$ (angles on a straight line add to $180°$)

The triangle is isosceles (has two equal sides and two equal angles) so angle RQP = x (angles at the base of an isosceles triangle are equal)

Therefore $2x + 30° = 180°$

Angles in a triangle add to $180°$

$2x = 150°$

$x = 75°$

Tip

Angle RQP means the angle between line RQ and line QP at Q.

Exercise 3.2B

Find the angles marked with letters. For the more difficult questions it is helpful to copy the diagram and write in the values of the angles as you find them.

1.

2.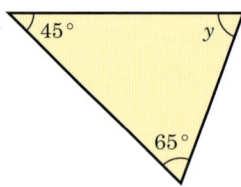

3.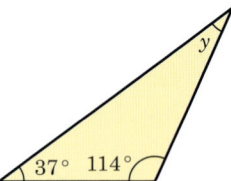

4.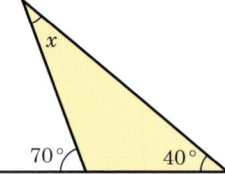

5.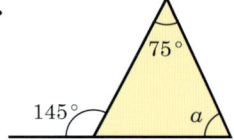

6.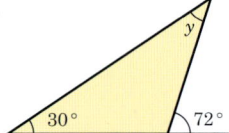

7.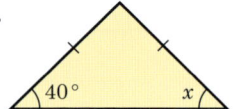

8.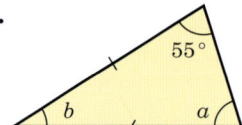

9.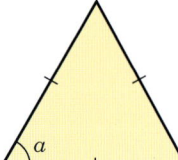

10.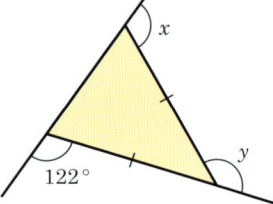

11.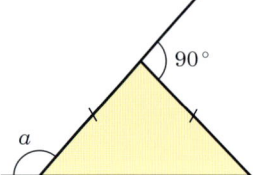

12.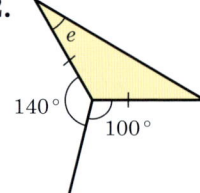

13.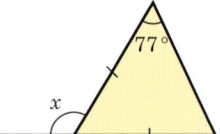

14.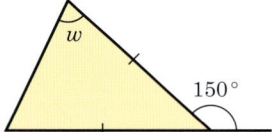

15.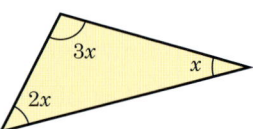

16.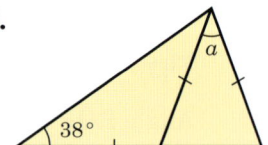

Parallel lines

When a line cuts a pair of parallel lines all the acute angles, a, are equal and all the obtuse angles, b, are equal.

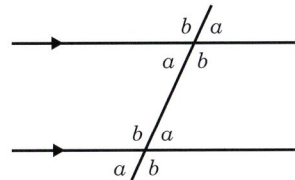

Corresponding angles are equal.

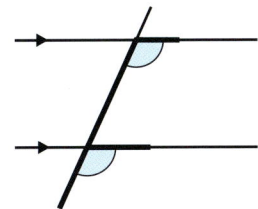

Alternate angles are equal.

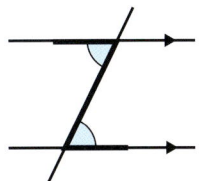

Co-interior (supplementary) angles sum to $180°$.

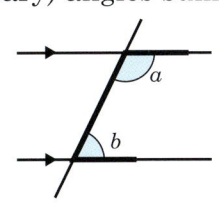

$$a + b = 180°$$

Example

Find the sizes of the unknown angles. Give geometrical reasons for your answers.

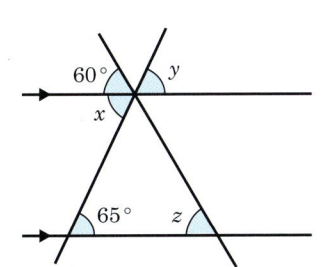

Angle $x = 65°$, alternate angles

Angle $y = 65°$, vertically opposite angles

Angle $z = 60°$, corresponding angles

Exercise 3.2C

Find the angles marked with letters. Give geometrical reasons for your answers.

1.
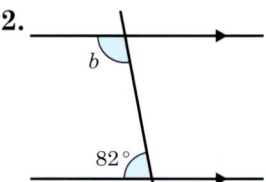

2.
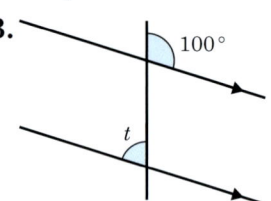

3.
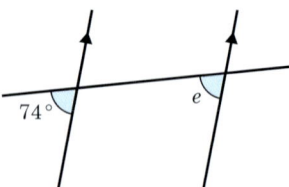

4.
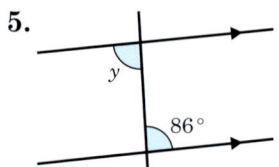

5.
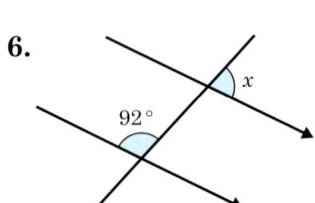

6.
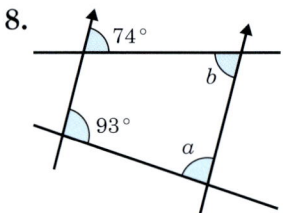

7.
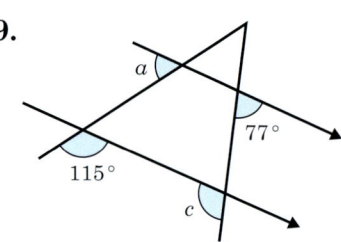

8.
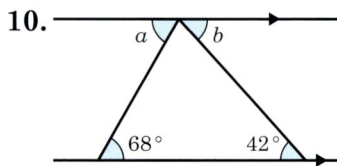

9.
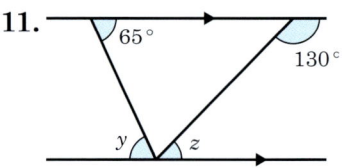

10.
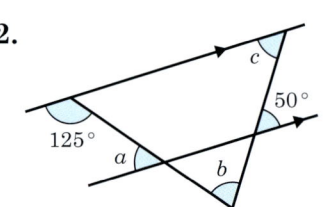

11.

12.

Quadrilaterals and regular polygons

A flat shape with three or more straight sides is called a **polygon**.

Two straight lines meet at a **vertex**.

A polygon with four sides is called a **quadrilateral**.

The sum of the angles in a quadrilateral is $360°$.

Proof: the quadrilateral PQRS has been split into two triangles.

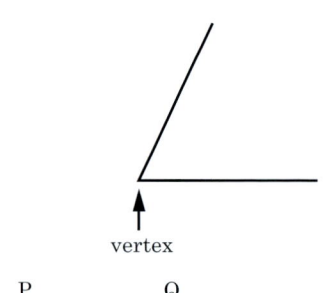

We know that $\qquad a + e + f = 180°$

and that $\qquad b + c + d = 180°$

Therefore $a + b + c + d + e + f = 360°$

But the angles of the quadrilateral are $(a + b)$, c, $(d + e)$ and f.

So, the sum of the angles in a quadrilateral is $360°$.

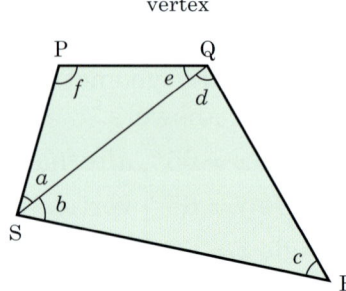

A **regular** polygon has equal sides and angles.

For example, a square is a regular quadrilateral.

Example

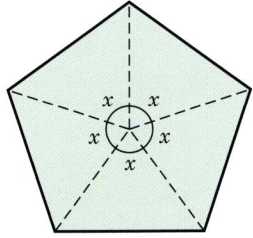

Regular pentagon:

$x + x + x + x + x = 360\degree$

Therefore angle at centre, $x = 72\degree$

Polygons	
Name	**Number of sides**
Quadrilateral	4
Pentagon	5
Hexagon	6
Heptagon	7
Octagon	8
Nonagon	9
Decagon	10

Exercise 3.2D

Find the angles marked with letters. Give geometrical reasons for your answers.

1.

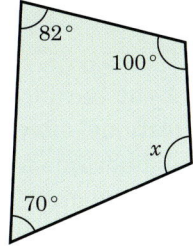

2.

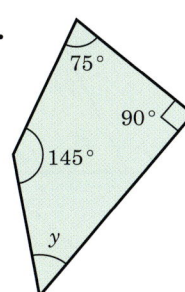

3.

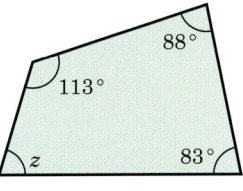

4.

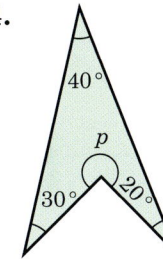

5.

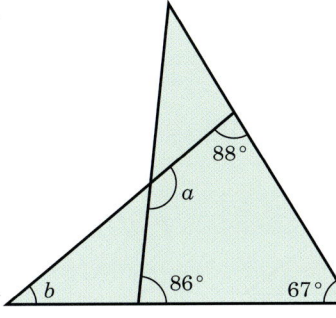

6.

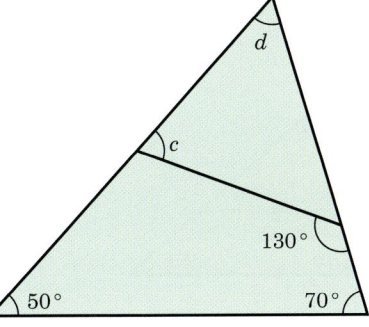

7.

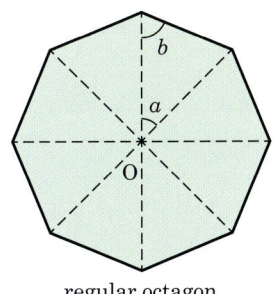

regular octagon
O is the centre

8. Copy and complete this table. The pentagon one has been done for you (as it is in the example before this exercise).

Polygons		
Name	**Number of sides**	**Angle at centre**
Quadrilateral	4	
Pentagon	5	72°
Hexagon	6	
Heptagon	7	
Octagon	8	
Nonagon	9	
Decagon	10	

Mixed questions

The next exercise contains questions which summarise the work of the last four exercises.

Exercise 3.2E

Find the angles marked with letters. Give reasons for your answers.

1.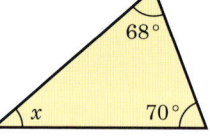

2.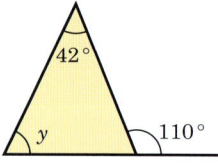

3.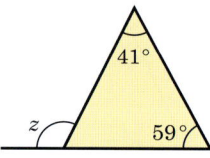

Tip

For the harder questions, copy the diagrams and fill in the angles as you find them.

4.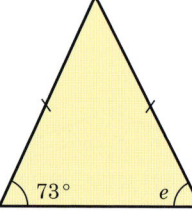

5.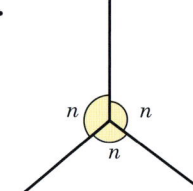

6.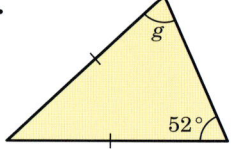

7.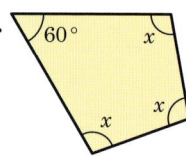

8.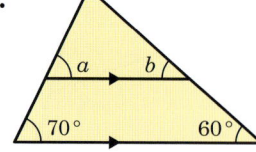

9.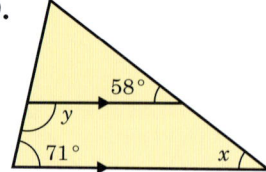

10.

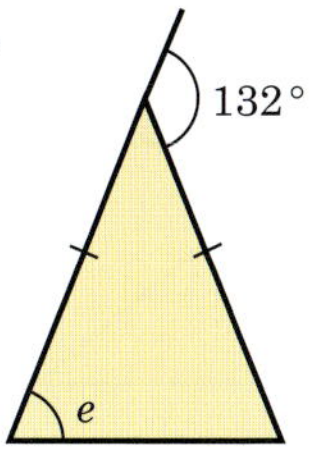

132°

e

11.

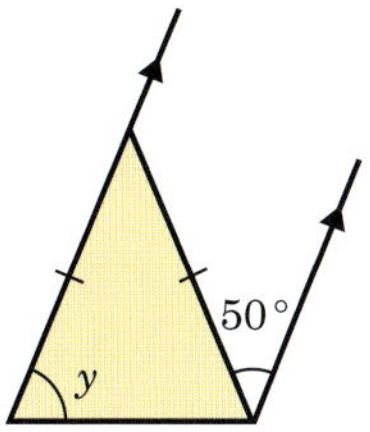

50°

y

12.

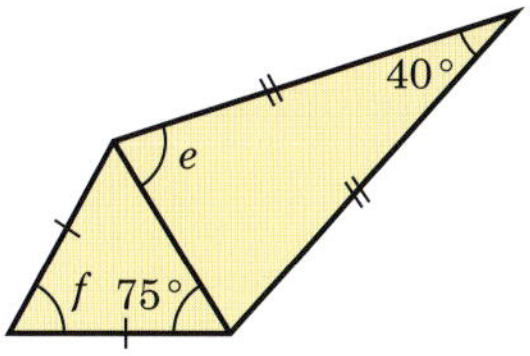

40°

e

f 75°

13.

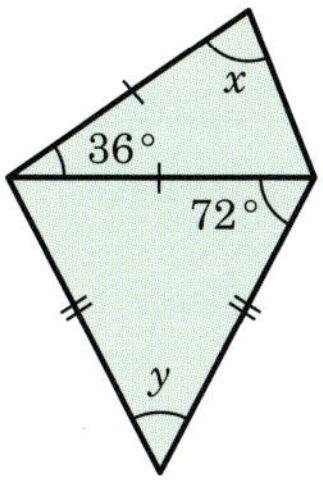

x

36°

72°

y

14.

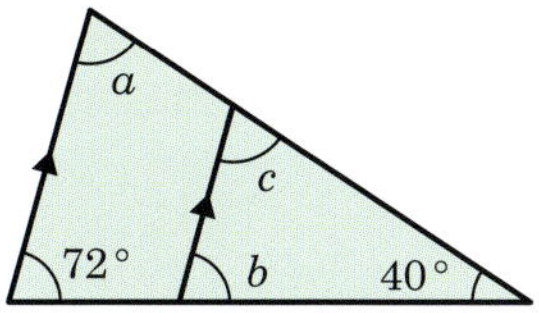

a

c

72°

b

40°

15.

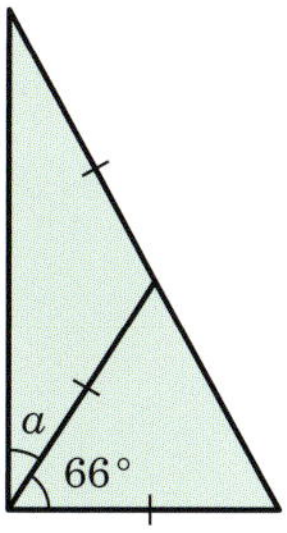

a

66°

16.

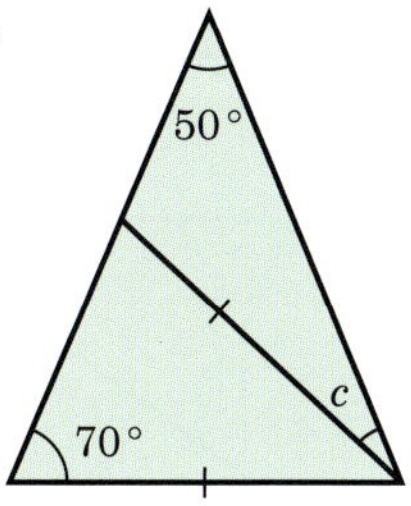

50°

c

70°

17.

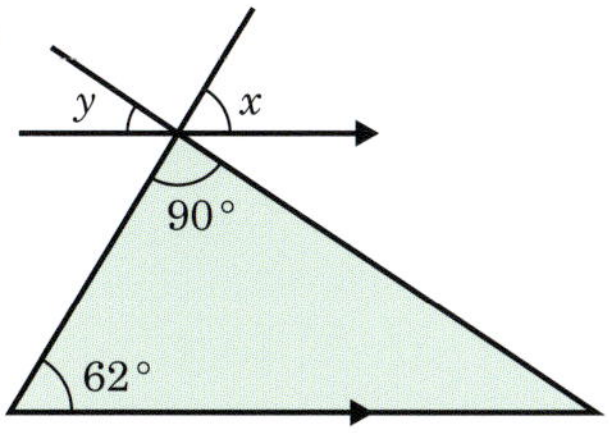

y   x

90°

62°

18.

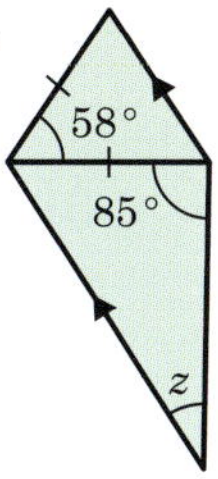

58°

85°

z

19.

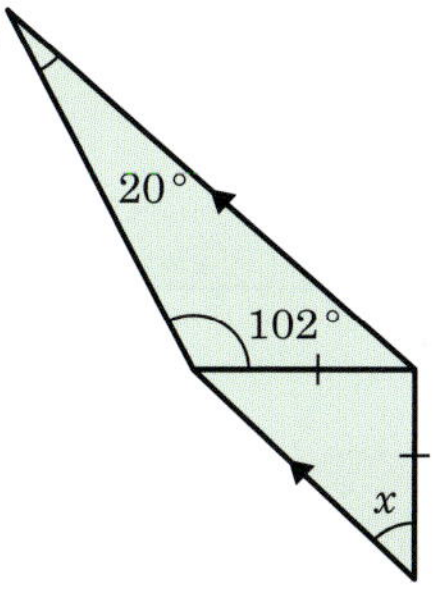

20°

102°

x

20.

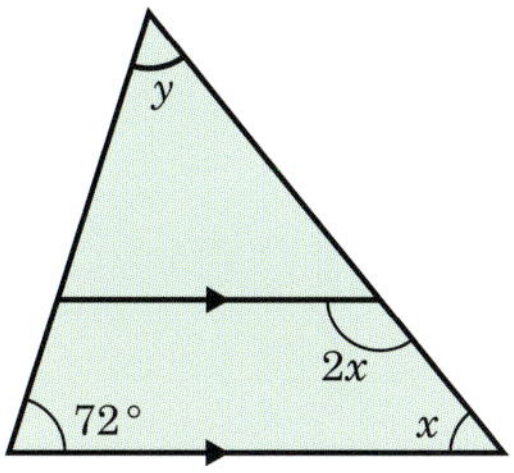

y

2x

72°   x

21.

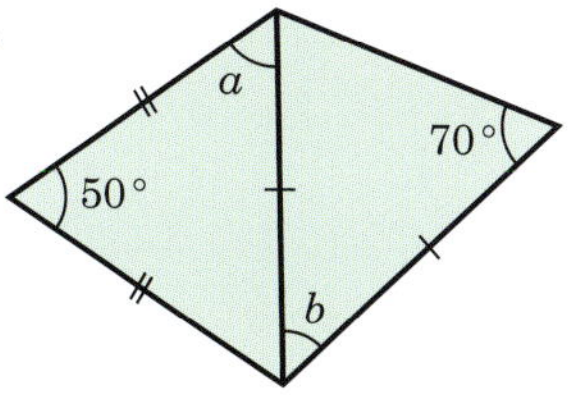

a

50°

70°

b

22.

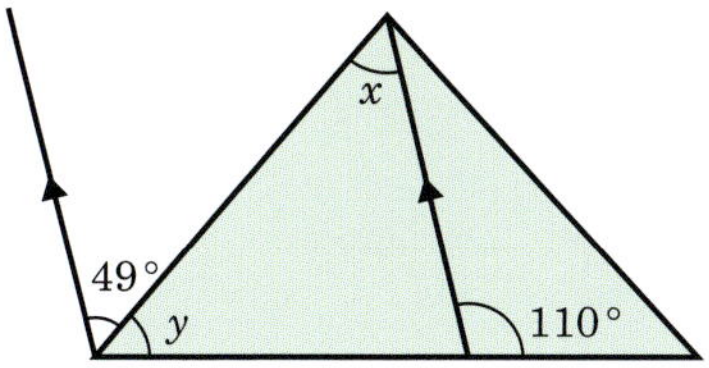

49°

x

y   110°

23.

b

100°

a

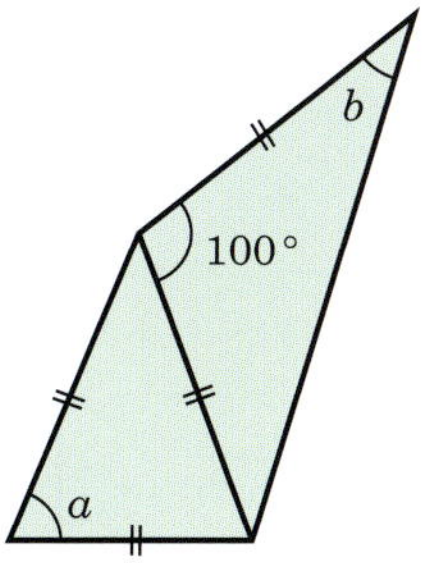

Questions **24** to **27** are more difficult.

24. The diagram shows two equal squares joined
to a triangle.

Find the angle x.

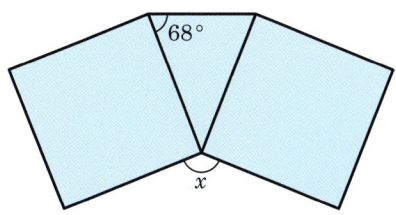

25. Find the angle a between the diagonals of the parallelogram.

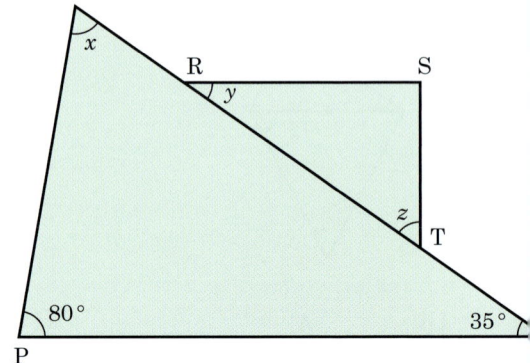

26. The diagram shows the cross-section of a roof. PQ and RS are horizontal and ST is vertical. Work out angles x, y and z.

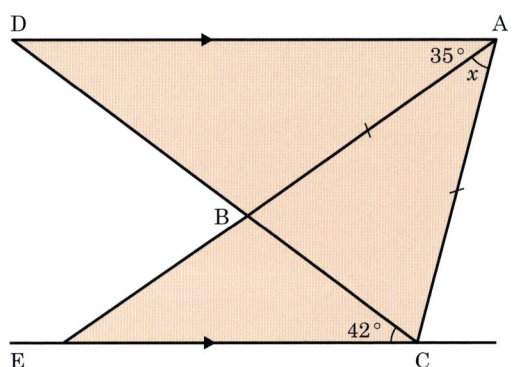

27. Given that AB = AC and DA is parallel to EC, find x.

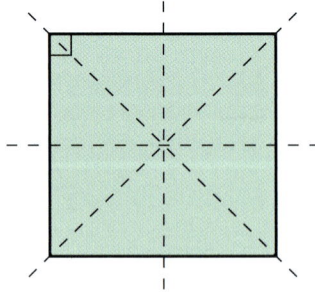

3.3 Quadrilaterals and other polygons

Properties of quadrilaterals

Square

Four equal sides
All angles 90°
Four lines of symmetry
Rotational symmetry
of order 4

Rectangle (not square)

Two pairs of equal and parallel sides

All angles 90°

Two lines of symmetry

Rotational symmetry of order 2

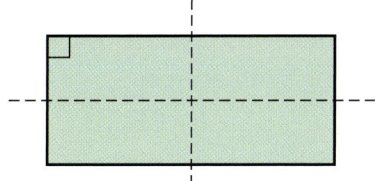

Rhombus

Four equal sides; opposite sides parallel

Diagonals bisect at right angles
(i.e. are perpendicular)

Diagonals bisect angles of rhombus

Two lines of symmetry

Rotational symmetry of order 2

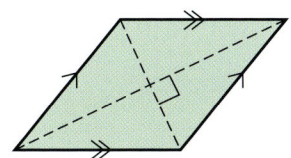

Parallelogram

Two pairs of equal and parallel sides

Opposite angles equal

No lines of symmetry (in general)

Rotational symmetry of order 2

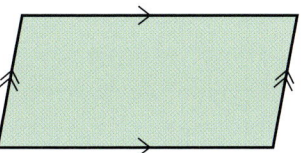

Trapezium

One pair of parallel sides

No rotational symmetry

Kite

AB = AD, CB = CD

Diagonals meet at 90° (i.e. are perpendicular)

One line of symmetry

No rotational symmetry

For all quadrilaterals the sum of the interior angles is 360°.

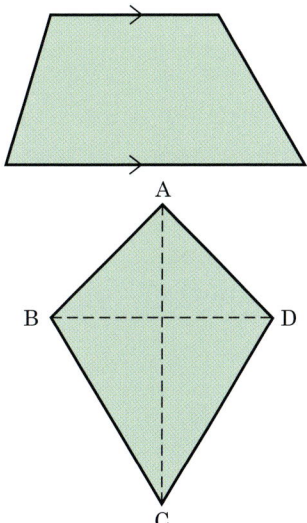

Exercise 3.3A

1. Name each of the following shapes:

   a) ABEH   **b)** EFGH   **c)** CDFE

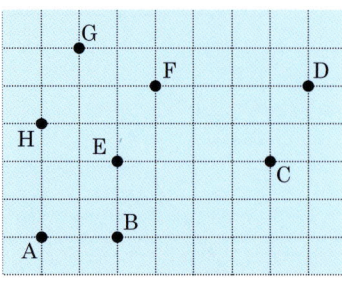

2. **a)** Write down the coordinates of point D if ABCD is a kite.

   b) Write down the coordinates of point E if ABCE is a parallelogram.

   c) Write down the coordinates of point G if BCGF is an arrowhead.

   [There is more than one possible answer to part **(c)**.]

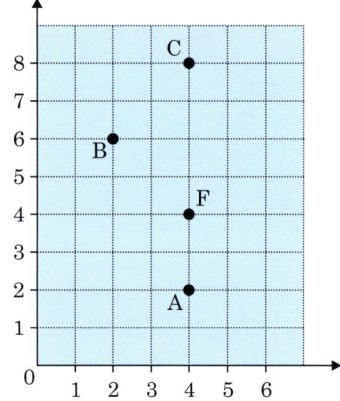

3. Copy the table and fill all the boxes with 'Yes', 'No' or a number.

	How many lines of symmetry?	How many pairs of opposite sides are parallel?	Diagonals always equal?	Diagonals are perpendicular?
Square				
Rectangle				
Kite				
Rhombus				
Parallelogram				
Arrowhead				

4. Find the value of angle x.

a)

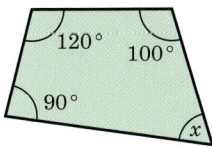

b)

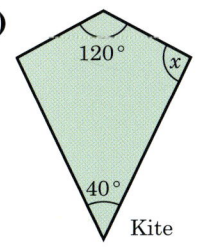

Kite

c)

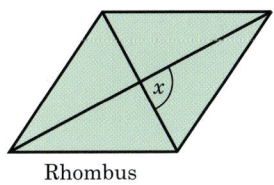

Rhombus

d)

Parallelogram

e)

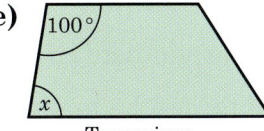

Trapezium

f)

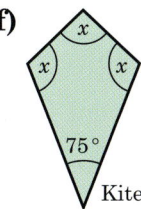

Kite

5. The diagram shows three vertices (corners) of a parallelogram. Copy the diagram and mark with crosses the *three* possible positions of the fourth vertex.

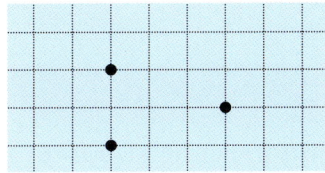

6. Line AC is one *diagonal* of a rhombus ABCD. Draw *two* possible rhombuses ABCD.

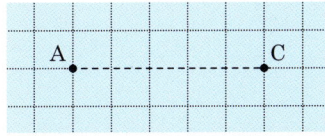

7. Suppose you cut along the diagonal of a rectangle to make two congruent triangles. Join the diagonals together in a different way. What shape is formed?

8. Suppose you have two identical isosceles triangles. Draw all the shapes that you can make by putting the equal sides together without overlapping. Name the shapes formed.

Tip

Congruent means exactly the same shape and size.

9. An equilateral triangle has vertices P, Q, R.

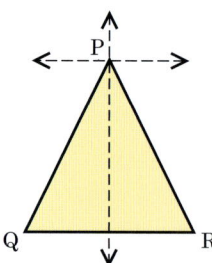

a) Suppose the vertex P moves perpendicular to QR.

What different types of triangle can be made?

Can you make: a right-angled triangle

an obtuse-angled triangle

a scalene triangle?

b) If the vertex P moves **parallel** to QR, what different types of triangle can be made?

Tip

A **scalene** triangle has no equal sides or angles.

Exercise 3.3B

1. ABCD is a rhombus whose diagonals intersect at M. Find the coordinates of C and D.

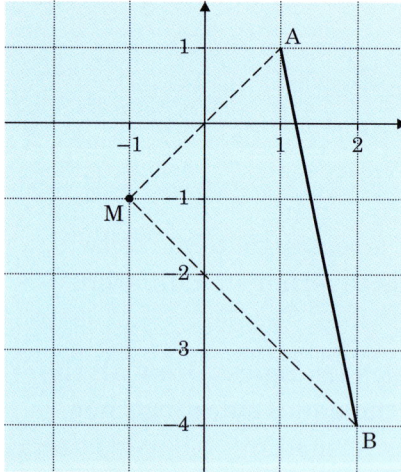

In Questions **2** to **12**, begin by drawing a diagram. Remember to put the letters around the shape in alphabetical order.

2. In a rectangle KLMN, angle LNM = 34°.

Calculate:

a) angle KLN   **b)** angle KML

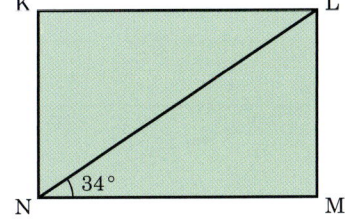

3. In a trapezium ABCD, angle ABD = 35°, angle BAD = 110° and AB is parallel to DC. Calculate:

a) angle ADB   **b)** angle BDC

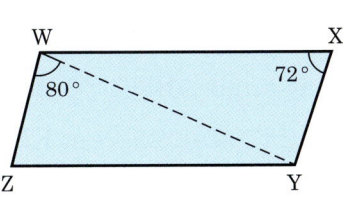

4. In a parallelogram WXYZ, angle WXY = 72°, angle ZWY = 80°. Calculate:

a) angle WZY   **b)** angle XWZ   **c)** angle WYX

5. In a kite ABCD, AB = AD, BC = CD, angle CAD = 40° and angle CBD = 60°. Calculate:

a) angle BAC   **b)** angle BCA   **c)** angle ADC

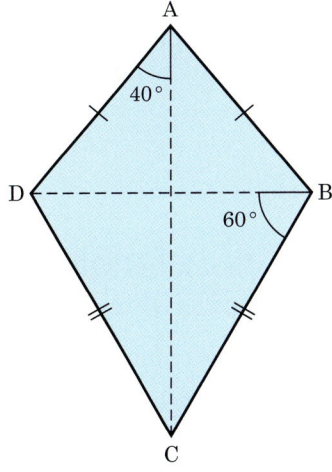

6. In a rhombus ABCD, angle ABC = 64°. Calculate:

a) angle BCD   **b)** angle ADB   **c)** angle BAC

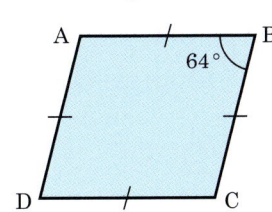

7. In a rectangle WXYZ, M is the midpoint of WX and angle ZMY = 70°.

Calculate:

a) angle MZY   **b)** angle YMX

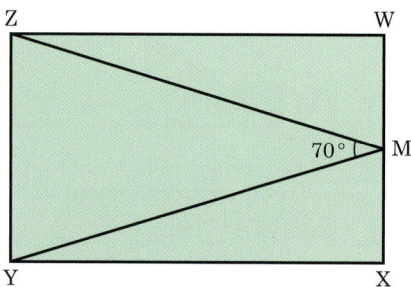

8. In a trapezium ABCD, AB is parallel to DC, AB = AD, BD = DC and angle BAD = 128°. Find:

a) angle ABD   **b)** angle BDC   **c)** angle BCD

9. In a parallelogram KLMN, KL = KM and angle KML = 64°.

Find:

a) angle MKL   **b)** angle KNM   **c)** angle LMN

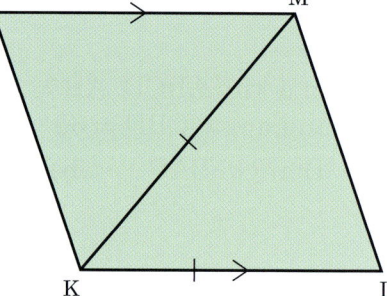

10. In a kite PQRS, PQ = PS and RQ = RS, angle QRS = 40° and angle QPS = 100°. Find angle PQR.

11. In a rhombus PQRS, angle RPQ = 54°. Find:

a) angle PRQ   **b)** angle PSR   **c)** angle RQS

12. In a kite PQRS, angle RPS is twice the angle PRS, PQ = QS = PS and QR = RS. Find:

a) angle QPS   **b)** angle PRS   **c)** angle QSR

Polygons

Remember that a polygon is a flat shape with three or more straight sides.

The table on page 85 shows the names of some common polygons.

A straight line drawn inside a polygon that joins two vertices that are not next to each other is called a **diagonal**.

> **Tip**
>
> Vertices is the plural of 'vertex'.

Exercise 3.3C

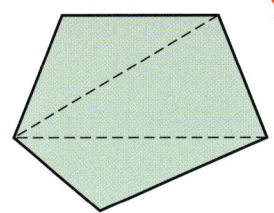

1. The diagram shows a pentagon in which some of the diagonals have been drawn. Draw your own diagram of a pentagon and find how many diagonals it has altogether.

2. Draw a hexagon and show that it has nine diagonals.

3. How many lines of symmetry has a regular pentagon?

4. How many lines of symmetry has a regular hexagon?

5. What is the name for a regular polygon with:
   a) four sides
   b) three sides?

6. This quadrilateral has one pair of parallel sides.
   Draw three **pentagons** with:
   a) one pair of parallel sides
   b) two pairs of parallel sides
   c) three right angles.

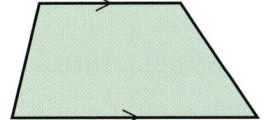

7. Sketch a diagram of a regular hexagon. Are all the diagonals the same length?

3.4 Angles in polygons and circles

Exterior angles of a polygon

The **exterior angle** of a polygon is the angle between a produced side and the adjacent side of the polygon. The word 'produced' in this context means 'extended'.

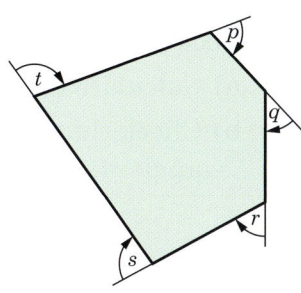

If you put all the exterior angles together, then you can see that the sum of the angles is 360°. This is true for any polygon.

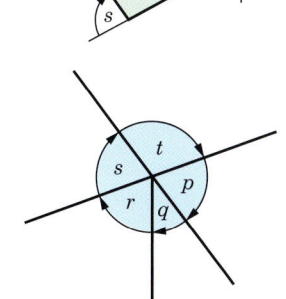

- A regular polygon has sides of equal length and all its **interior angles** are equal.
- The sum of the exterior angles of any polygon = 360°.
- All exterior angles of a regular polygon are equal.
- For a regular polygon with n sides, each exterior angle = $\dfrac{360°}{n}$.

Example

The diagram shows a regular octagon (8 sides).

a) Calculate the size of each exterior angle (marked e).

b) Calculate the size of each interior angle (marked i).

c) Calculate the sum of the interior angles of the regular octagon.

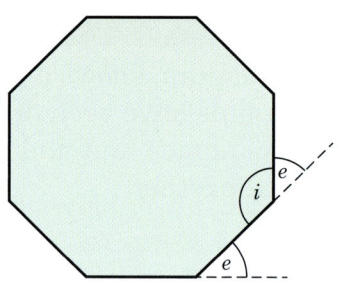

a) There are 8 exterior angles and the sum of these angles is $360°$.

Therefore angle $e = \dfrac{360°}{8} = 45°$

b) $e + i = 180°$ (angles on a straight line sum to $180°$)

∴ $i = 135°$

c) There are 8 interior angles of $135°$ so the sum of the interior angles is $8 \times 135° = 1080°$

Note: this can be done without working out the interior angle first.

Split the regular octagon into 6 triangles.

The angles in each triangle add up to $180°$ and $6 \times 180° = 1080°$

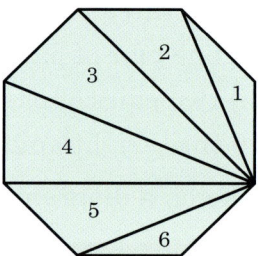

Exercise 3.4A

1. Look at the polygon shown.

   a) Calculate the size of each exterior angle.

   b) Check that the total of the exterior angles is $360°$.

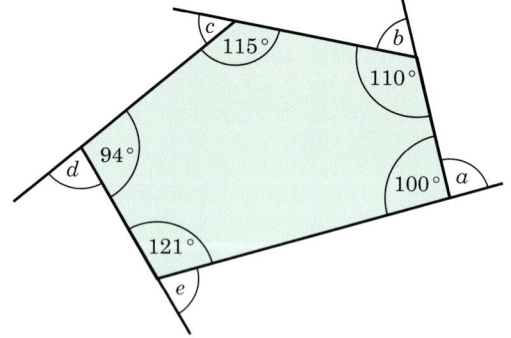

2. The diagram shows a regular decagon.

　a) Calculate the size of angle a.

　b) Calculate the size of each interior angle of a regular decagon.

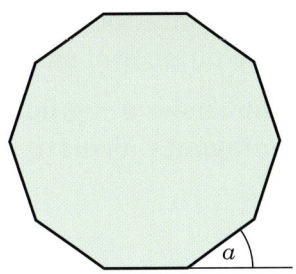

3. Find:

　a) the size of each exterior angle

　b) the size of each interior angle of a regular polygon with:

　　i) 9 sides　　**ii)** 18 sides　　**iii)** 45 sides　　**iv)** 60 sides.

4. Find the sizes of the angles marked with letters.

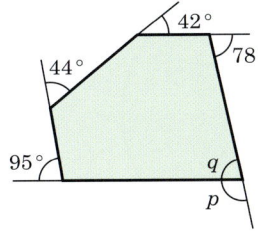

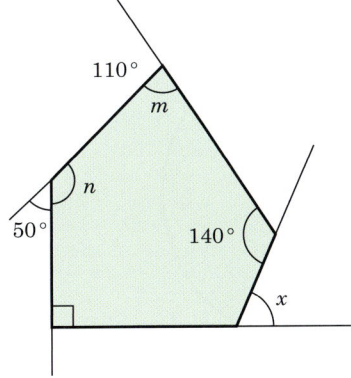

5. Each exterior angle of a regular polygon is $15°$.
How many sides has the polygon?

6. Each interior angle of a regular polygon is $140°$.
How many sides has the polygon?

7. Each exterior angle of a regular polygon is $18°$.
How many sides has the polygon?

8. The sides of a regular polygon subtend angles
of $18°$ at the centre of the polygon.
How many sides has the polygon?

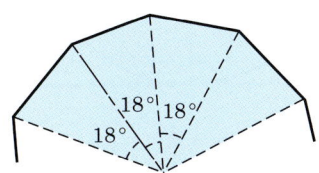

9. Calculate the sum of the interior angles of:

　a) a regular hexagon

　b) a regular decagon.

10. The sum of the interior angles of a regular polygon is 2160°.
Work out the number of sides of the polygon.

11. The diagram shows a regular hexagon joined to a
regular pentagon. Calculate the size of angle x.

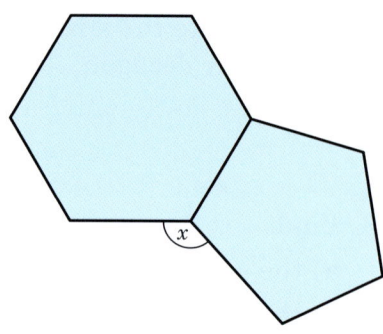

Parts of a circle

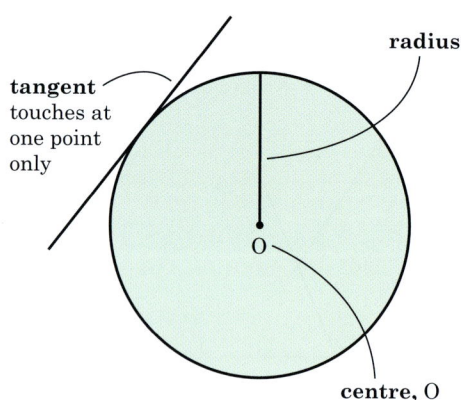

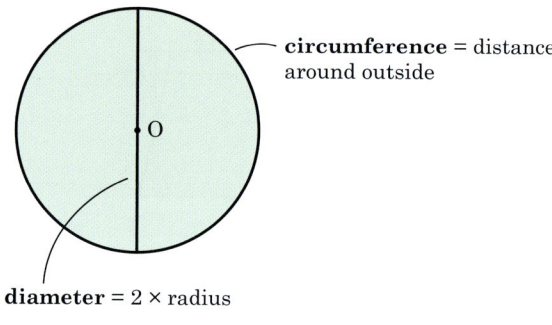

Angles in circles

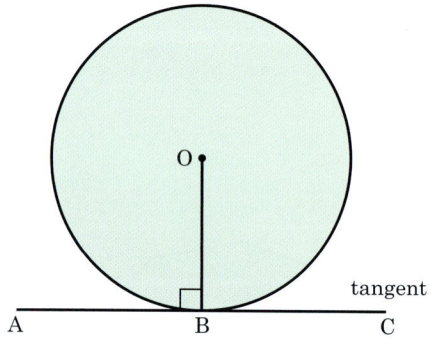

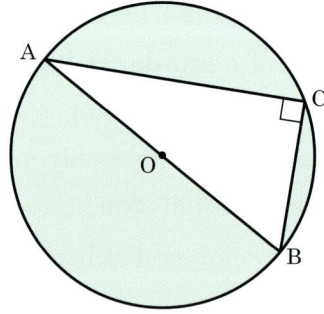

The **tangent** ABC touches the circle at B.

OB is a radius of the circle.

Angle OBA = 90°.

AB is a diameter.

The angle at the circumference,
angle ACB, is 90°.

Example

Calculate the value of each angle marked x in these circles with centre O.

In the second diagram the two straight lines are tangents to the circle.

a)

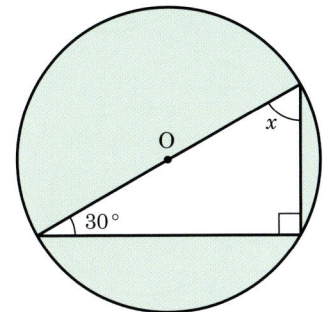

b)

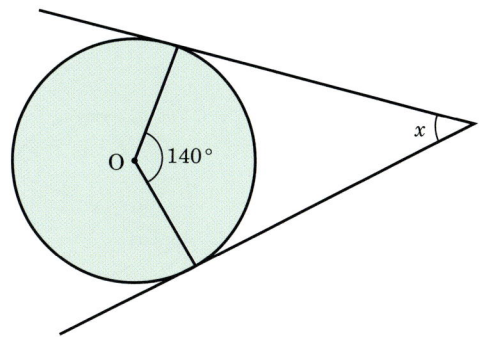

a) The triangle is a right-angled triangle because the angle in a semi-circle is $90°$.

$x = 180° - 90° - 30°$ (angles in a triangle add up to $180°$)

$x = 60°$

b) The two unmarked angles in the quadrilateral containing $140°$ and x are both right angles because the angle between a tangent and a radius is $90°$.

$x = 360° - 90° - 90° - 140°$ (angles in a quadrilateral add up to $360°$)

$x = 40°$

Exercise 3.4B

1. a) Draw a circle with radius 5 cm and draw any diameter AB.

b) Draw triangles ABC, ABD, ABE and measure the angles at the circumference. What do you notice?

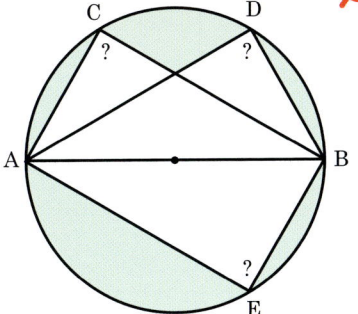

In Questions **2** to **13** find the angles marked with letters. Point O is the centre of the circle. Give geometrical reasons for your answers.

2.

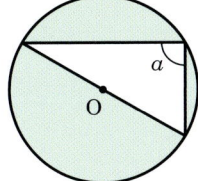

3.

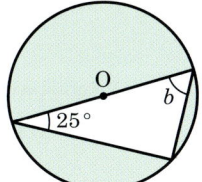

4.

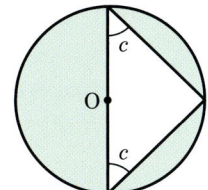

5.

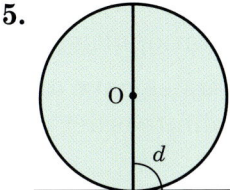

6.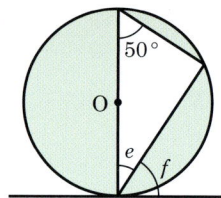

7.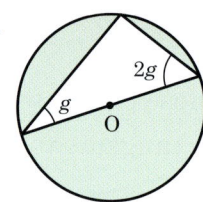

8.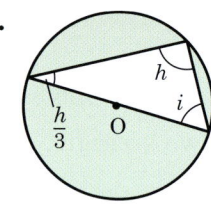

9.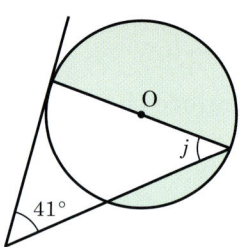

10.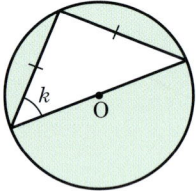

11.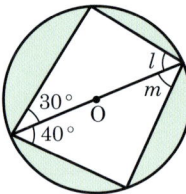

12.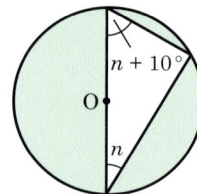

13.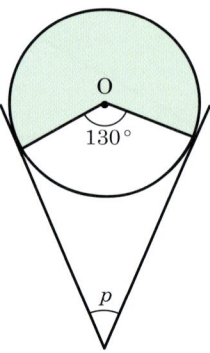

3.5 Symmetry

a) Line symmetry

The letter M has one line of symmetry, shown dotted.

b) Rotational symmetry

The shape may be turned about O into three identical positions in one complete rotation. It has rotational symmetry of order 3.

Note:

- Isosceles triangles have one line of symmetry but no rotational symmetry.
- Equilateral triangles have three lines of symmetry and rotational symmetry of order 3.
- Circles have an infinite number of lines of symmetry and an infinite order of rotational symmetry.

Exercise 3.5A

For each shape state:

a) the number of lines of symmetry

b) the order of rotational symmetry.

1. 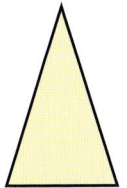   2. 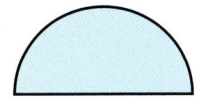   3.    4.

5.    6.    7. 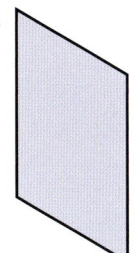   8.

9. 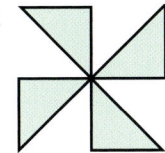   10.    11.    12.

13. Make three copies of this grid.

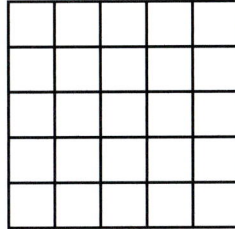

Shade in some of the squares in each grid so that the diagram has

a) 1 line of symmetry but no rotational symmetry

b) rotational symmetry but no lines of symmetry

c) 2 lines of symmetry and rotational symmetry of order 4.

Exercise 3.5B

In Questions **1** to **8**, the dashed lines are axes of symmetry. In each question, only *part of the shape* is given. Copy what is given onto squared paper and then carefully complete the shape.

1.
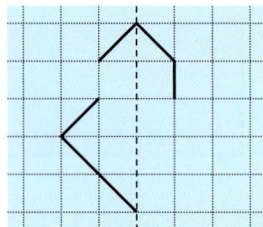

2.
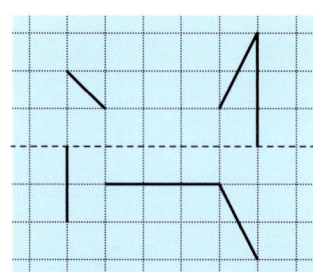

3.
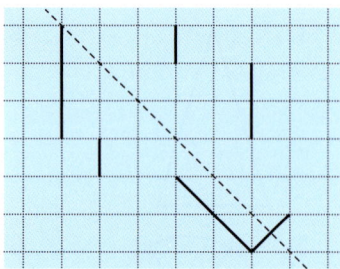

4.
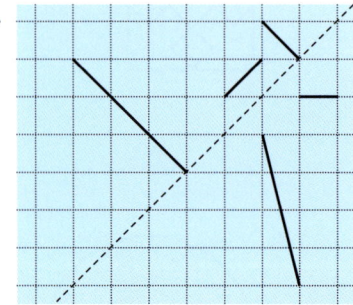

5.
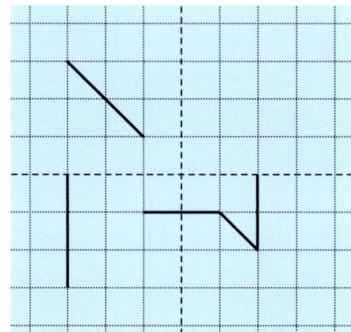

6.
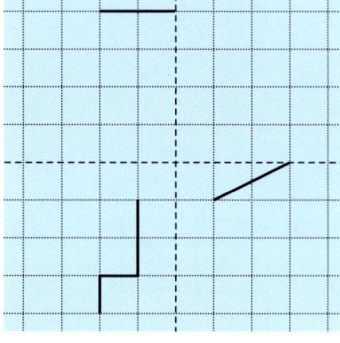

7.
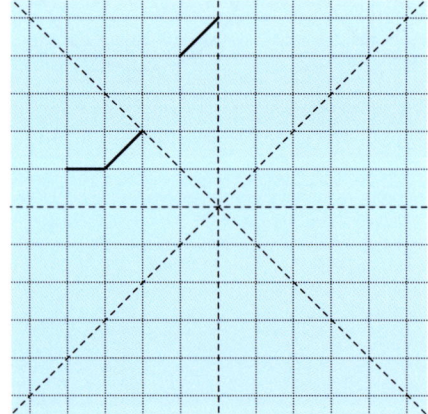

8.
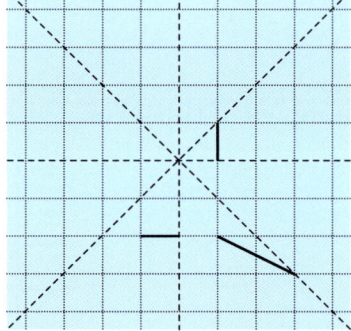

9. Fold a piece of paper twice and cut out any shape from the corner. What are the number of lines of symmetry and the order of rotational symmetry of your shape?

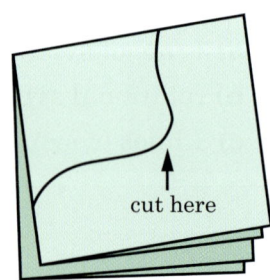

cut here

3.6 Scale drawing

A **scale drawing** is a drawing that shows a real object with accurate sizes reduced or enlarged by a certain amount, called the scale. Use a ruler for all straight edges.

Example

Draw an accurate scale drawing of this rectangle using the scale 1 cm represents 2 m.

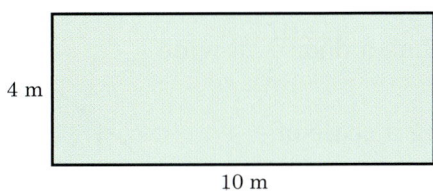

4 m

10 m

Divide each length in metres by 2 to find the number of centimetres.

$10 \div 2 = 5$ and $4 \div 2 = 2$

Then draw a rectangle that is 5 cm long and 2 cm wide.

Exercise 3.6A

Draw an accurate scale drawing of each shape below using the scale shown.

1. Scale: 1 cm for 1 m.

   Measure and write down the length of AB (in cm).

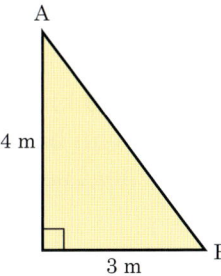

A

4 m

3 m

B

2. Scale: 1 cm for every 3 m.

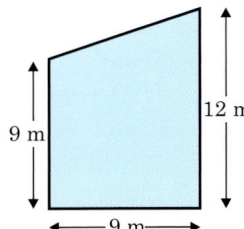

9 m

12 m

9 m

3. Scale: 2 cm for every 1 m.

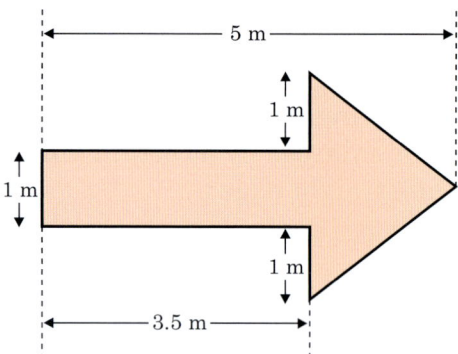

4. A hut is 3 m high and $1\frac{1}{2}$ m wide. It has a door $\frac{1}{2}$ m wide and 2 m high.

Make an accurate scale drawing using a scale of 2 cm for every 1 m.

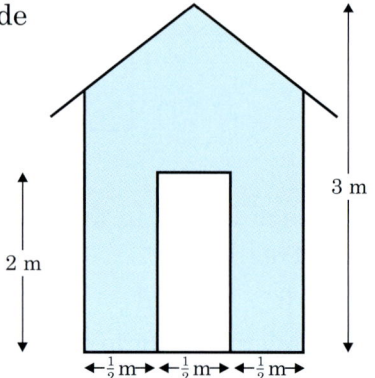

5. A bird leaves its nest and makes three flights.

a) Make a scale drawing to show the journey.

b) How far does the bird have to fly to return to its nest?

	Direction	Distance
1st stage	west	5 km
2nd stage	south-east	6 km
3rd stage	east	12 km

6. The diagram shows ports P and Q, where P is 10 km west of Q.
A boat A is 9 km north-east of P. A lighthouse S is 4 km north-west of Q.
The light can be seen up to 4 km from the lighthouse.
Will the boat see the light?

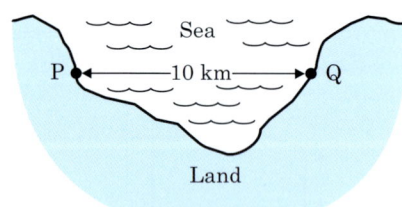

7. A pirate has buried treasure in the field shown.

Make a scale drawing of the field and shade the region where you think the treasure lies. Use a scale of 1 cm to 100 m.

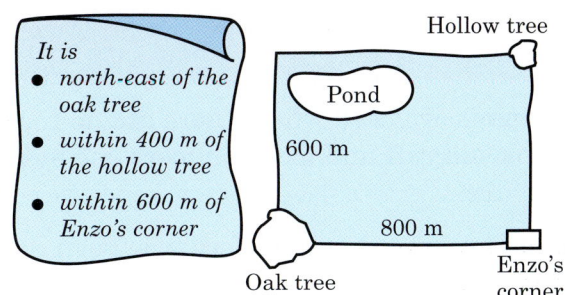

It is
• north-east of the oak tree
• within 400 m of the hollow tree
• within 600 m of Enzo's corner

Hollow tree
Pond
600 m
800 m
Oak tree
Enzo's corner

8. Here is a sketch of a logo which is to be painted onto a ship.

Make a scale drawing of the logo with a scale of 1 cm to 1 m and find the height of the logo.

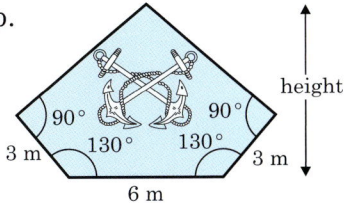

height
90° 90°
3 m 130° 130° 3 m
6 m

9. Make a scale drawing of a room in your home. Design a layout for the furniture you would like to have in the room.

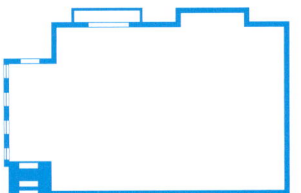

10. This is a plan of a house and gardens. It has been drawn to a scale of 1 cm for every 2 m.

a) How wide is:

 i) the front garden

 ii) the drive

 iii) the bay window?

b) How long is flower bed 2?

c) How wide is flower bed 1?

d) If the fish pond is 4 m wide, what size should it be on the plan?

e) Measure carefully the width of the patio on the plan. How many *metres* wide is the real patio?

f) What is the real *area* of the double garage?

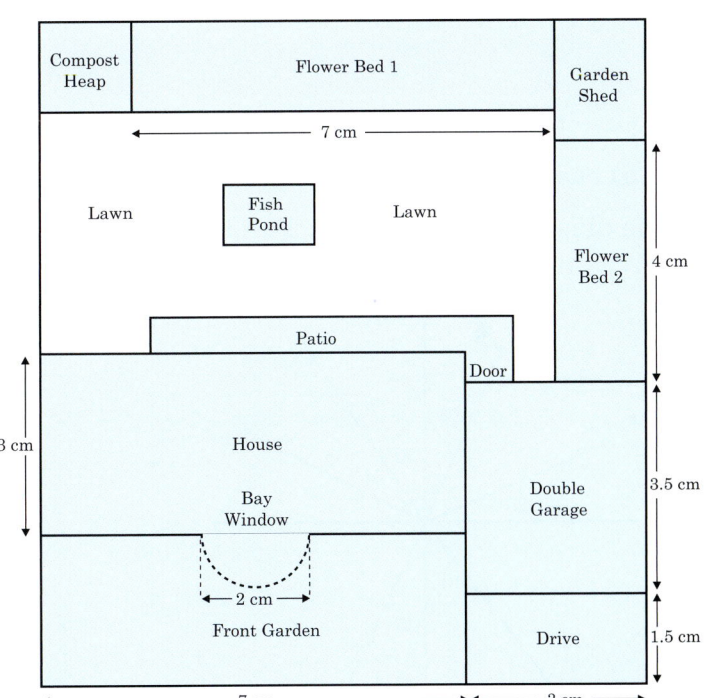

Compost Heap
Flower Bed 1
Garden Shed
7 cm
Lawn
Fish Pond
Lawn
Flower Bed 2 4 cm
Patio
Door
3 cm
House
Double Garage 3.5 cm
Bay Window
2 cm
Front Garden
Drive 1.5 cm
7 cm
3 cm

3.7 Bearings

Bearings are used where there are no roads to guide the way. Ships, aircraft and mountaineers use bearings to work out where they are.

Bearings are measured clockwise from north and have 3 figures from 000° to 360°.

Example

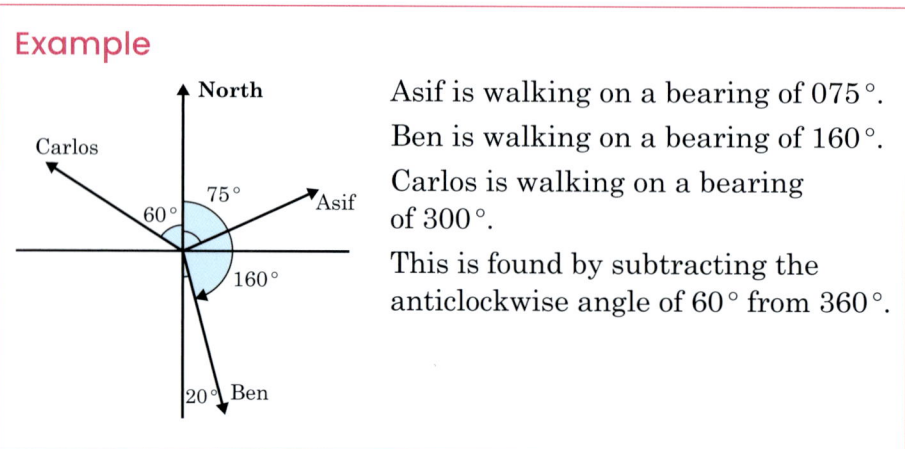

Asif is walking on a bearing of 075°.

Ben is walking on a bearing of 160°.

Carlos is walking on a bearing of 300°.

This is found by subtracting the anticlockwise angle of 60° from 360°.

Exercise 3.7A

1. Give the bearing of:

   a) east   **b)** west   **c)** south   **d)** north   **e)** south-west   **f)** north-east.

The diagrams in Questions **2** and **3** show the directions in which several people are travelling. Work out the bearing for each person.

2.

3.

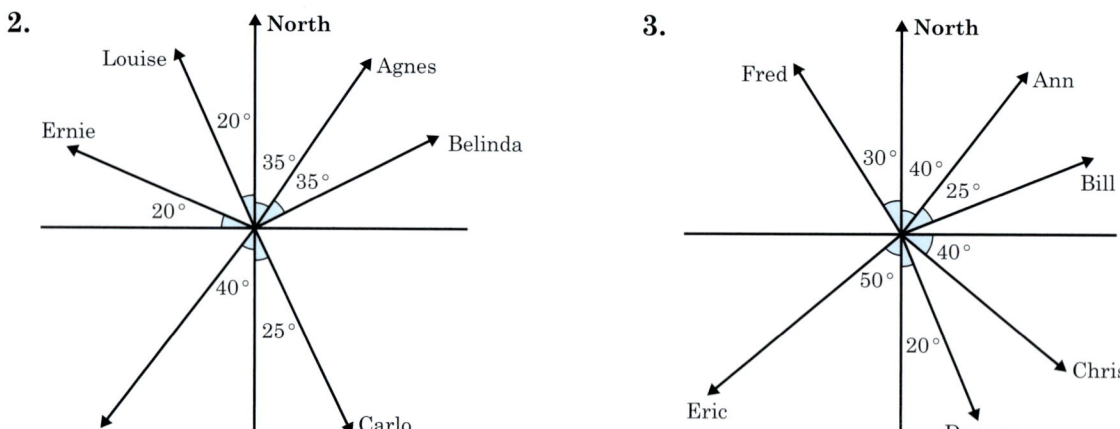

4. Debbie is facing south-east. She turns through 90° anticlockwise and then walks forward in a straight line. On what bearing does she walk?

Relative bearings

The bearing of A from B is the direction in which you travel to get to A from B.

It helps to show the journey with an arrow, as in the example below.

Example 1

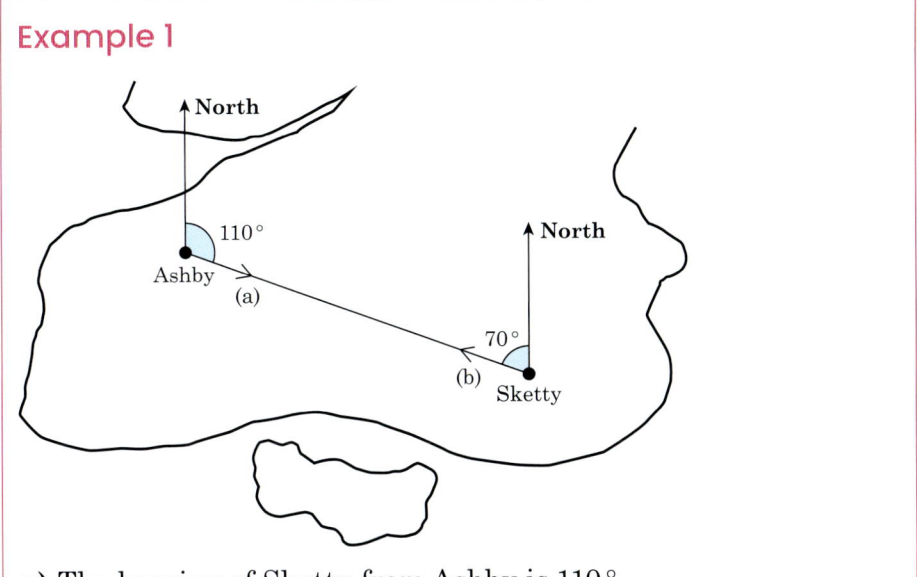

a) The bearing of Sketty from Ashby is 110°.

b) The bearing of Ashby from Sketty is 290°.

This is found by subtracting 70° from 360°.

Example 2

The bearing of A from B is 075°. Find the bearing of B from A.

Draw a sketch. The two north lines are parallel.

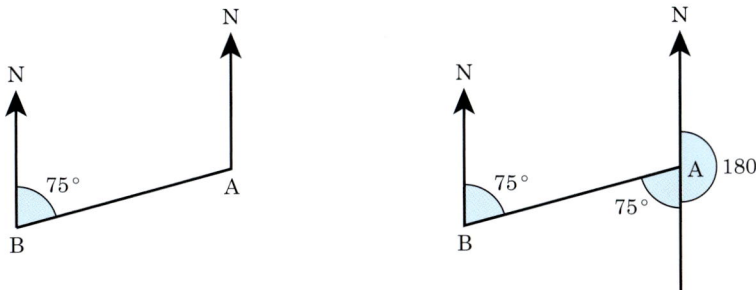

Use alternate angles to find another angle equal to 75°.

The bearing of B from A is 180° + 75° = 255°.

Exercise 3.7B

The map of part of North America shows six radar tracking stations,
A, B, C, D, E, F.

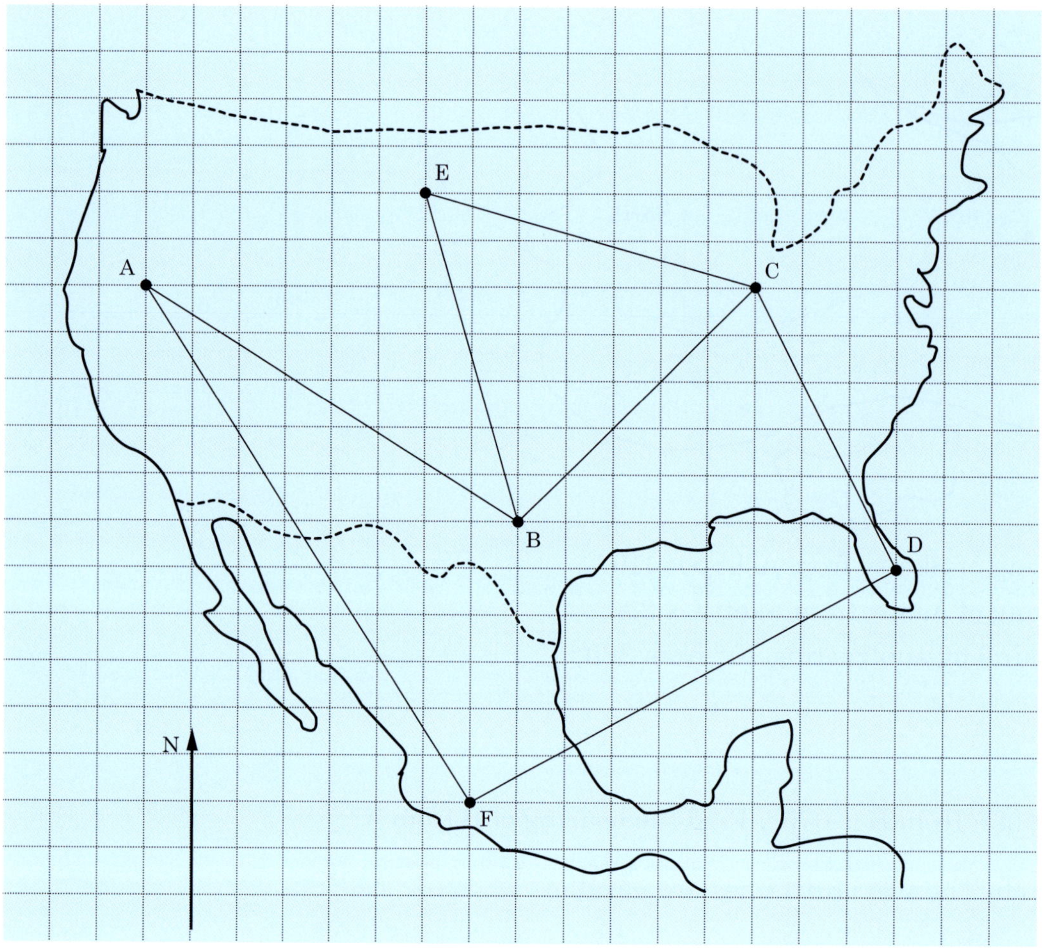

1. From A, measure the bearing of   **a)** F   **b)** B   **c)** C.

2. From C, measure the bearing of   **a)** E   **b)** B   **c)** D.

3. From F, measure the bearing of   **a)** D   **b)** A.

4. From B, measure the bearing of   **a)** A   **b)** E   **c)** C.

5. Find the bearing of a station G from A if the bearing of A from G is:

   a) 079°   **b)** 152°   **c)** 248°.

Ships or aircraft can be located when their bearings from two places are known.

Example

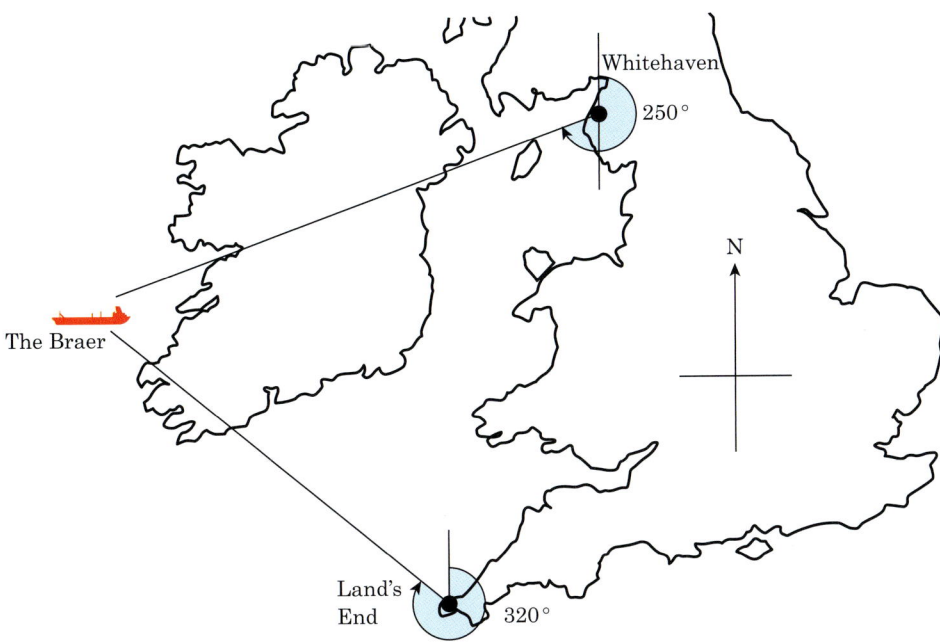

On the map the tanker *Braer* is on a bearing 320° from Land's End.

From Whitehaven, the *Braer* is on a bearing of 250°.

There is only one place where the tanker can be.

Exercise 3.7C

Draw the points P and Q as shown in the middle of a clean page of squared paper. Mark the points A, B, C, D and E accurately, using the information given.

1. A is on a bearing of 040° from P and 015° from Q.

2. B is on a bearing of 076° from P and 067° from Q.

3. C is on a bearing of 114° from P and 127° from Q.

4. D is on a bearing of 325° from P and 308° from Q.

5. E is on a bearing of 180° from P and 208° from Q.

Exercise 3.7D

Draw the points X and Y as shown in the middle of a piece of squared paper. Mark the points K, L, M, N and O accurately using the information given.

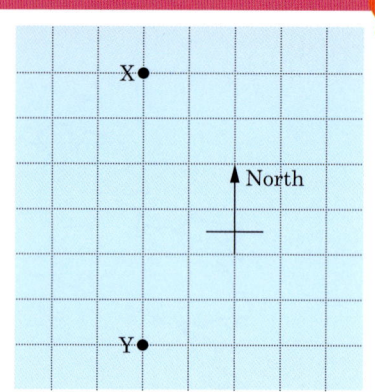

1. K is on a bearing of 041° from X and 025° from Y.
2. L is on a bearing of 090° from X and 058° from Y.
3. M is on a bearing of 123° from X and 090° from Y.
4. N is on a bearing of 203° from X and 215° from Y.
5. O is on a bearing of 288° from X and 319° from Y.

Example

A ship sails 6 km west and then changes direction and sails a further 3 km on a bearing 205°.

Draw a scale drawing to find how far the ship is now from its starting point.

Use a scale of 1 cm for 1 km.

Start with a 6 cm line west.

West ⟵ ———————————• Start
          6 cm

Then draw a 3 cm line on a bearing of 205°.

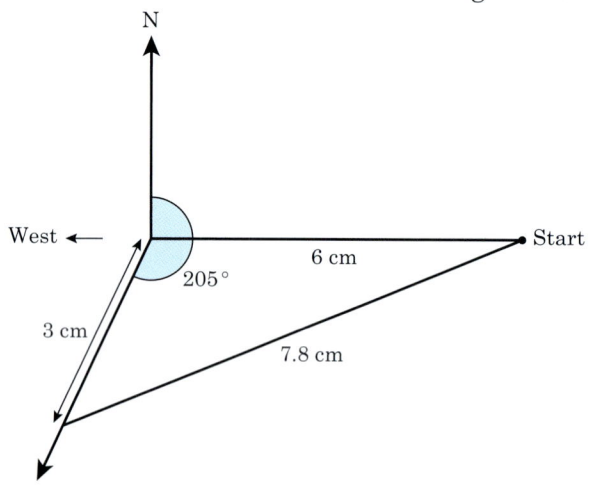

Tip

If you do not have a 360° protractor, then you can measure anticlockwise 155° on a 180° protractor.

Measure the distance the ship is now from its starting position, 7.8 cm.

Convert this to kilometres: the real distance is 7.8 km.

Exercise 3.7E

Make accurate scale drawings with a scale of 1 cm to 1 km, unless told otherwise. Use squared paper and begin each question by drawing a small sketch of the journey.

1. A ship sails 8 km due north and then a further 7 km on a bearing 080°, as in the diagram (which is not drawn to scale).

   How far is the ship now from its starting point?

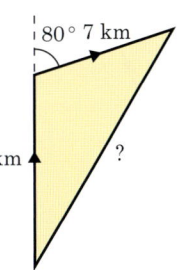

2. A ship sails 9 km on a bearing 090° and then a further 6 km on a bearing 050°, as shown in the diagram.

   How far is the ship now from its starting point?

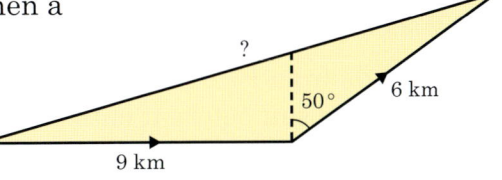

3. A ship sails 6 km on a bearing 160° and then a further 10 km on a bearing 240°, as shown.

   a) How far is the ship from its starting point?

   b) On what bearing must the ship sail so that it returns to its starting point?

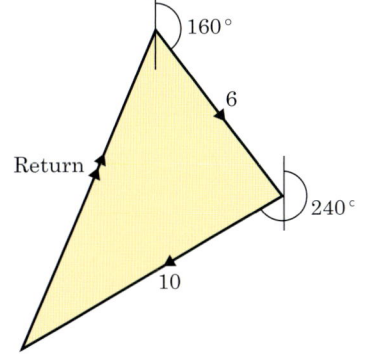

4. A ship sails 5 km on a bearing 030°, then 3 km on a bearing 090° and finally 4 km on a bearing 160°. How far is the ship now from its starting point?

5. Point B is 8 km from A on a bearing 140° from A.

   Point C is 9 km from A on a bearing 200° from A.

   a) How far is B from C?

   b) What is the bearing of B from C?

6. Point Q is 10 km from P on a bearing 052° from P. Point R is 4 km from P on a bearing 107° from P.

   a) How far is Q from R?

   b) What is the bearing of Q from R?

7. A laser beam L is 120 km from P on a bearing 068°.

The laser beam is directed on a bearing 270° from L.

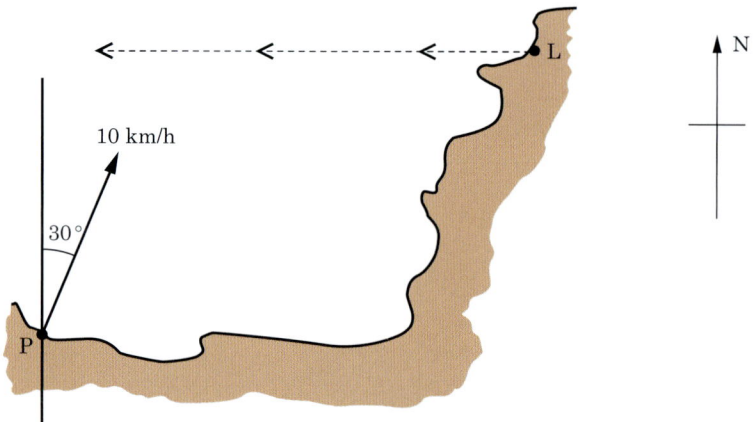

a) Draw a diagram, with a scale of 1 cm to 10 km, to show the positions of P and L.

b) A ship sails from P at a speed of 10 km/h on a bearing 030°. For how long does the ship sail before it cuts the beam?

8. Robinson Crusoe is on a tiny island R. There is an airport at A which is 150 km from R on a bearing 295° from R.

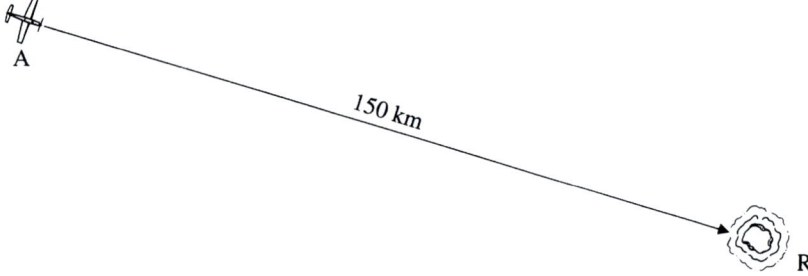

An aircraft flies from A. If the aircraft gets within 40 km of R, the pilot will see Robinson's bonfire and he will be saved. Will he be saved if the plane flies on a bearing of 098°?

Revision exercise 3

1. Which of the nets below can be used to make a cube?

a)

b)

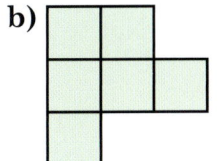

c)

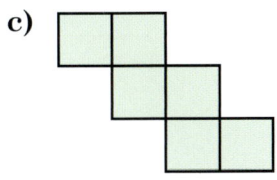

2. In the quadrilateral PQRS, PQ = QS = QR; PS is parallel to QR and angle QRS = 70°. Calculate

a) angle RQS

b) angle PQS.

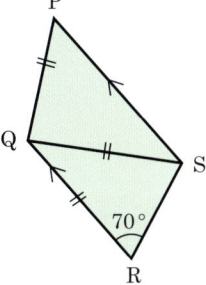

3. Point B is on a bearing 120° from point A. The distance from A to B is 110 km.

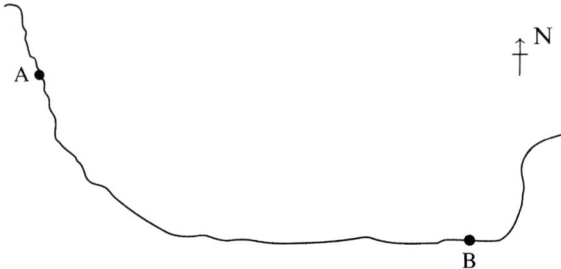

a) Draw a diagram showing the positions of A and B. Use a scale of 1 cm to 10 km.

b) Ship S is on a bearing 072° from A. Ship S is on a bearing 325° from B.

Show S on your diagram and find the distance from S to B.

4. a) A lies on a bearing of 040° from B. Calculate the bearing of B from A.

b) The bearing of X from Y is 115°. Calculate the bearing of Y from X.

5. Use a ruler and pair of compasses to draw triangle ABC with AB = 8 cm, BC = 6 cm and AC = 5 cm.

Measure each of the angles in the triangle.

6. Copy the pattern onto squared paper and shade a further 5 squares so that it has rotational symmetry of order 2.

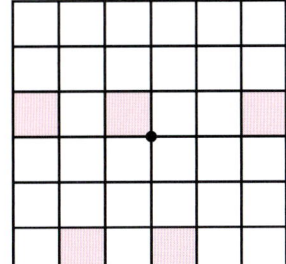

7. Each interior angle of a regular polygon is 156°.

How many sides does the polygon have?

8. Find the sizes of the angles marked with letters in this diagram. Give reasons for your answers.

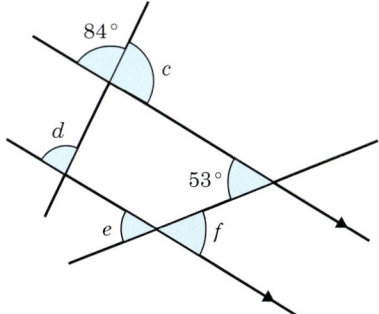

NON-CALCULATOR

1. In the diagram AB is parallel to CD. Calculate the value of b. [2]

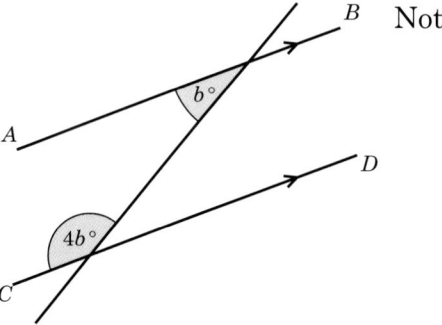

Not to scale

2.

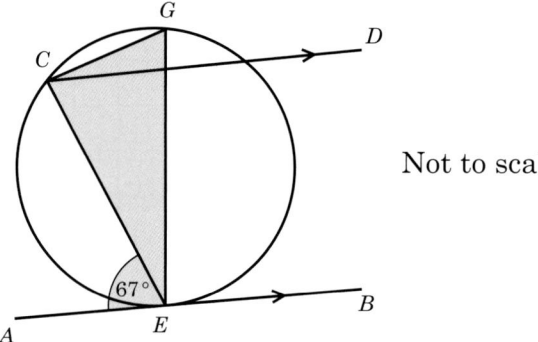

Not to scale

EG is a diameter of the circle through E, C and G. The tangent AEB is parallel to CD and angle $AEC = 67°$. Calculate the size of the following angles and give a reason for each answer.

  a) Angle ECD [2]

  b) Angle CEG [2]

  c) Angle ECG [2]

  d) Angle CGE [2]

3.

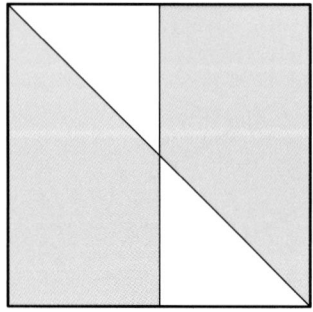

Write down the order of rotational symmetry of the diagram. [1]

4. a) Calculate the size of one interior angle of a regular nonagon
(nine-sided polygon). [2]

b)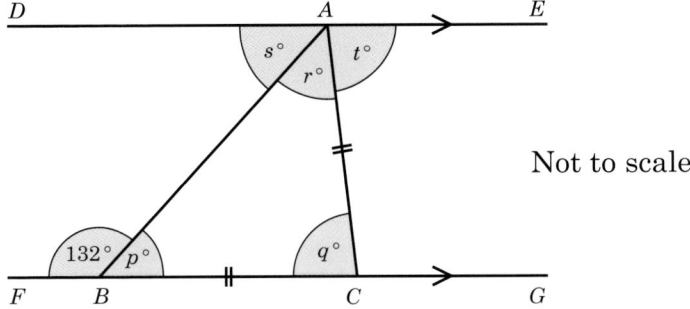

Not to scale

In the diagram, *DAE* and *FBCG* are parallel lines. *AC* = *BC* and
angle *FBA* = 132°.

Work out the values of *p*, *q*, *r*, *s* and *t*. [5]

c) *P*, *Q* and *R* lie on a circle, centre *O*.

QOR is a straight line and angle *PQR* = 63°.

Find the value of *x*. [2]

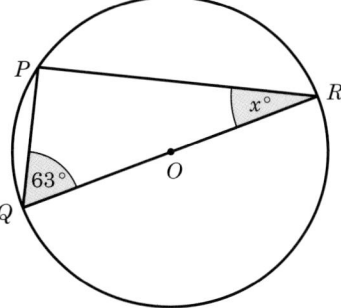

5.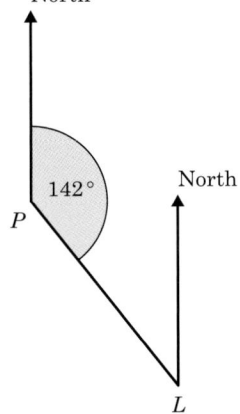

The bearing of Lily, *L*, from Petra, *P*, is 142°.

Find the bearing of Petra from Lily. [2]

6. The diagram shows the route of an aeroplane.

The plane flies from A to B on a bearing of $120°$ and then from B to C on a bearing of $085°$. $AB = BC$

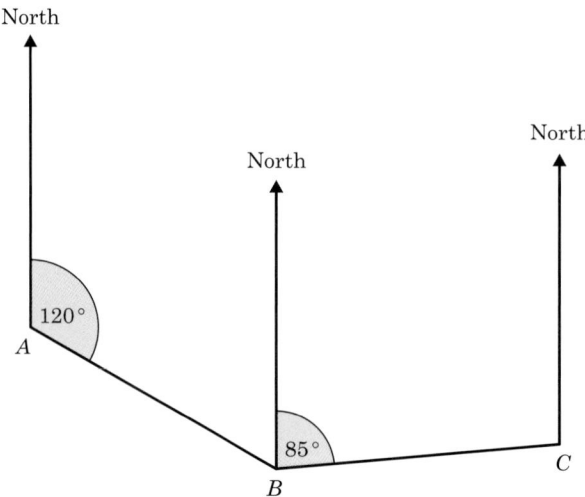

a) Show that angle $ABC = 145°$. [2]

b) Work out the bearing of B from C. [2]

William Shockley (1910–1989), along with two other scientists, was awarded the Nobel Prize in Physics in 1956 for inventing the transistor. This is a good example of how mathematics can be used to solve practical problems.

Every time you use a calculator, you are making use of integrated circuits that were developed from the first transistors.

The first electronic computers did not make use of transistors or integrated circuits and they were so big that they occupied entire rooms. The modern computers we use today are a direct result of Shockley's work.

Katherine Johnson (1918–2020) was an American mathematician who worked for NASA. Her orbital calculations were critical to the success of many spaceflights, including the Apollo missions to the Moon in the 1960s. She was one of the first black American women to work as a scientist at NASA. The story of her life was made into a film called *Hidden Figures* and, in 2015, President Barack Obama awarded her the Presidential Medal of Freedom.

- Round to a given number of decimal places or significant figures and using rounded values to estimate calculations, including choosing an appropriate degree of accuracy when a question is in context.
- Understand and use bounds of numbers.
- Effective calculator use including entering values correctly and interpreting values correctly, e.g. for time and money.

4.1 Approximations

Consider the calculation $158 \div 3.5$.

The answer on a calculator display is 45.142 857 14

It is not sensible to give all these figures in the answer, so you can approximate the number by rounding.

You can approximate in two ways:

a) you can give **significant figures** (s.f.)

b) you can give **decimal places** (d.p.)

Each type of approximation is described below.

Significant figures

Significant figures are the figures in a number after any leading zeros. Zeros that are 'trapped' by other digits *are* significant.

Consider 45 607

- The first significant figure is the '4'.
- The second significant figure is the '5'.
- The third significant figure is the '6'.
- The fourth significant figure is the '0'.
- The fifth significant figure is the '7'.

Now consider 0.00542

- The first significant figure is the '5'.
- The second significant figure is the '4'.
- The third significant figure is the '2'.

The zeros before the 5 are not significant.

To round a number to a given number of significant figures, first look at the number which is in the place you are asked for, and then look at the number to its right (marked with an arrow in the example below) to see if it is '5 or more'. If it is 5 or more, then round the appropriate significant figure UP.

Example

Write the following numbers correct to three significant figures (3 s.f.).

a) $2.65\underset{\uparrow}{8}2 = 2.66$ (to 3 s.f.)   The third significant figure is the 5. Since $8 \geqslant 5$, you round up.

---

b) $0.514\underset{\uparrow}{2} = 0.514$ (to 3 s.f.)   The third significant figure is the 4. Since 2 < 5, you don't round up.

c) $84\,6\underset{\uparrow}{6}0 = 84\,700$ (to 3 s.f.)   The third significant figure is the first 6. Since $6 \geqslant 5$, you round up.

d) $0.040\underset{\uparrow}{3}1 = 0.0403$ (to 3 s.f.)   The third significant figure is the 3. Since 1 < 5, you don't round up.

Exercise 4.1A

For Questions **1** to **8**, write the numbers correct to three significant figures.

1. 2.3462   **2.** 0.814 38   **3.** 26.241

4. 35.55   **5.** 112.74   **6.** 210.82

7. 0.8254   **8.** 0.031 162

For Questions **9** to **16**, write the numbers correct to two significant figures.

9. 5.894   **10.** 1.232   **11.** 0.5456

12. 0.7163   **13.** 0.1443   **14.** 1.831

15. 24.83   **16.** 31.37

For Questions **17** to **24**, write the numbers correct to four significant figures.

17. 486.72   **18.** 500.36   **19.** 2.8888

20. 3.1125   **21.** 0.071 542   **22.** 3.0405

23. 2463.5   **24.** 488 852

For Questions **25** to **36**, write the numbers to the number of significant figures indicated.

25. 0.5126 (3 s.f.)   **26.** 5.821 (2 s.f.)   **27.** 65.89 (2 s.f.)

28. 587.55 (4 s.f.)   **29.** 0.581 (1 s.f.)   **30.** 0.0713 (1 s.f.)

31. 5.8354 (3 s.f.)   **32.** 87.84 (2 s.f.)   **33.** 2482 (2 s.f.)

34. 52 666 (3 s.f.)   **35.** 0.0058 (1 s.f.)   **36.** 4.397 (3 s.f.)

Decimal places

Decimal places are different to significant figures since you count all of the digits that come after the decimal point, whether they are zero or not.

The process for rounding to a given number of decimal places is the same as for significant figures. You look at the number which is in the place you are asked for, and then look at the number marked with the arrow to see if it is '5 or more'. If it is 5 or more, then round the appropriate decimal place UP.

Example

Write the following numbers correct to two decimal places (2 d.p.).

a) 8.358 = 8.36 (to 2 d.p.)
↑

The second decimal place is the 5. Since 8 ⩾ 5, you round up.

b) 0.0328 = 0.03 (to 2 d.p.)
↑

The second decimal place is the 3. Since 2 < 5, you don't round up.

c) 74.355 = 74.36 (to 2 d.p.)
↑

The second decimal place is the first 5. Since 5 ⩾ 5, you round up.

Exercise 4.1B

For Questions **1** to **8**, write the numbers correct to two decimal places (2 d.p.).

1. 5.381
2. 11.0482
3. 0.414
4. 0.3666
5. 8.015
6. 87.044
7. 9.0062
8. 0.0724

For Questions **9** to **16**, write the numbers correct to one decimal place.

9. 8.424
10. 0.7413
11. 0.382
12. 0.095
13. 6.083
14. 19.53
15. 8.111
16. 7.071

For Questions **17** to **28**, write the numbers to the number of decimal places indicated.

17. 8.155 (2 d.p.)
18. 3.042 (1 d.p.)
19. 0.5454 (3 d.p.)
20. 0.005 55 (4 d.p.)
21. 0.7071 (2 d.p.)
22. 6.8271 (2 d.p.)
23. 0.8413 (1 d.p.)
24. 19.646 (2 d.p.)
25. 0.071 35 (4 d.p.)
26. 60.051 (1 d.p.)
27. −7.30 (1 d.p.)
28. 3.799 (2 d.p.)

29. Use a ruler to measure the dimensions of the rectangles below.

    a) Write down the length and width in cm correct to 1 d.p.

    b) Work out the area of each rectangle and give the answer in cm² correct to 1 d.p.

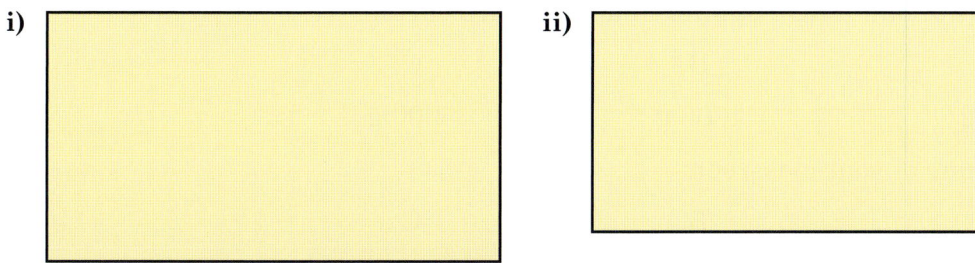

    i)   **ii)**

4.2 Estimating

In some circumstances it is unrealistic to work out the exact answer to a problem. It might be quite satisfactory to give an estimate for the answer.

For example, a builder does not know *exactly* how many bricks a new garage will require. She may estimate that she needs 2500 bricks and place an order for that number. In practice, she may need only 2237.

You also do not need to know the exact number of sheep in the world. Just over 1 billion sheep will be adequate as an approximation.

When you go into the supermarket, you can estimate the total value of the items in your shopping basket (e.g. by rounding each up to the nearest dollar), so you know that you have enough money to pay for the goods at the checkout.

Exercise 4.2A

For Questions **1** to **14**, estimate which value is closest to the actual value.

1. The height of a double-decker bus:

A	B	C
3 m	6 m	10 m

2. The height of the tallest player in the Olympic basketball competition:

A	B	C
1.8 m	3.0 m	2.2 m

3. The height of the Eiffel Tower:

A	B	C
30 m	300 m	1000 m

4. The mass of half a litre of milk in a cardboard carton:

A	B	C
500 g	1000 g	5000 g

5. The volume of your classroom:

A	B	C
100 m³	1000 m³	10 000 m³

6. The top speed of a Grand Prix racing car:

A	B	C
600 km/h	80 km/h	300 km/h

7. The number of times your heart beats in one day (24 h):

A	B	C
10 000	100 000	1 000 000

8. The thickness of one page in this book:

A	B	C
0.01 cm	0.001 cm	0.0001 cm

9. The number of cars in a traffic jam 10 km long on a 3-lane motorway:

A	B	C
4000	40 000	200 000

10. The time it takes to walk $1\frac{1}{2}$ km:

A	B	C
10 minutes	20 minutes	60 minutes

11. The area of a standard postcard:

A	B	C
150 cm²	1000 cm²	0.1 m²

12. The mass of an ordinary apple:

A	B	C
100 g	250 g	400 g

13. The telephone charge to Australia is 70c per minute. The number of words you will be able to say in a call costing $4 is:

A	B	C
120	500	1200

14. The largest tree in the world has a diameter of 11 m. Estimate the number of 'average' 15-year-olds, standing with their arms stretched out, that are required to circle the tree so that they form an unbroken chain around the trunk.

Use the fact that the circumference of the tree is approximately 3 × diameter and assume that the trunk has a circular cross-section.

15. When you multiply by a number greater than 1, you make the original number bigger.

When you multiply by a number less than 1, you make the original number smaller.

so   **5.3** × 1.03 > **5.3**   and   **6.75** × 0.89 < **6.75**

When you divide by a number greater than 1, you make the original number smaller.

When you divide by a number less than 1, you make the original number bigger.

so   **8.9** ÷ 1.13 < **8.9**   and   **11.2** ÷ 0.73 > **11.2**

State whether each statement is true or false:

a) 3.72 × 1.3 > 3.72

b) 253 × 0.91 < 253

c) 0.92 × 1.04 > 0.92

d) 8.5 ÷ 1.4 > 8.5

e) 113 ÷ 0.73 < 113

f) 17.4 ÷ 2.2 < 17.4

g) 0.73 × 0.73 < 0.73

h) 2511 ÷ 0.042 < 2511

i) 614 × 0.993 < 614

Example

Estimate the answers to the following questions by rounding each number in the calculation to 1 s.f.:

a) $9.7 \times 3.1 \approx 10 \times 3$. About 30.

b) $81.4 \times 98.2 \approx 80 \times 100$. About 8000.

c) $19.2 \times 49.1 \approx 20 \times 50$. About 1000.

d) $102.7 \div 19.6 \approx 100 \div 20$. About 5.

e) $\dfrac{41.2}{9.83 \times 0.791} \approx \dfrac{40}{10 \times 1}$. About 4.

Sometimes rounding to one significant figure is not the easiest option.

If you want to estimate the answer to $23.47 \div 2.98$ then $20 \div 3$ is not as easy to calculate as $24 \div 3$.

Make sure you read the question carefully to see whether or not you are expected to use one significant figure.

Exercise 4.2B

Write down each question and decide (by estimating) which answer is correct. Do not work out the exact answers to the calculations.

	Question	Answer A	Answer B	Answer C
1.	$551.1 \div 11$	6.92	50.1	5623
2.	$207.1 + 11.65$	310.75	23.75	218.75
3.	664×0.51	256.2	338.64	828.62
4.	$(5.6 - 0.21) \times 39$	389.21	210.21	20.51
5.	$\dfrac{17.5 \times 42}{2.5}$	294	504	86
6.	$(906 + 4.1) \times 0.31$	473.21	282.131	29.561
7.	$\dfrac{543 + 472}{18.1 + 10.9}$	65	35	85
8.	$\dfrac{112.2 \times 75.9}{6.9 \times 5.1}$	242	20.4	25.2

9. The petrol consumption of a car is 4 km per litre and petrol costs $1.52 per litre.

   Jasper estimates that the petrol costs of a round trip of about 1200 km will be $150. Is this a reasonable estimate? Explain your answer.

10. 44 teachers buy 19 books at $24.20 each. They share the cost equally between them.

    The headteacher used a calculator to work out the cost per teacher and got an answer of $10.50 to the nearest cent. *Without* using a calculator, work out an estimate for the answer to check whether or not the headteacher got it right. Show your working.

11. Each year about 150 million trees are cut down to make paper. One tree is enough to make about 650 kg of paper.

    a) Find the mass of several newspapers (large and small) and estimate the number of newspapers which can be made from one tree.

    b) Estimate the number of newspapers which could be made from all the trees cut down each year.

    c) Find the mass of some of the exercise books you use at school. Estimate the number of books your class will use in a whole year and estimate the number of trees required to supply the paper for your class for one year.

12. A hockey team are going on tour and need to buy 18 airline tickets. The cost of each ticket is $205. The sponsor of the team estimates that he will need to give them $4000 to pay for the tickets.

    Is this a reasonable estimate? Explain your answer.

13. Bilal goes into the supermarket and buys 11 apples each costing 85 cents, 14 oranges each costing $1.02 and 19 pomegranates each costing $2.53.

    Bilal says that he will need about $75 dollars to make sure he has enough money at the checkout.

    Without using a calculator, show that Bilal is correct.

4.3 Errors in measurement

Whenever a quantity is measured, the measurement is never *exact*. If you measure the thickness of a wire with a ruler, you might read the thickness as 2 mm. If you use a more accurate device for measuring you might read the thickness as 2.3 mm. An even more accurate device might read the thickness as 2.31 mm. None of these figures is precise.

They are all **approximations** to the actual thickness. This means that there is always an error in making any kind of measurement such as length, mass, time, temperature and so on. An error of this kind is not the same as making a mistake in a calculation!

Bounds of accuracy

a) Suppose the length of a book is measured at 22 cm to the nearest cm. The actual length could be from 21.5 to *almost* 22.5. You say 'almost' 22.5 because a length of 22.4999999... would be rounded to 22 cm. The number 22.499999... is effectively 22.5 and we take 22.5 as the **upper bound**. The **lower bound** is 21.5.

Lower bound   Upper bound

$$21.5 \text{ cm} \leqslant \text{length} < 22.5 \text{ cm}$$

Lower bound = 22 – 0.5 = 21.5   Upper bound = 22 + 0.5 = 22.5

b) Using a ruler, the length of a nail is measured as 3.8 cm to the nearest 0.1 cm. In this case:

Lower bound   Upper bound

$$3.75 \text{ cm} \leqslant \text{length} < 3.85 \text{ cm}$$

Lower bound = 3.8 – 0.05 = 3.75   Upper bound = 3.8 + 0.05 = 3.85

c) Sometimes measurements are given 'to the nearest 10, 100, etc.' Suppose the length of a lake is measured as 4200 m to the nearest 100 m. The bounds of accuracy are:

Lower bound   Upper bound

$$4150 \text{ m} \leqslant \text{length} < 4250 \text{ m}$$

Lower bound = 4200 – 50 = 4150   Upper bound = 4200 + 50 = 4250

When the bounds of accuracy are written using inequality notation, it is called an **error interval**.

In **(a)**, **(b)** and **(c)** above the maximum possible error is always half of the level of accuracy.

In part **(b)** the level of accuracy is the nearest 0.1 cm. The maximum possible error is 0.05 cm.

d) Here are some further examples:

i) The mass of an apple is 43 g to the nearest gram.

ii) The temperature of a room is 22.9 °C to one decimal place.

iii) The length of a road is 780 m to the nearest 10 m.

iv) The capacity of a mug is 115 ml to the nearest 5 ml.

v) The mass of a lorry is 23 000 kg to two significant figures.

Value	Lower bound	Upper bound
43 g	42.5 g	43.5 g
22.9 °C	22.85 °C	22.95 °C
780 m	775 m	785 m
115 ml	112.5 ml	117.5 ml
23 000 kg	22 500 kg	23 500 kg

Exercise 4.3A

1. Copy and complete each statement. Part **(a)** is done as an example.

   a) A length d is 42 m, to the nearest m, so $41.5 \leqslant d < 42.5$.

   b) A volume V is 8 m³, to the nearest cubic metre, so $7.5 \leqslant V < \boxed{}$

   c) A mass m is 72 kg, to the nearest kg, so $\boxed{} \leqslant m < \boxed{}$

   d) A time t is 3.2 h, to the nearest 0.1 h, so $\boxed{} \leqslant t < 3.25$

   e) A radius r is 5.8 cm, to the nearest 0.1 cm, so $\boxed{} \leqslant r < \boxed{}$

2. The height of a table is measured at 84 cm to the nearest cm. Write down the lower bound for the height of the table.

3. A man measures the mass of a parcel to be 5.2 kg to the nearest 0.1 kg. Write down the upper bound for the mass of the parcel.

4. The length and width of a rectangle are measured to the nearest 0.1 cm, as shown.

   a) Write down the upper bound for the length of the rectangle.

   b) Write down the lower bound for the width of the rectangle.

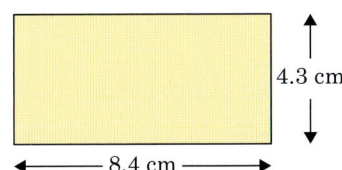

5. The height of a man is measured at 173 cm, to the nearest cm. Write down the upper bound for the height of the man.

6. A scientist finds the mass of a sea shell to be 3.7 g, correct to one decimal place. What is the least possible mass of the shell?

7. A book states that the distance from the Earth to the Sun is 93 million miles correct to two significant figures. What is the shortest possible distance?

8. In a 200 m race a sprinter is timed at 20.63 seconds to the nearest 0.01 seconds. Write down the least possible time.

9. The mass of an egg is 17.8 g, correct to one decimal place. What is the greatest possible mass of the egg?

10. Copy and complete the table:

a) length of nail = 5.6 cm, to the nearest mm

b) height of lighthouse = 37 m, to the nearest m

c) mass of insect = 0.27 mg, to 2 d.p.

d) temperature in oven = 230 °C to the nearest 10 °C

e) length of oil pipeline = 315 km, to the nearest km

	Lower bound	Upper bound
a)		
b)		
c)		
d)		
e)		

4.4 Using a calculator

All calculators work differently so it is important that you are confident using yours. Calculators cannot think for themselves. *You* have to decide in which order the buttons have to be pressed.

You should familiarise yourself with some of the most often used buttons on your calculator, for example the button for squaring, typically x^2, the button for square rooting, typically $\sqrt{\square}$, and the fraction button, typically $\frac{\square}{\square}$.

Example

a) Work out $8.43 + \dfrac{9.72}{3.3}$ to 4 s.f.

Type $8.43 + \dfrac{9.72}{3.3}$ into your calculator and press $=$

Use the fraction button for the second part.

Answer = 11.37545455 = 11.38 (to 4 s.f.)

Write down all the digits and then round.

b) Work out $(5.2 - 4.737)^2$ to 4 s.f.

Type $(5.2 - 4.737)\ \boxed{x^2}$ into your calculator and press $\boxed{=}$

Use the squaring button once you have entered your brackets.

Answer $= 0.214369 = 0.2144$ (to 4 s.f.)

Write down all the digits and then round.

c) Work out $\sqrt{17^2 - 15^2} + 2.5$

Press $\boxed{\sqrt{\ }}$ first and then enter

$17\ \boxed{x^2} - 15\ \boxed{x^2}$ inside the box

Then you can add the 2.5

Make sure you use the arrow buttons to exit the square root box first.

Answer $= 10.5$

Exercise 4.4A

Use a calculator to work out:

1. $3 \times 5 - 28 \div 4$

2. $7 \times 4 + 2 \times 2$

3. $30 \div 3 + 5 \times 4$

4. $20 \div 2 - 3 \times 2$

5. $8 \div 8 - 1 \times 1$

6. $\dfrac{27 + 3 \times 3}{3 \times 2}$

7. $\dfrac{6 + 8 \times 3}{8 \times 2 - 10}$

8. $\dfrac{13 - 12 \div 4}{4 + 3 \times 2}$

9. $\dfrac{11 + 6 \times 6}{5 - 8 \div 2}$

10. $\dfrac{12 + 3 \times 6}{4 + 3 \div 3}$

11. $\dfrac{24 - 18 \div 3}{1.5 + 4.5}$

Work out, correct to four significant figures:

12. 85.3×21.7

13. $18.6 \div 2.7$

14. $10.074 \div 8.3$

15. 0.982×6.74

16. $\dfrac{8.3 + 2.94}{3.4}$

17. $\dfrac{6.1 - 4.35}{0.76}$

18. $\dfrac{19.7 + 21.4}{0.985}$

19. $7.3 + \dfrac{8.2}{9.5}$

20. $\dfrac{6.04}{18.7} - 0.214$

21. $\dfrac{2.4 \times 0.871}{4.18}$

22. $19.3 + \dfrac{2.6}{1.95}$

23. $6.41 + \dfrac{9.58}{2.6}$

24. $6.71^2 + 0.64$

25. $3.45^3 + 11.8$

26. $2.93^3 - 2.641$

27. $\dfrac{7.2^2 - 4.5}{8.64}$

28. $\dfrac{13.9 + 2.97^2}{4.31}$

29. $(3.3 - 2.84)^2$

30. $(8.7 - 5.95)^4$

31. $\sqrt{68.4} + 11.63$

32. $9.45 - \sqrt{8.248}$

33. $3.4^2 - \sqrt{1.962}$

34. $\dfrac{3.54 + 2.4}{8.47^2}$

35. $2065 - \sqrt{44\,000}$

36. $\sqrt{5.69 - 0.0852}$

37. $\sqrt{0.976 + 1.03}$

38. $\sqrt{\dfrac{17.4}{2.16 - 1.83}}$

39. $\dfrac{\sqrt{4.79} + 1.6}{9.63}$

40. $\dfrac{0.761^2 - \sqrt{4.22}}{1.96}$

41. $\sqrt[3]{\dfrac{1.74 \times 0.761}{0.0896}}$

42. $\left(\dfrac{8.6 \times 1.71}{0.43}\right)^3$

43. $\dfrac{\sqrt[3]{86.6}}{\sqrt[4]{4.71}}$

Rounding an answer to a calculation

Sometimes, when you are asked to round the answer to a given degree of accuracy (significant figures, or decimal places), you might be tempted to round the numbers in the calculation first.

For example, when calculating 0.26147×1.9782, correct to 2 d.p., you might enter 0.26×1.98 into your calculator but *this is incorrect.*

You should always use the given values within a calculation, then write down all of the digits on your calculator display before then rounding the final answer.

So $0.26147 \times 1.9782 = 0.517239954 = 0.52$ (to 2 d.p.).

Try the questions in this exercise to practise this.

Exercise 4.4B

Use a calculator to work out each of these. Write down all of the digits on your calculator display and then round the answers to the degree of accuracy indicated.

1. 0.153×3.74 (2 d.p.)

2. $18.09 \div 5.24$ (3 s.f.)

3. 184×2.342 (3 s.f.)

4. $17.2 \div 0.89$ (1 d.p.)

5. $58 \div 261$ (2 s.f.)

6. 88.8×44.4 (1 d.p.)

7. $(8.4 - 1.32) \times 7.5$ (2 s.f.)

8. $(121 + 3758) \div 211$ (3 s.f.)

9. $(1.24 - 1.144) \times 0.61$ (3 d.p.)

10. $1 \div 0.935$ (1 d.p.)

11. 78.3524^2 (3 s.f.)

12. $(18.25 - 6.941)^2$ (2 d.p.)

13. $9.245^2 - 65.2$ (1 d.p.)

14. $(2 - 0.666) \div 0.028$ (3 s.f.)

15. 8.43^3 (1 d.p.)

16. $0.924^2 - 0.835^2$ (2 d.p.)

Time and money

You can use your calculator to work out problems involving time and money.

Money can just be treated as a decimal written correct to 2 decimal places.

When working with time, remember that there are 60 minutes in an hour, not 100.

This means that 2 hours and 25 minutes is not written as 2.25 hours, because 2.25 hours means $2\frac{1}{4}$ hours, which is 2 hours and 15 minutes.

If you want to enter 2 hours and 25 minutes into your calculator, the easiest way is to use the 'degrees, minutes and seconds' button, which usually looks like ⌗°'".

To enter 2 hours and 25 minutes, press 2 ⌗°'" 25 ⌗°'".

If the time you want to enter also contains seconds, you can press the button again.

For example, to enter 3 hours 14 minutes and 8 seconds you would press

3 ⌗°'" 14 ⌗°'" 8 ⌗°'".

If you then want to convert this time into a decimal, you can do this using the ⌗S⇔D button.

The ⌗°'" button can also be used in reverse to convert a decimal into hours, minutes and seconds.

For example, if you type in 2.18 and press ⌗°'" ⌗⊠, the calculator will display 2°10'48", which means that 2.18 hours is the same as 2 hours, 10 minutes and 48 seconds.

> ### Example
> How many hours, minutes and seconds is 3.61 hours?
>
> 3.61 ⌗°'" ⌗⊠ 3°36'36"
> So 3.61 hours is 3 hours 36 minutes and 36 seconds.

Exercise 4.4C

1. Convert into a decimal number of hours:

   a) 2 hours, 13 minutes and 6 seconds

   b) 5 hours, 42 minutes and 38 seconds

   c) 1 hour, 6 minutes and 50 seconds

2. Convert into hours, minutes and seconds:

   a) 2.64 hours   **b)** 3.88 hours   **c)** 8.29 hours

3. Work out in hours, minutes and seconds:

   a) 1.54 hours + 2.83 hours

   b) 9.82 hours + 4.72 hours

4. How many cents are there in 5.3 dollars?

5. Eric buys 34 cakes at $1.42 each and 28 drinks at $0.89 each.

   How much does he spend in total?

6. Eight people share $215 000 equally between them. How much do they each get?

7. Four people split a restaurant bill of $34.40 equally between them. How much do they each have to pay?

8. Three singers book 6.75 hours of studio time that they share equally.

   How many hours and minutes do they each get?

Revision exercise 4

1. Work out on a calculator, correct to 4 s.f.

   a) $3.61 - (1.6 \times 0.951)$

   b) $\dfrac{4.65 + 1.09}{3.6 - 1.714}$

2. Evaluate the following and give the answers to three significant figures:

   a) $\sqrt[3]{9.61 \times 0.0041}$

   b) $\left(\dfrac{1}{9.5} - \dfrac{1}{11.2} \right)^3$

   c) $\dfrac{15.6 \times 0.714}{0.0143 \times 12}$

   d) $\sqrt[4]{\dfrac{1}{5 \times 10^3}}$

3. Estimate the answer correct to one significant figure.

   a) $(612 \times 52) \div 49.2$

   b) $(11.7 + 997.1) \times 9.2$

   c) $\sqrt{\dfrac{91.3}{10.1}}$

   d) $\pi\sqrt{5.2^2 + 18.2^2}$

4. The mass of the planet Jupiter is about 350 times the mass of the Earth. The mass of the Earth is approximately 6.03×10^{21} tonnes. Give an estimate correct to two significant figures for the mass of Jupiter.

5. 📱 Evaluate the following using a calculator and give the answers to four significant figures:

a) $\dfrac{0.74}{0.81 \times 1.631}$

b) $\sqrt{\dfrac{9.61}{8.34 - 7.41}}$

c) $\left(\dfrac{0.741}{0.8364}\right)^4$

d) $\dfrac{8.4 - 7.642}{3.333 - 1.735}$

6. A wedding cake with a mass of 9.2 kg is shared between 107 guests. About how much cake, in grams, does each person get?

7. The total mass of 95 000 marbles is 308 kg. What is the approximate mass of each marble?

8. Here are nine calculations and nine answers. Write down each calculation and choose, by estimating, the correct answer from the list given.

a) 1.8×10.4 **b)** 9.8×9.1

c) 7.9×8.1 **d)** 76.2×1.9

e) 3.8×8.2 **f)** 8.15×5.92

g) $36.96 \div 4$ **h)** $9.6 \div 5$

i) $0.11 + 3.97$

> Answers: 63.99, 18.72, 31.16, 4.08, 1.92, 9.24, 144.78, 89.18, 48.248.

9. A piece of wire is measured as 45 cm, correct to the nearest centimetre.

a) Write down the lower bound for the length of the wire.

b) Write down the upper bound for the length of the wire.

10. Safa's reaction time to a light coming on is timed as 0.287 seconds, correct to three decimal places.

a) Write down the lower bound for Safa's reaction time.

b) Write down the upper bound for Safa's reaction time.

CALCULATOR

1. The diameter of the Sun is 1 392 532 kilometres.

   Write this value correct to three significant figures. [1]

2. A bridge has a height limit of 4.4 metres, measured to the nearest 10 cm.

   A lorry has a height of 4.38 metres, to the nearest cm.

   Will it be safe for the lorry to drive under the bridge? Explain your answer. [2]

3. The distance, d kilometres, between Christchurch and Kyoto is 9500 km, correct to the nearest 100 kilometres.

   Complete the statement about the value of d.

   $$\_\_\_\_ \leqslant d < \_\_\_\_$$ [2]

4. **a)** Work out $\dfrac{13.48 \times 0.073}{\sqrt{6} + 5.52}$

   Write down all the figures on your calculator display. [1]

   b) Write your answer to part **(a)** correct to two significant figures. [1]

5. Work out an estimate for $\dfrac{8.19}{3.98} \times 21.31$ [3]

6. Terence buys 1012 books at a cost of $8.95 each.

   Estimate the total amount of money that Terence spends on the books. [2]

7. **a)** Work out $\dfrac{85.41 + 17.93}{45 \times \left(2 + 3.1^2\right)}$ [1]

   b) Write your answer to part **(a)** correct to three significant figures. [1]

8. The mass, m, of a honey badger is 7.3 kg, correct to 1 decimal place.

   Copy and complete the error interval for the mass of the honey badger.

   $$\_\_\_\_ \leqslant m < \_\_\_\_$$ [2]

Florence Nightingale (1820–1910), was a famous nurse during the Crimean War, which took place between 1853 and 1856. She is generally considered to have founded the modern nursing profession, by establishing her own nursing school at St Thomas' Hospital in London in 1860. She was also a gifted mathematician, and used statistical diagrams to illustrate the conditions that existed in the hospitals where she worked. Although she did not invent the pie chart, she popularised its use, along with other diagrams such as the rose diagram, which is like a circular histogram. In 1859, she was elected the first female member of the Royal Statistical Society.

- Use tables for statistical data.
- Read, interpret, draw inferences from and compare data using tables, graphs and statistical diagrams and measures, including an appreciation of limits on any conclusions made.
- Find the mean, median, mode and range and choose the most appropriate statistical measures.
- Draw and interpret the following statistical diagrams:
  - bar charts
  - pie charts
  - pictograms
  - stem-and-leaf diagrams
  - simple frequency distributions
  - scatter diagrams (including understanding correlation and drawing a line of best fit).

5.1 Averages

If you have a set of data, say exam marks or heights, and are told to find the 'average', what are you trying to find?

The answer is: a single number which can be used to represent the entire set of data.

This could be done in three different ways.

a) The median

To find the **median**, first arrange the data in order from smallest to largest. Then select the middle number.

For example, for the data set 5, 7, 11, 15, 16, the median is 11.

If there are two 'middle' numbers, the median is halfway between these two numbers.

For example, for the data set 4, 5, 7, 9, 15, 17, the median is halfway between the 7 and the 9, so it is 8.

b) The mean

To find the **mean**, add all the data values up and then divide the total by the number of items. This is equivalent to sharing out all the data evenly.

For example, for the data set 6, 7, 11, 12, 19, the mean is

$$\frac{6 + 7 + 11 + 12 + 19}{5} = \frac{55}{5} = 11$$

c) The mode

The **mode** is the most commonly occurring value (from the French 'à la mode' meaning 'fashionable').

For example, in the data set 4, 5, 5, 6, 8, the mode is 5.

Each average has its purpose and sometimes one is preferable to the others.

The median is fairly easy to find and has the advantage of being hardly affected by atypical values, such as very large or very small values that sometimes occur at the ends of the distribution.

Consider these exam marks:

20, 21, 21, 22, 23, 23, 25, 27, 27, 27, 29, 98, 98

The median (shown by the arrow) is 25.

The mean is calculated as

$$\frac{20 + 21 + 21 + 22 + 23 + 23 + 25 + 27 + 27 + 27 + 29 + 98 + 98}{13} = 35.5$$

The median gives a truer picture of the centre of the distribution since it is not affected by the two atypical scores of 98. Since the mean uses *all* the data, these atypical scores **skew** the mean value.

The mode of this data is 27. It is easy to find and it eliminates some of the effects of extreme values. However, it does have disadvantages, particularly in data which has two or more 'most frequent' values, so it is not widely used.

> **Tip**
>
> Data sets with two modes are called **bimodal**.

Range

In addition to knowing the centre of a distribution, it is useful to know the range or spread of the data.

range = (largest value) − (smallest value)

For the examination marks, range = 98 − 20 = 78.

Example

Find **a)** the median **b)** the mean **c)** the mode and **d)** the range of this set of 10 numbers.

7, 4, 10, 3, 3, 4, 7, 4, 7, 5

a) Arrange the numbers in order of size to find the median.

3, 3, 4, 4, 4, 5, 7, 7, 7, 10

The median is the value halfway between the 4 and 5.
Median = 4.5

b) Mean $= \dfrac{7 + 4 + 10 + 3 + 3 + 4 + 7 + 4 + 7 + 5}{10} = \dfrac{54}{10} = 5.4$

c) Mode = 4 and 7 since there are more 4s and 7s (3 of each) than any other number.

d) Range = 10 − 3 = 7

Frequency tables

A frequency table shows a number x, for example a mark or a score, against the frequency f, which is the number of times that x occurs.

The symbol Σ (or sigma) means 'the sum of'.

The next example shows how these symbols are used in calculating the mean, the median and the mode.

Example 1

The table shows the marks obtained by 100 students in a test.

Mark (x)	0	1	2	3	4
Frequency (f)	4	19	25	29	23

Find:

a) the mean mark

b) the median mark

c) the modal mark.

a) Mean $= \frac{\Sigma xf}{\Sigma f}$

where Σxf means Σ (number × frequency), or 'the sum of each number multiplied by its frequency' and Σf means 'the sum of the frequencies'.

Mean $= \frac{(0 \times 4) + (1 \times 19) + (2 \times 25) + (3 \times 29) + (4 \times 23)}{100}$

$= \frac{248}{100} = 2.48$

b) The median mark is the number halfway between the 50th and 51st data values. The 50th data value is 3 and the 51st data value is also 3, so the median is 3.

Mark (x)	0	1	2	3	4
Frequency (f)	4	19	25	29	23
Position of data value	1–4	5–23	24–48	49–77	78–100

c) The modal mark $= 3$ since there are 29 students who scored 3.

> **Tip**
>
> $4 + 19 + 25 + 29 + 23 = 100$, which is the number of students who took the test.

> **Tip**
>
> The position of the data value is the position it would be in if the scores were written out individually, e.g. 0, 0, 0, 0, 1, 1, 1, 1, ...

You can also find the combined mean of two data sets.

> ### Example 2
> Some students take a chemistry test.
> The mean mark in group A, which contains 12 students, is 13.5.
> The mean mark in group B, which contains 8 students, is 12.25.
> Calculate the mean mark of all 20 students.
>
> | | |
> |---|---|
> | $12 \times 13.5 = 162$ | Find the total mark of group A. |
> | $8 \times 12.25 = 98$ | Find the total mark of group B. |
> | $162 + 98 = 260$ | Find the total mark for all the students. |
>
> The mean mark for all 20 students is $\frac{260}{20} = 13$

Exercise 5.1A

1. Find the mean, median and mode of the following sets of numbers:

   a) 3, 12, 4, 6, 8, 5, 4

   b) 7, 21, 2, 17, 3, 13, 7, 4, 9, 7, 9

   c) 12, 1, 10, 1, 9, 3, 4, 9, 7, 9

   d) 8, 0, 3, 3, 1, 7, 4, 1, 4, 4.

2. The temperature in °C on 17 days was:

   1, 0, 2, 2, 0, 4, 1, 3, 2, 1, 2, 3, 4, 5, 4, 5, 5.

   What was the modal temperature?

3. A dice was thrown 14 times and landed as follows:

   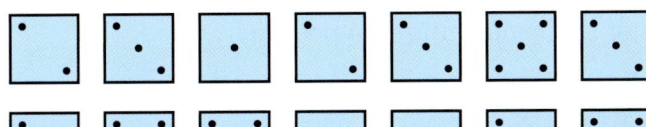

   a) What was the modal score?

   b) What was the median score?

4. Write down five numbers so that:

   the mean is 6

   the median is 5

   the mode is 4.

5. Louise claims that she is better at maths than her sister Petra.

Louise's last five marks were 63, 72, 58, 84 and 75 and Petra's last four marks were 69, 73, 81 and 70. Find the mean mark for Louise and for Petra. Is Louise better at maths than Petra?

6. The bar chart shows the marks scored in a test. What was the modal mark?

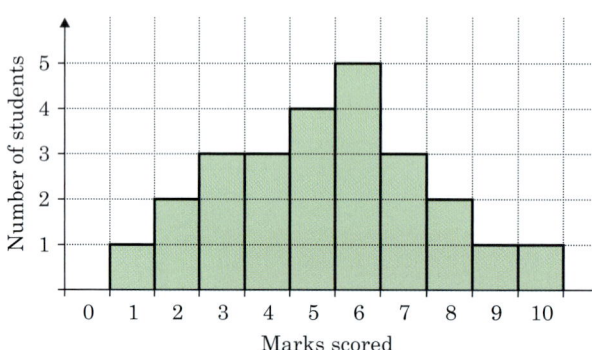

7. Six boys have heights of 1.53 m, 1.49 m, 1.60 m, 1.65 m, 1.90 m and 1.43 m.

a) Find the mean height of the six boys.

b) The shortest boy leaves the group. Find the mean height of the remaining five boys.

8. Seven people in a lift have masses of 44 kg, 51 kg, 57 kg, 63 kg, 48 kg, 49 kg and 45 kg.

a) Find the mean mass of the seven people.

b) The people with the highest and lowest mass leave the lift. Find the mean mass of the remaining five people.

9. In a maths test, the marks for group A were 9, 7, 8, 7, 5 and the marks for group B were 6, 3, 9, 8, 2, 2.

a) Find the mean mark for group A.

b) Find the mean mark for group B.

c) Find the mean mark for all the students.

10. The following are the salaries of 5 people:

A $22 500   B $17 900   C $21 400

D $22 500   E $155 300.

a) Find the mean and the median of their salaries.

b) Which does *not* give a fair 'average'? Write a sentence to explain why.

11. A farmer has 32 sheep to sell. The masses of the sheep in kg are:

81	81	82	82	83	84	84	85
85	86	86	87	87	88	89	91
91	92	93	94	96	150	152	153
154	320	370	375	376	380	381	390

[Total mass = 5028 kg]

On the telephone to a potential buyer, the farmer says that the 'average' mass of the sheep is 'over 157 kg'.

a) Find the mean mass and the median mass of the sheep.

b) Which 'average' has the farmer used to describe his sheep? Does this average describe the sheep fairly?

12. A company sells seeds and claims that the average height of the plants after one year's growth will be 85 cm. A sample of 24 of the plants was measured after one year and gave the following results (in cm):

6	7	7	9	34	56	85	89
89	90	90	91	91	92	93	93
93	94	95	95	96	97	97	99

[The sum of the heights is 1788 cm.]

a) Find the mean and the median height of the sample of plants.

b) Is the company's claim about average height justified? Explain your answer.

13. A group of 50 adults were asked how many books they had read in the previous year. The results are shown in the frequency table below.

Number of books	0	1	2	3	4	5	6	7	8
Frequency	5	5	6	9	11	7	4	2	1

a) Calculate the mean number of books read per adult.

b) Calculate the range of the number of books read.

A group of children was asked the same question and it was found that the mean number of books read was 6 with a range of 11.

c) i) Explain why this information supports the view that the children had read, on average, more books than the adults.

ii) Explain why this information supports the view that the number of books read by the children was more varied than the number read by the adults.

14. A number of people were asked how many coins they had in their pockets. The results are shown below.

Number of coins	0	1	2	3	4	5	6	7
Frequency	3	6	4	7	5	8	5	2

a) Calculate the mean number of coins per person.

b) Calculate the range of the number of coins.

The same people were asked how many bank notes they had in their wallets. It was found that the mean number of bank notes was 2 with a range of 5.

c) Write two sentences comparing the sets of data.

15. The following tables give the distribution of marks obtained by different classes in various tests. For each table, find the mean, median and mode.

a)

Mark	0	1	2	3	4	5	6
Frequency	3	5	8	9	5	7	3

b)

Mark	15	16	17	18	19	20
Frequency	1	3	7	1	5	3

c)

Mark	0	1	2	3	4	5	6
Frequency	10	11	8	15	25	20	11

16. One hundred golfers play the same hole and the number of shots they took are summarised below.

Number of shots	2	3	4	5	6	7	8
Number of players	2	7	24	31	18	11	7

Find:

a) the mean number of shots

b) the median number of shots.

17. The mean mass of five men is 76 kg. The masses of four of the men are 72 kg, 74 kg, 75 kg and 81 kg. What is the mass of the fifth man?

18. The mean length of 6 rods is 44.2 cm. The mean length of 5 of them is 46 cm. How long is the sixth rod?

19. a) The mean of 3, 7, 8, 10 and x is 6. Find x.

b) The mean of 3, 3, 7, 8, 10, x and x is 7. Find x.

20. The mean height of 12 men is 1.70 m, and the mean height of 8 women is 1.60 m. Find:

a) the total height of the 12 men

b) the total height of the 8 women

c) the mean height of the 20 men and women.

21. The total mass of 6 rugby players is 540 kg and the mean mass of 14 ballet dancers is 40 kg. Find the mean mass of the group of 20 rugby players and ballet dancers.

22. The mean mass of 8 boys is 55 kg and the mean mass of a group of girls is 52 kg. The mean mass of all the children is 53.2 kg. How many girls are there?

23. For the set of numbers below, find the mean and the median.

1, 3, 3, 3, 4, 6, 99

Which average best describes the set of numbers?

24. In a history test, Andrew got 62%. For the whole class, the mean mark was 64% and the median mark was 59%. Which 'average' tells Andrew whether he is in the 'top' half or the 'bottom' half of the class?

25. The mean age of three people is 22 and their median age is 20. The range of their ages is 16. How old is each person?

5.2 Displaying data

Raw numerical data can be collected by carrying out surveys or experiments. This sort of information is often much easier to understand when displayed in a diagram.

Pictograms

In a pictogram, you represent the frequency by a simple visual symbol that is repeated.

For example, this pictogram shows how many pizzas were sold on four days.

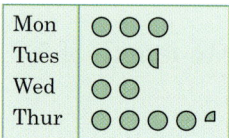

Key: ◯ represents 4 pizzas

In the pictogram:

- 12 pizzas were sold on Monday
- 10 pizzas were sold on Tuesday
- 8 pizzas were sold on Wednesday
- 17 pizzas were sold on Thursday.

A disadvantage of a pictogram is that it is not always easy to show fractions of the symbol and this can sometimes only be approximate.

Exercise 5.2A

1. The pictogram shows the money spent in a café by four students.

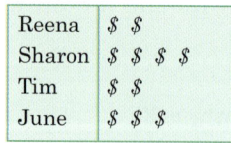

Key: $ represent $1

a) Who spent most?

b) How much was spent altogether?

c) How would you show that someone spent 50c?

2. The pictogram shows the make of cars in a car park.

a) How many cars does the represent?

Make	Number of cars	
Ford	20	🚗 🚗
Renault		🚗 🚗 🚗
Toyota	25	
Audi		🚗 🚗

b) Copy and complete the pictogram.

3. The frequency table shows the number of letters posted to six houses during 1 week. Draw a pictogram to represent these data.

House 1	House 2	House 3	House 4	House 5	House 6
25	32	20	17	14	27

Stem-and-leaf diagrams

Data can be displayed in groups using a **stem-and-leaf diagram**. Here are the ages of 20 people who attended a concert:

25	65	43	16	28	32	57	21	17	61
21	43	36	21	14	35	22	44	52	47

We start by choosing a sensible way to group the data. Here we can use the tens digit of the ages, giving us the groups 10–19, 20–29, 30–39, 40–49, 50–59, 60–69.

We then use the tens digit as the 'stem' and the ones digit as the 'leaf'.

Stem	Leaf					
1	6	7	4			
2	5	8	1	1	1	2
3	2	6	5			
4	3	3	4	7		
5	7	2				
6	5	1				

The highlighted number represents the first data value, 25.

You complete the diagram using the data in the order it appears in the list. This called an **unordered stem-and-leaf diagram**.

You can then make a second diagram, putting the data in numerical order. This is called an **ordered stem-and-leaf diagram**. You also need to include a key, containing a sample piece of data, so that people know how to interpret your diagram.

Stem	Leaf					
1	4	6	7			
2	1	1	1	2	5	8
3	2	5	6			
4	3	3	4	7		
5	2	7				
6	1	5				

Key

1 | 4 means age 14

From this diagram, it is easy to find the mode, the median and the range of the data.

Example

For the data in the stem-and-leaf diagram above, find the mode, median and range.

Stem	Leaf					
1	4	6	7			
2	1	1	1	2	5	8
3	2	5	6			
4	3	3	4	7		
5	2	7				
6	1	5				

The mode is most frequent value, in this case 21.

There are 20 data values, so the median is halfway between the 10th and the 11th values.

The median is 33.5

The range is the smallest value subtracted from the largest value.

The range is $65 - 14 = 51$.

Back-to-back stem-and-leaf diagrams

Two sets of data can be compared using a back-to-back stem-and-leaf diagram.

10 students who received coaching and 10 students who did not receive coaching all took part in a mathematics competition. Their scores are shown in this back-to-back stem-and-leaf diagram.

Coached			Stem	Not coached			
			0	4	8		
9	5	4	1	0	2	5	9
7	6	0	2	4	8		
6 5	4	1	3	2	4		

Key

9 | 1 | 2 means a score of 19 for coached and a score of 12 for uncoached.

Because the two sets of data share the same stem, you can use the stem-and-leaf diagram to compare the data.

> ## Example
>
> Use the stem-and-leaf diagram above to compare the performance of the two groups of students.
>
> The median score for the coached students is 26.5.
>
> The range of scores for the coached students is $36 - 14 = 22$.
>
> The median score for the uncoached students is 17.
>
> The range of scores for the uncoached students is $34 - 4 = 30$.
>
> The coached students did better on average in the competition (higher median), and the scores for the coached students are more consistent (lower range).

Exercise 5.2B

1. The marks scored by 25 students in a history test are as follows:

62	45	53	76	60	45	33	64	53	36
71	42	26	48	62	66	29	37	21	74
48	56	52	68	62					

   a) Draw a stem-and-leaf diagram to display this data.

   b) What was the median score for the students?

   c) Write down the range of the scores.

2. Here is a stem-and-leaf diagram showing the times taken for a group of amateur athletes to run 100 metres, measured to the nearest tenth of a second.

Stem	Leaf			
12	9			
13	0	1	5	8
14	1	6	8	
15	2	2	6	7

Key

13 | 5 means 13.5 seconds

a) How many athletes' scores were recorded?

b) Find the median time taken.

c) Find the range of the times recorded.

d) Write down the modal time.

3. A group containing girls and boys measured their hand spans in centimetres. Here are the results:

Girls   17.4   19.4   18.8   16.7   16.1   21.0   19.3   16.5   20.8   18.5
Boys   17.7   21.0   21.9   18.2   23.1   22.2   18.8   22.7   17.5   19.3

a) Copy and complete the following back-to-back stem-and-leaf diagram to display this data.

Girls	Stem	Boys
	16	
	17	
	18	
	19	
	20	
	21	
	22	
	23	

Key

4 | 17 | 7 means 17.4 cm
for the girls and 17.7 cm
for the boys

b) What are the median hand spans for the girls and for the boys?

c) What are the ranges of the hand spans for the girls and for the boys?

4. The lengths, to the nearest minute, of 10 horror films and 10 action films are collected:

Horror	99	90	94	85	105	92	88	95	89	100
Action	110	88	99	90	119	100	121	106	93	110

a) Display this data in a back-to-back stem-and-leaf diagram.

b) Find the median length of each type of film.

c) Based on this sample, which type of film is, on average, longer?

5.3 Pie charts

Example

The pie chart shows the holiday intentions of 600 people.

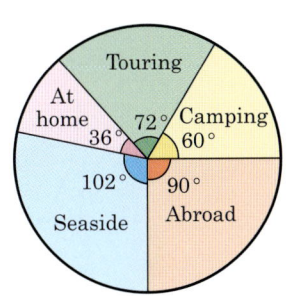

a) Number of people camping $= \dfrac{60°}{360°} \times 600 = 100$

b) Number of people touring $= \dfrac{72°}{360°} \times 600 = 120$

c) Number of people at seaside $= \dfrac{102°}{360°} \times 600 = 170$

A pie chart shows the relative proportion of the data in each sector, not the actual number of data items.

To find the actual number of data items represented by each sector, you need to know the total number of data values, as in the example above.

Exercise 5.3A

1. The total cost of a holiday was \$900. The pie chart shows how this cost was made up.

a) How much was spent on food?

b) How much was spent on travel?

c) How much was spent on the hotel?

d) How much was spent on other items?

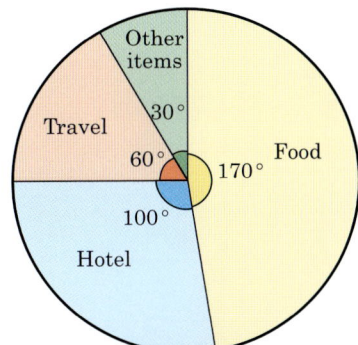

2. Qing-nian has an income of $60 000. The pie chart shows how he uses the money.

How much does he spend on:

a) food

b) rent

c) savings

d) entertainment

e) travel?

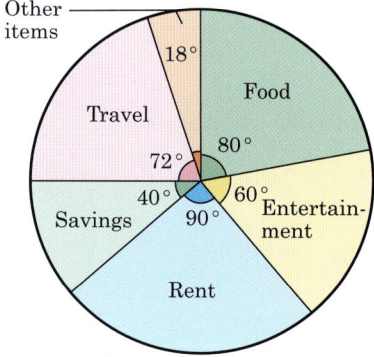

3. The total expenditure of a council is $36 000 000.

The pie chart shows how the money was spent.

a) How much was spent on:

   i) education   **ii)** health care?

b) What is the angle representing expenditure on highways?

c) How much was spent on highways?

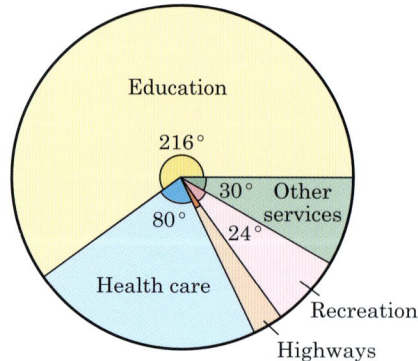

4. The pie chart shows how a student spends her time in a maths lesson which lasts 60 minutes.

   a) How much time does she spend:

   i) getting ready to work

   ii) working

   iii) listening to the teacher?

   b) She spends 3 minutes talking. What is the angle on the pie chart for the time spent talking?

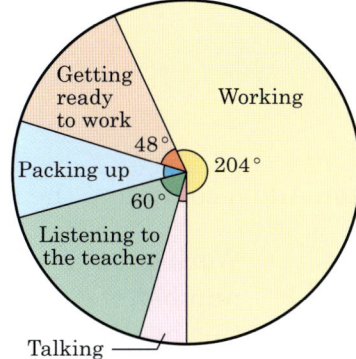

5. Between finishing their dinner and going to bed, Zac and Lucy had four hours. These pie charts show how they spent their time.

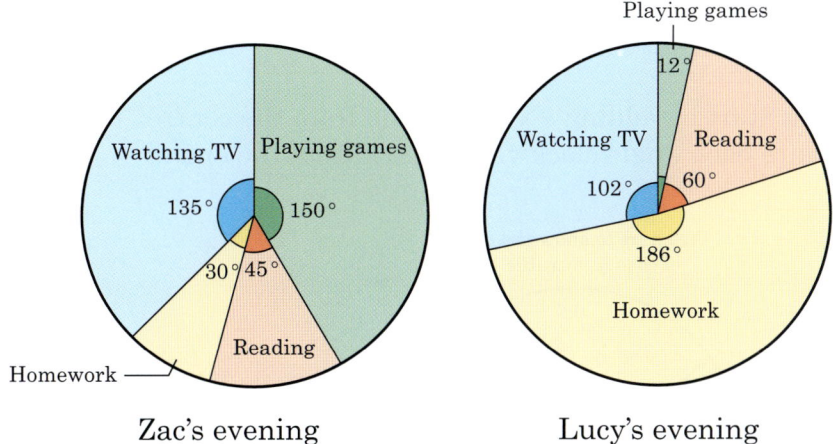

Zac's evening   Lucy's evening

   a) Which one of them spent the most time reading?

   b) How many minutes did Zac spend doing his homework?

   c) How many more minutes did Zac spend watching TV than Lucy?

Drawing pie charts

If you are asked to draw a pie chart, you need to find the angle of the sector representing each item in the table.

Example

The colours of 40 cars in a traffic jam are recorded.

Colour	Frequency
Red	7
White	12
Grey	15
Black	6

Draw a pie chart to show this data.

You need to work out the angle for each sector. Add a column to the table:

Colour	Frequency	Angle
Red	7	$\dfrac{7}{40} \times 360° = 63°$
White	12	$\dfrac{12}{40} \times 360° = 108°$
Grey	15	$\dfrac{15}{40} \times 360° = 135°$
Black	6	$\dfrac{6}{40} \times 360° = 54°$

Work out 7 as a fraction of the total number of cars and then multiply this by $360°$.

Repeat this process for each of the other rows in the table.

When drawing the pie chart, start by adding a vertical line from the centre going upwards. Then measure each angle clockwise in turn from the end of the previous sector.

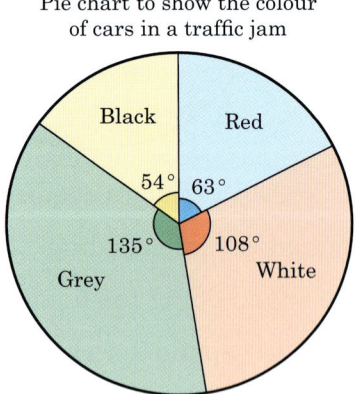

Pie chart to show the colour of cars in a traffic jam

Exercise 5.3B

1. At the semi-final stage of the T20 Cricket World Cup in 2022, 72 neutral commentators were asked to predict who they thought would win. Their answers were:

Pakistan	9	England	22
India	40	New Zealand	1

   a) Work out:

   i) $\dfrac{9}{72}$ of $360°$ **ii)** $\dfrac{40}{72}$ of $360°$ **iii)** $\dfrac{22}{72}$ of $360°$ **iv)** $\dfrac{1}{72}$ of $360°$

   b) Draw an accurate pie chart to display the predictions of the 72 commentators.

2. A survey was carried out to find what 400 students did when they left school:

   120 went into employment

   160 went to university

   80 went to college in another country

   40 were unemployed.

   a) Simplify the following fractions: $\dfrac{120}{400}$; $\dfrac{160}{400}$; $\dfrac{80}{400}$; $\dfrac{40}{400}$.

   b) Draw an accurate pie chart to show the information above.

3. In a survey on washing powder 180 people were asked to state which brand they preferred. 45 chose Brand A.

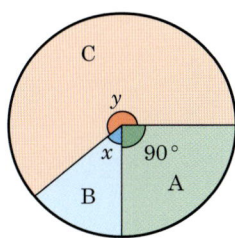

   If 30 people chose brand B and 105 chose Brand C, calculate the angles x and y.

4. A packet of breakfast cereal with a mass of 600 g contains four ingredients as follows:

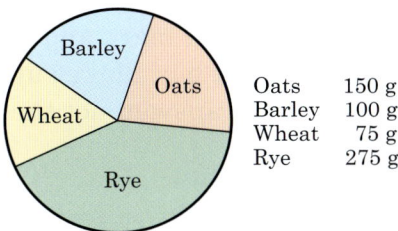

Oats	150 g
Barley	100 g
Wheat	75 g
Rye	275 g

   Calculate the angles on the pie chart shown and draw an accurate diagram.

5. The table below shows the share of car sales achieved by four companies in one year.

Company	A	B	C	D
Share of sales	50%	10%	25%	15%

In a pie chart to show this information, find the angle of the sectors representing:

a) Company A   **b)** Company B   **c)** Company C   **d)** Company D.

5.4 Tally charts, two-way tables and frequency diagrams

Example

The marks obtained by 36 students in a test were as follows:

1	3	2	3	4	2	1	3	0
5	3	0	1	4	0	4	4	3
3	4	3	1	3	4	3	1	2
1	3	4	0	4	3	2	5	3

Show the data:

a) on a tally chart   **b)** on a frequency diagram.   **c)** What was the modal mark?

a)

Mark	Tally	Frequency
0	||||	4
1	⊥⊥⊥⊥ |	6
2	||||	4
3	⊥⊥⊥⊥ ⊥⊥⊥⊥ ||	12
4	⊥⊥⊥⊥ |||	8
5	||	2

The tally marks are grouped into 5s.

b)

The modal mark is the most frequently occurring mark and corresponds to the highest bar. Make sure you give the *mark* as the answer, not the frequency.

c) The modal mark is 3.

Exercise 5.4A

1. In a survey, the number of occupants in the cars passing a school was recorded.

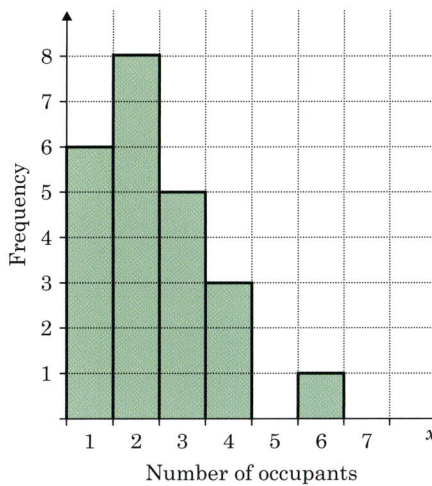

a) How many cars had 3 occupants?

b) How many cars had fewer than 4 occupants?

c) How many cars were in the survey?

d) What was the total number of occupants in all the cars in the survey?

e) What fraction of the cars had only one occupant?

f) Write down the modal number of occupants.

g) Work out the range of the number of occupants.

2. In an experiment, two dice were thrown sixty times and the total score showing was recorded.

2	3	5	4	8	6	4	7	5	10
7	8	7	6	12	11	8	11	7	6
6	5	7	7	8	6	7	3	6	7
12	3	10	4	3	7	2	11	8	5
7	10	7	5	7	5	10	11	7	10
4	8	6	4	6	11	6	12	11	5

a) Draw a tally chart to show the results of the experiment. The tally chart has been started below.

Score	Tally	Frequency
2	||	2
3	||||	4
4		
.		
.		

b) Draw a frequency diagram to illustrate the results. Plot the frequency on the vertical axis.

3. The bar chart shows the profit/loss made by a toy shop from September 2021 to April 2022.

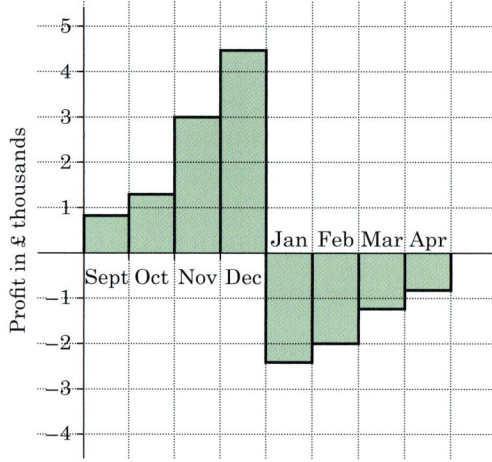

a) Estimate the total profit in this period.

b) Describe what is happening to the shop's profits in this period.
Hint: think of an explanation for the shape of the bar chart.

Two-way tables, dual bar charts and composite bar charts

Sometimes the data collected is subdivided into two (or more) groups, for example boys and girls, or adults and children.

In cases like this, it can be displayed in a **two-way table.**

Here is an example of a two-way table, showing the favourite sports of some men and some women.

	Cricket	Hockey	Basketball	Football
Men	12	14	7	3
Women	8	16	9	5

Data of this type can be displayed in two different ways, depending on exactly what you are trying to illustrate.

If you want to compare the number of men and women who like each sport, you can use a **dual** (or side-by-side) bar chart.

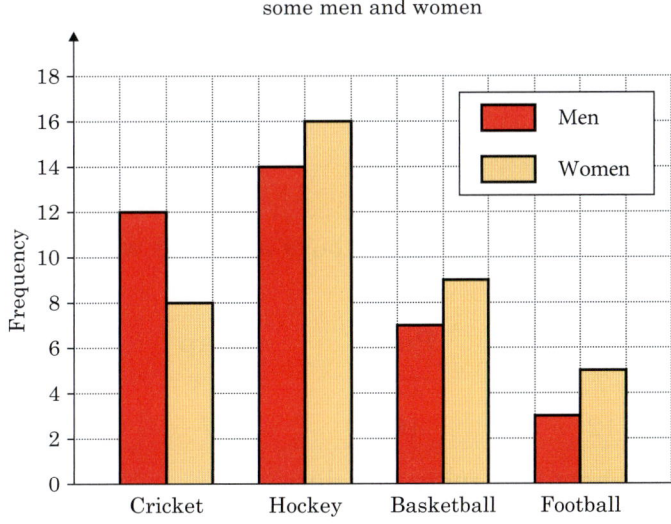

Dual bar chart showing the favourite sports of some men and women

From this chart, you can clearly see that more men than women like cricket, and that more women than men like basketball.

However, it is difficult to see the total number of men and women from this type of chart since you would need to add up all the different bar heights. But if you display the data using a **composite** (or stacked) bar chart, then this information is clearly shown.

You can compare the number who like each sport by looking at the relative height of each part of the bar. For example in the next chart, the yellow part is bigger for the women than for the men, so more women than men in the survey like football.

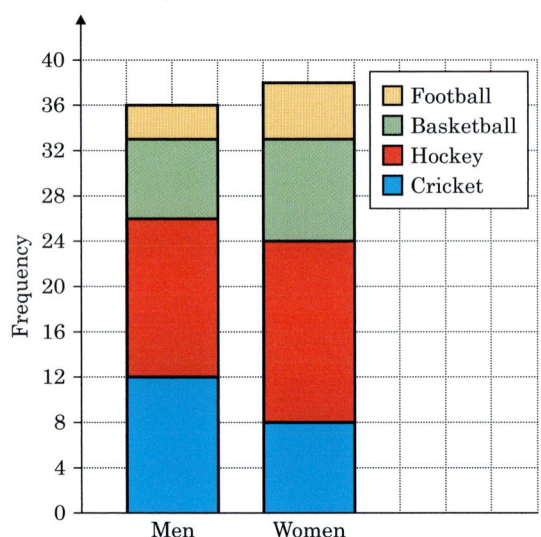

Composite bar chart showing the favourite sports of some men and women

You can see that there are more women than men in this survey.

Compare the size of each part to see whether more men or more women liked each sport.

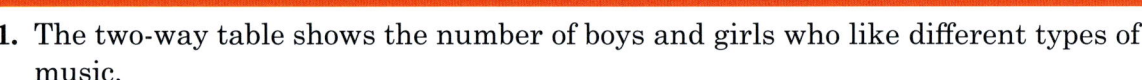

Exercise 5.4B

1. The two-way table shows the number of boys and girls who like different types of music.

	Rock	Pop	Rap	Dance hall
Boys	7	12	13	9
Girls	12	16	7	10

a) Draw a dual bar chart to display this data.

b) Draw a composite bar chart to show this data.

c) Which diagram is best for comparing the total number of boys and girls?

2. In a survey, some children and some adults are asked to name their favourite food.

12 of the adults surveyed say their favourite food is curry, 15 say it is kebabs and 8 say it is dhal.

Of the children surveyed, 9 say their favourite food is curry, 12 say it is kebabs and 16 say it is dhal.

a) Show this information in a two-way table.

b) Draw a dual bar chart to show this data.

c) Draw a composite bar chart to show this data.

d) Which diagram is best for comparing the number of adults and children that like kebabs?

3. Two shops record their daily sales of ice cream for a week. The results are shown in the two-way table.

	Monday	Tuesday	Wednesday	Thursday	Friday
Shakee's	80	70	55	60	80
Bargie's	60	55	20	65	70

a) Draw a dual bar chart to show this data.

b) Bargie's ran out of ice cream and had to close early on one of the days. Use your dual bar chart to identify the day.

c) Draw a composite bar chart to show this data.

d) Which ice cream shop had the highest weekly sales?

4. A survey was carried out to find out the favourite type of chocolate of some children and some adults. A dual bar chart was drawn.

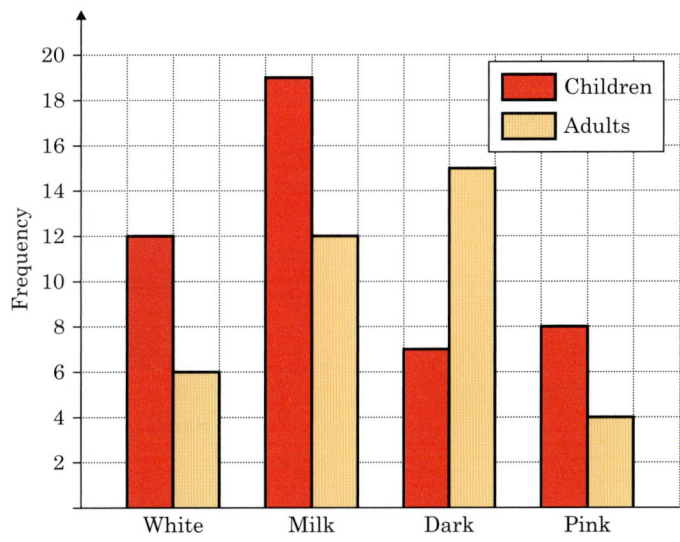

Dual bar chart showing favourite types of chocolate

a) Draw a composite bar chart to show this data.

b) Were there more adults or more children in the survey? How many more?

Grouped data

Sometimes the data to be displayed can take a wide range of values. In such cases, it is convenient to put the data into groups before drawing a tally chart and frequency diagram.

Example

The hand spans of 21 children were measured as follows:

14.8	20.0	16.9	20.7	18.1	17.5	18.7
19.0	19.8	17.8	14.3	19.2	21.7	17.4
16.0	15.9	18.5	19.3	16.6	21.2	18.4

Group the data and draw up a tally chart and frequency diagram.

The smallest value is 14.3 cm and the largest is 21.7 cm. The data can be grouped as shown in the table.

Inequalities are used to make sure that each value is included in only one interval.

For example, 20.0 goes into the last group $20 \leqslant s < 22$.

Then the frequency diagram can be drawn.

Notice that there are no gaps between the bars as this is a continuous scale.

Class interval	Tally
$14 \leqslant s < 16$	|||
$16 \leqslant s < 18$	ⵀ|	
$18 \leqslant s < 20$	ⵀ |||
$20 \leqslant s < 22$	||||

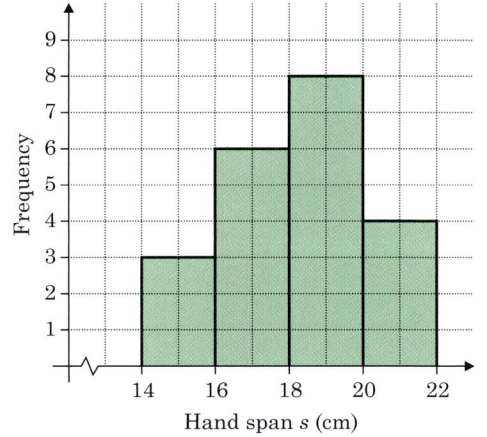

Exercise 5.4C

1. The frequency diagrams show the heights of two groups of people. One group is high school students and the other is adults.

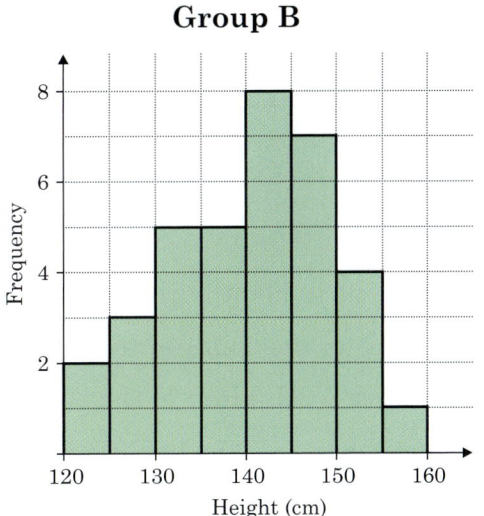

a) Which of the diagrams represents the high school students?

b) How many students were over 150 cm tall?

c) How many students were in the group?

d) How many adults were there?

e) What fraction of the adults were under 150 cm tall?

2. In a survey, the heights of children aged 15 were measured in four countries around the world. A random sample of children was chosen, not necessarily the same number from each country.

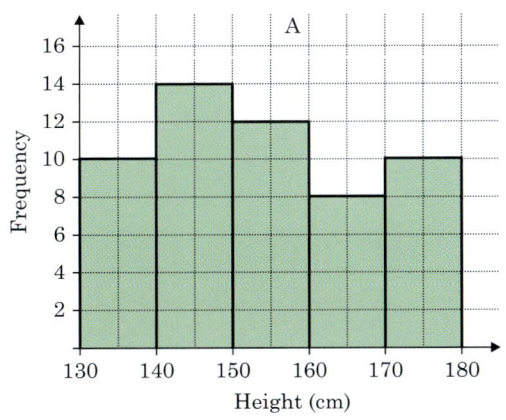

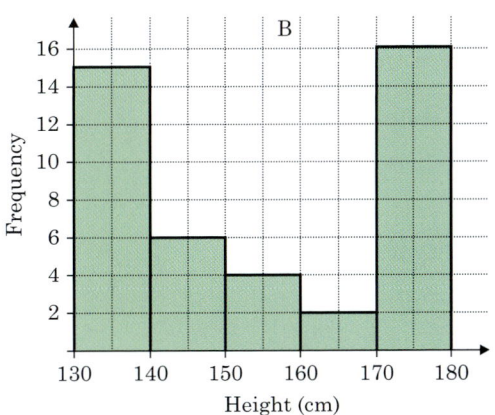

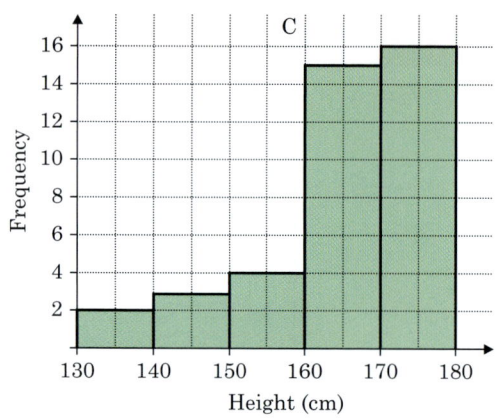

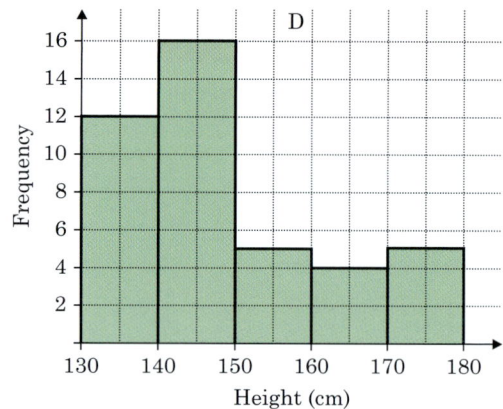

Use the graphs to identify the country in each of the statements below.

a) Two-thirds of the children from Country _____ were less than 150 cm tall.

b) There were 54 children in the sample from Country _____ .

c) In Country _____ the heights were spread fairly evenly across the range 130 to 180 cm.

d) The smallest sample of children came from Country _____ .

e) In Country _____ three-quarters of the children were either tall or short.

3. A farmer grew carrots in two adjacent fields A and B and treated one of the fields with fertiliser. A random sample of 50 carrots was taken from each field and the mass found. Here are the results for Field A (all in grams).

118	91	82	105	72	92	103	95	73	109
63	111	102	116	101	104	107	119	111	108
112	97	100	75	85	94	76	67	93	112
70	116	118	103	65	107	87	98	105	117
114	106	82	90	77	88	66	99	95	103

Make a tally chart using the groups given.

Mass	Tally	Frequency
$60 \leqslant m < 70$		
$70 \leqslant m < 80$		
$80 \leqslant m < 90$		
$90 \leqslant m < 100$		
$100 \leqslant m < 110$		
$110 \leqslant m < 120$		

The frequency diagram for Field B is shown below.

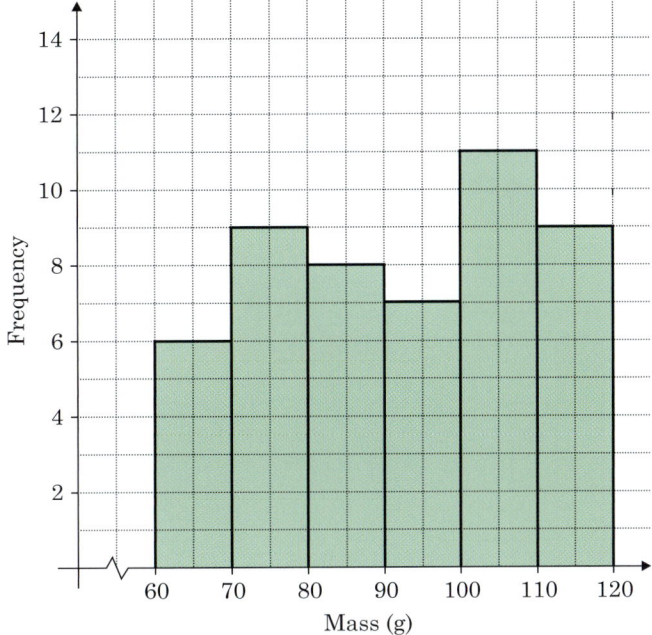

Mass (g)

Copy the graph above and, in a different colour, draw the graph for Field A. Which field do you think was treated with the fertiliser?

4. In an experiment, 52 children took an IQ test before and after a course of vitamin pills. Here are the results.

Before:

81	107	93	104	103	96	101	102	93	105	82	106	97
108	94	111	92	86	109	95	116	92	94	101	117	102
95	108	112	107	106	124	125	103	127	118	113	91	113
113	114	109	128	115	86	106	91	85	119	129	99	98

After:

93	110	92	125	99	127	114	98	107	128	103	91	104
103	83	125	91	104	99	102	116	98	115	92	117	97
126	100	112	113	85	108	97	101	125	93	102	107	116
94	117	95	108	117	96	102	87	107	94	103	95	96

a) Put the scores into convenient groups between 80 and 130.

b) Draw two frequency diagrams to display the results.

c) Write a conclusion. Did the vitamin pills make a significant difference to the children's scores in the test?

5.5 Frequency polygons

You saw on page 154 how a frequency distribution can be displayed in a bar chart or frequency diagram.

The number of peas in 40 pea pods is shown below. Frequency goes on the vertical axis.

A **frequency polygon** is formed by joining the midpoints of the tops of the bars in a frequency diagram using straight lines.

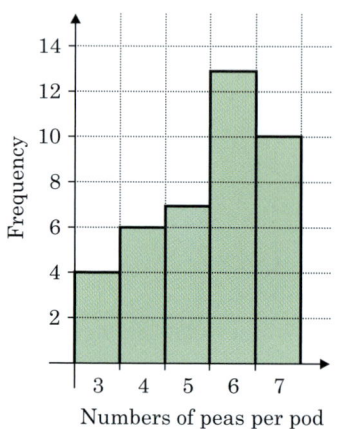

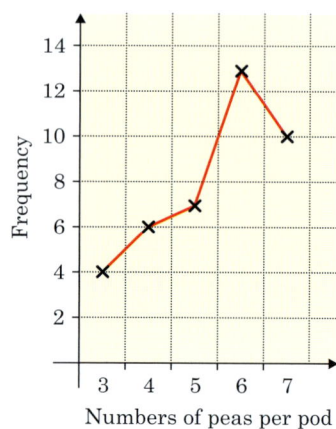

Discrete and continuous data

The data that we record can be either **discrete** or **continuous**.
Discrete data can take only certain values:

- the number of peas in a pod
- the number of children in a class
- shoe sizes.

Continuous data comes from measuring and can take any value:

- height of a child
- mass of an apple
- time taken to boil a kettle.

Class boundaries

The lengths of 36 pea pods were measured and rounded to the nearest mm. So a pea pod which is actually 59.2 mm long is rounded off to 59 mm.

52	80	65	82	77	60	72	83	63
78	84	75	53	73	70	86	55	88
85	59	76	86	73	89	91	76	92
66	93	84	62	79	90	73	68	71

This data can be put into a grouped frequency table.

Length (mm)	Tally	Frequency
$50 \leqslant l < 60$	||||	4
$60 \leqslant l < 70$	|||| |	6
$70 \leqslant l < 80$	ⵍⵍⵍ ⵍⵍⵍ ||	12
$80 \leqslant l < 90$	ⵍⵍⵍ ⵍⵍⵍ	10
$90 \leqslant l < 100$	||||	4

For the class $50 \leqslant l < 60$, the class boundaries are 50 and 60. The bar will go from 50 to 60 mm.

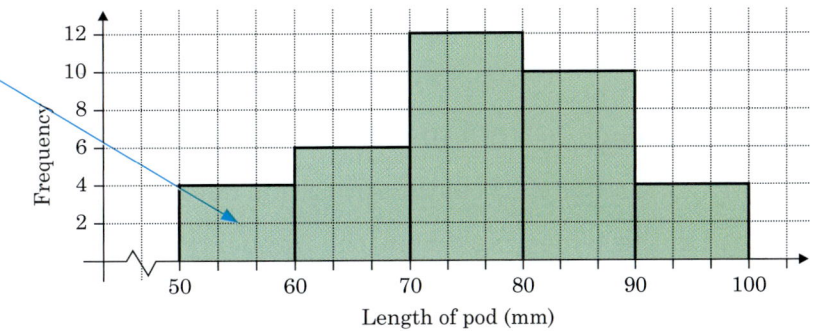

The frequency polygon for this data can be drawn in the same way as with discrete data. Note that you can draw the frequency polygon *without* drawing a bar chart. First, you must calculate the midpoint of each group.

For the $50 \leqslant l < 60$ group:

$$\text{midpoint} = \frac{50 + 60}{2} = 55$$

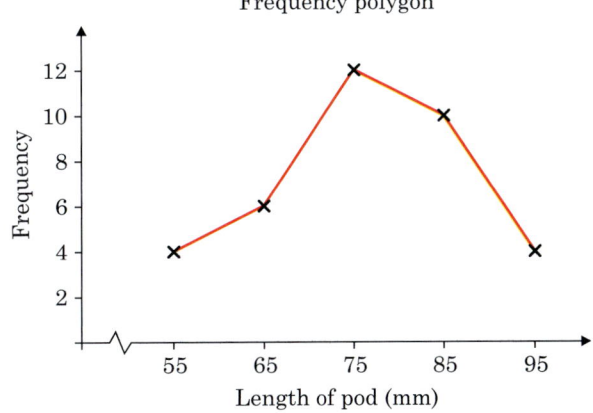

Frequency polygon

Midpoints

The midpoints of other groups can be calculated as follows:

a)

Mark	Midpoint
0–9	4.5
10–19	14.5

$$\left(\frac{10 + 19}{2}\right)$$

$$\left(\frac{0 + 9}{2}\right)$$

b)

Height	Midpoint
$150 \leqslant h < 155$	152.5
$155 \leqslant h < 160$	157.5

$$\left(\frac{155 + 160}{2}\right)$$

$$\left(\frac{150 + 155}{2}\right)$$

Exercise 5.5A

1. In a survey, the number of people in 100 cars passing a set of traffic lights was counted. Here are the results.

Number of people in car	0	1	2	3	4	5	6
Frequency	0	10	35	25	20	10	0

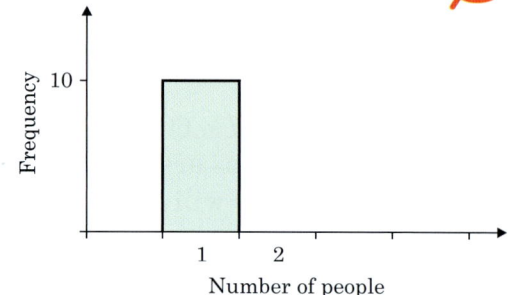

   a) Draw a bar chart to illustrate this data.

   b) On the same graph draw the frequency polygon.

   The bar chart has been started for you. For the frequency, use a scale of 1 cm for 5 units.

2. The frequency polygon shows the marks obtained by students in a maths test.

   a) How many students obtained 7 marks?

   b) How many students altogether took the test?

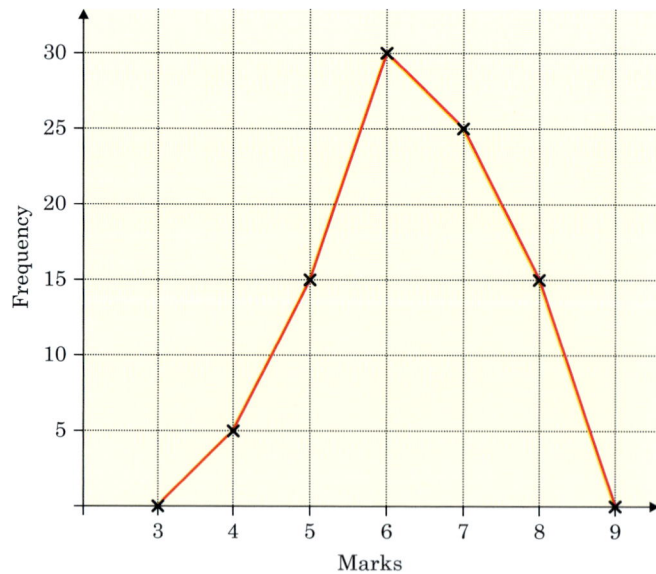

3. The heights of the members of several professional basketball teams were measured. The results are shown in the table

Draw a frequency diagram and a frequency polygon to illustrate this data.

Height	Frequency
$180 \leqslant h < 185$	5
$185 \leqslant h < 190$	8
$190 \leqslant h < 195$	15
$195 \leqslant h < 200$	11
$200 \leqslant h < 205$	6
$205 \leqslant h < 210$	2

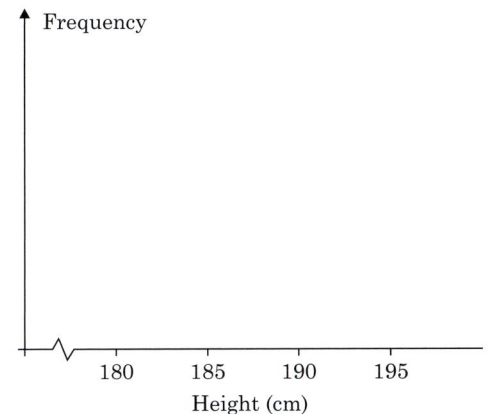

4. The two frequency polygons show the distribution of the masses of players in two different sports, A and B.

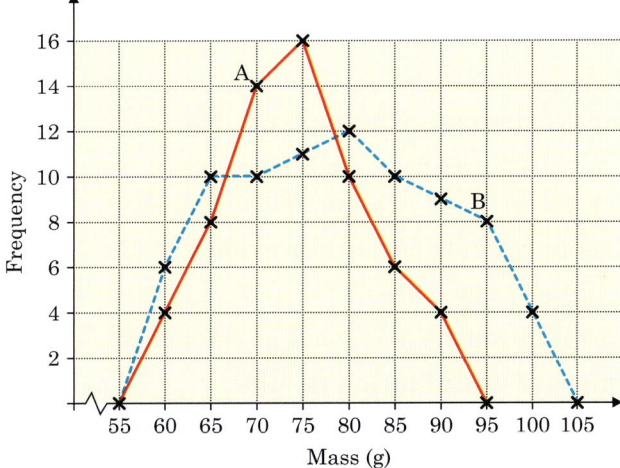

a) How many people played sport A?

b) Give two differences between the two frequency polygons.

c) For either A or for B, suggest a sport where you would expect a frequency polygon of masses to have this shape.

Explain in one sentence why you have chosen that sport.

5. A scientist measures the heights of some raspberry plants and also the total mass of fruit collected. She does this for two sets of plants: one with fertiliser and one without it. Here are the frequency polygons.

a) What effect did the fertiliser have on the heights of the plants?

b) What effect was there on the mass of fruit collected?

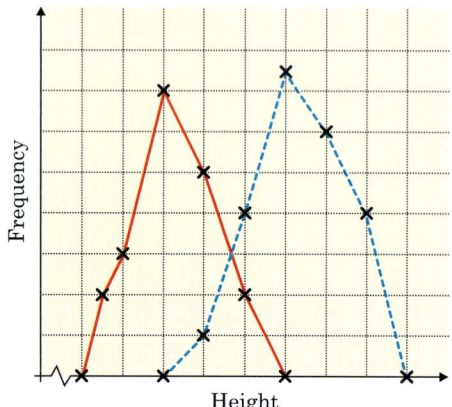
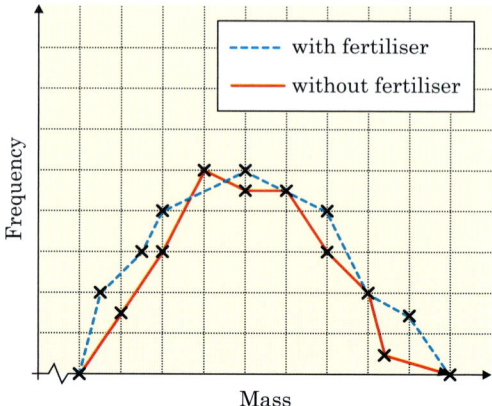

5.6 Scatter diagrams

Scatter diagrams

Sometimes it is important to discover whether there is a connection or relationship between two sets of data.

Examples:

- Are more ice creams sold when the weather is hot?
- Do tall people have higher pulse rates?
- Are people who are good at maths also good at science?
- Does watching TV improve examination results?

If there is a relationship, it will be easy to spot if your data is plotted on a **scatter diagram** – this is a graph in which one set of data is plotted on the horizontal axis and the other on the vertical axis.

Here is a scatter diagram showing the price of pears and the quantity sold.

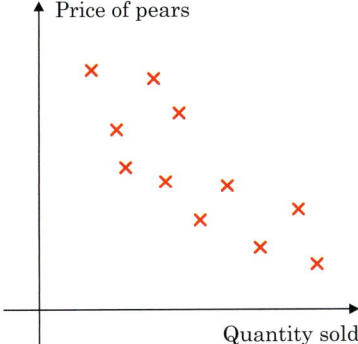

We can see a *connection* – when the price was high the sales were low and when the price went down the sales increased.

This scatter diagram shows the sales of a newspaper and the temperature. We can see there is *no connection* between the two variables.

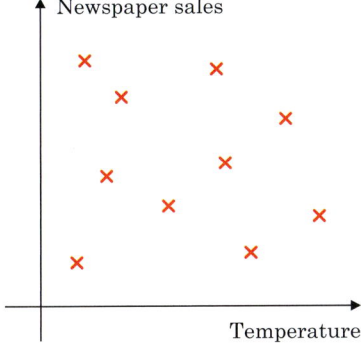

Correlation

The word **correlation** describes how things *co-relate*. There is correlation between two sets of data if there is a connection or relationship.

The correlation between two sets of data can be positive or negative and it can be strong or weak as indicated by these scatter diagrams.

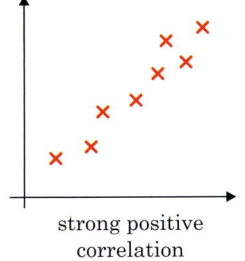

strong positive correlation

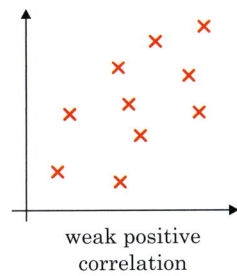

weak positive correlation

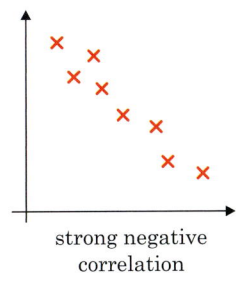

strong negative correlation

When the correlation is positive, the points approximately form a line that slopes upwards to the right. When the correlation is negative, the 'line' slopes downwards to the right.

When the correlation is strong, the points are bunched close to a line through their middle. When the correlation is weak, the points are more scattered.

It is important to realise that often there is *no* correlation between two sets of data.

If, for example, we take a group of students and plot their maths test results against their time to run 800 m, the diagram might look like the one on the right. While it looks, at first glance, that there is a positive correlation, a closer inspection of the two data sets indicates that there is almost certainly *no* correlation. A common mistake in this topic is to 'see' a correlation on a scatter diagram where none exists.

Time to run 800 m

Maths results

There is also *no* correlation in these two scatter diagrams.

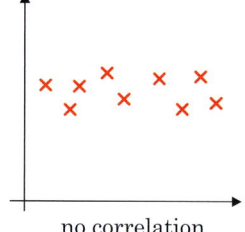

no correlation

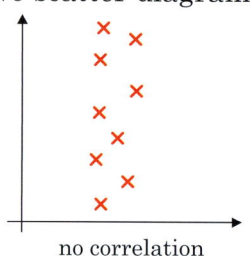

no correlation

Exercise 5.6A

1. Plot the points given on a scatter diagram, with *s* across the page and *p* up the page. Draw axes with values from 0 to 20.

   Describe the correlation, if any, between the values of *s* and *p* [i.e. 'strong negative', 'weak positive', etc.].

   a)

s	7	16	4	12	18	6	20	4	10	13
p	8	15	6	12	17	9	18	7	10	14

   b)

s	3	8	12	15	16	5	6	17	9
p	4	2	10	17	5	10	17	11	15

   c)

s	11	1	16	7	2	19	8	4	13	18
p	5	12	7	14	17	1	11	8	11	5

2. Describe the correlation, if any, in these scatter diagrams.

a)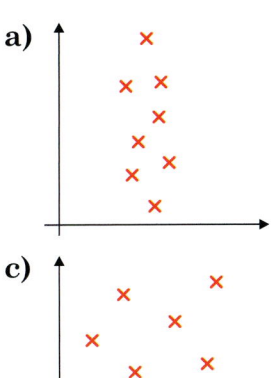

b)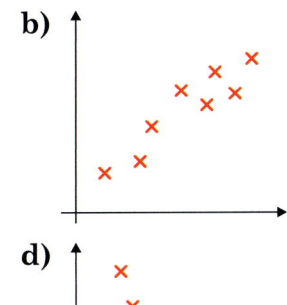

c)

d)

3. The scatter diagram shows the marks scored by some students in a chemistry and a physics test.

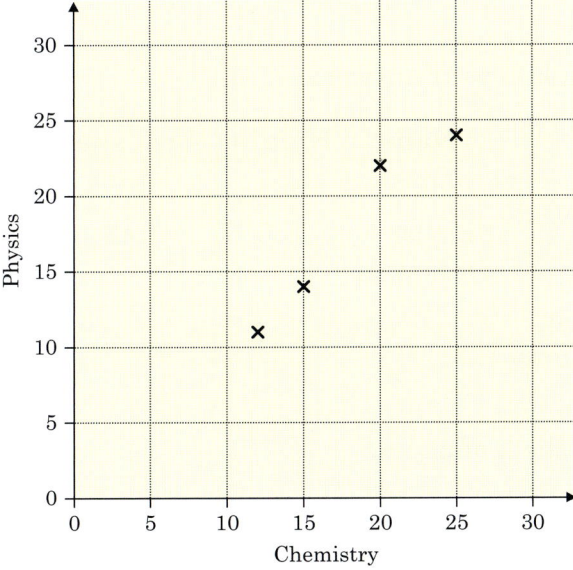

Alicia says: 'The points nearly lie in a straight line so there is evidence of a strong positive correlation'.

With reference to the scatter diagram, give a reason why Alicia's inference may be incorrect.

4. The scatter diagram shows the marks obtained by some students in a biology test and their hand span, in centimetres.

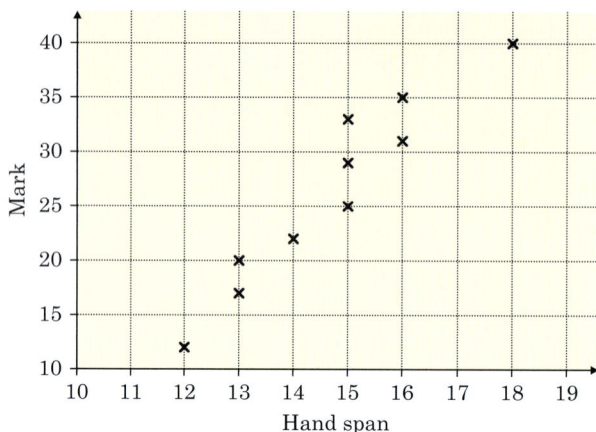

Mishthi says: 'The points lie close to a straight line so there is evidence that the bigger their hand span, the better the student does in biology'.

Is Mishthi's inference correct? Explain your answer.

5. Make the following measurements for everyone in your class:

Name	Height	Arm span	Head	
Ravi	161	165	56	
Layla	150	148	49	
Govinder				

height   (nearest cm)
arm span   (nearest cm)
head circumference   (nearest cm)
hand span   (nearest cm)
pulse rate   (beats/minute)

Enter all the measurements in a table, either on the board or on a sheet of paper.

a) Draw the scatter diagrams shown below:

i)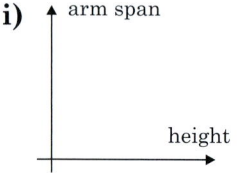

ii)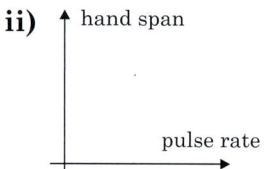

b) Describe the correlation, if any, in the scatter diagrams you drew in part **(a)**.

c) **i)** Draw a scatter diagram of two measurements where you think there might be positive correlation.

ii) Was there really a positive correlation?

Line of best fit

When a scatter diagram shows either positive or negative correlation, a **line of best fit** can be drawn. The sums of the distances to points on either side of the line are equal and there should be roughly an equal number of points on each side of the line. The line is easier to draw when a transparent ruler is used.

Here are the marks obtained in two tests by 9 students.

Student	A	B	C	D	E	F	G	H	I
Maths mark	28	22	9	40	37	35	30	22	?
Physics mark	48	45	35	57	50	55	53	45	52

A line of best fit can be drawn as there is a strong positive correlation between the two sets of marks.

The line of best fit can be used to estimate the maths result of student I, who missed the maths test but scored 52 in the physics test.

You can estimate that student I would have scored about 33 in the maths test by drawing a line across from 52 on the physics axis, to the line of best fit and then down to the maths axis.

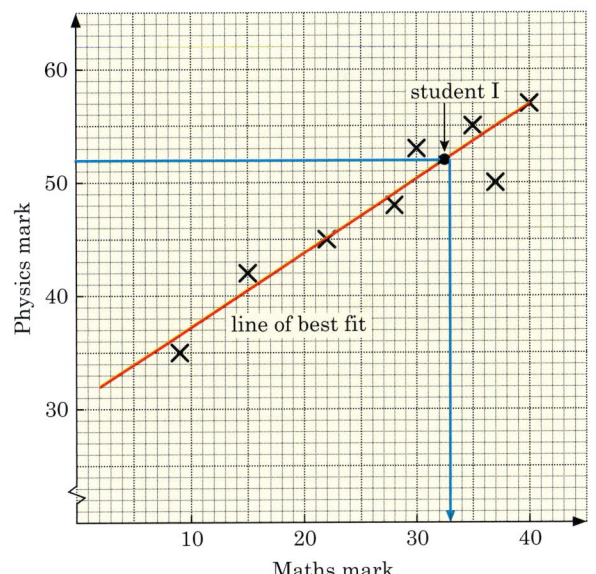

Exercise 5.6B

In Questions **1**, **2** and **3** plot the points given on a scatter diagram, with *s* across the page and *p* up the page.

Draw axes with the values from 0 to 20.

If possible draw a line of best fit on the diagram.

Where possible, estimate the value of *p* on the line of best fit where *s* = 10.

1.

s	2	14	14	4	12	18	12	6
p	5	15	16	6	12	18	13	7

2.

s	2	15	17	3	20	3	6
p	13	7	5	12	4	13	11

3.

s	4	10	15	18	19	4	19	5
p	19	16	11	19	15	3	1	9

4. The following data gives the marks of 11 students in a French test and in a German test.

French	15	36	36	22	23	27	43	22	43	40	26
German	6	28	35	18	28	28	37	9	41	45	17

a) Plot this data on a scatter diagram, with French marks on the horizontal axis.

b) Draw the line of best fit.

c) Estimate the German mark of a student who got 30 in French.

d) Estimate the French mark of a student who got 45 in German.

5. The data below gives some information about cars with the same size engine, when driven at different speeds.

Speed (m.p.h.)	30	62	40	80	70	55	75
Petrol consumption (m.p.g.)	38	25	35	20	26	34	22

a) Plot a scatter diagram and draw a line of best fit.

b) Estimate the petrol consumption of a car travelling at 45 m.p.h.

c) Estimate the speed of a car whose petrol consumption is 27 m.p.g.

6. The table shows the marks of 7 students in the two papers of a science examination.

Paper 1	35	10	60	17	43	55	49
Paper 2	26	15	40	15	30	34	35

a) Plot the marks on a scatter diagram, using a scale of 1 cm to 5 marks and draw a line of best fit.

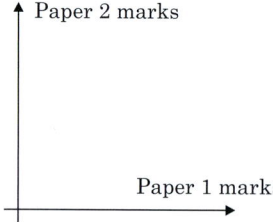

Paper 2 marks

Paper 1 marks

b) A student gained 25 marks on paper 1 but missed paper 2. What mark would you expect them to get on paper 2?

Revision exercise 5

1. The marks of nine students in a test were 7, 5, 2, 7, 4, 9, 7, 6, 6. Find

   a) the mean mark

   b) the median mark

   c) the modal mark.

2. The mean height of 10 boys is 1.60 m and the mean height of 15 girls is 1.52 m. Find the mean height of the 25 boys and girls.

3. a) The mean mass of 10 boys in a class is 56 kg.

   i) Calculate the total mass of these 10 boys.

   ii) Another boy, whose mass is 67 kg, joins the group. Calculate the mean mass of the 11 boys.

   b) A group of 10 boys whose mean mass is 56 kg joins a group of 20 girls whose mean mass is 47 kg. Calculate the mean mass of the 30 children.

4. The mean of four numbers is 21.

   a) Calculate the sum of the four numbers.

   Six other numbers have a mean of 18.

   b) Calculate the mean of the ten numbers.

5. The pie chart shows the after-school activities of 200 students.

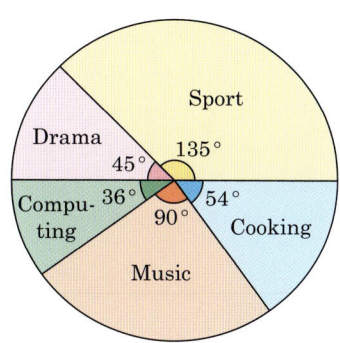

a) How many students do drama?

b) How many students do sport?

c) How many students do computing?

6. Forty teenagers were asked to name their favourite holiday destinations. Here are the results.

   Dubai 12   Egypt 5   Greece 10

   Turkey 4   India 9

   Display this information on a pie chart, showing the angles corresponding to each country.

7. The frequency diagram shows the rainfall recorded in a village in one month.

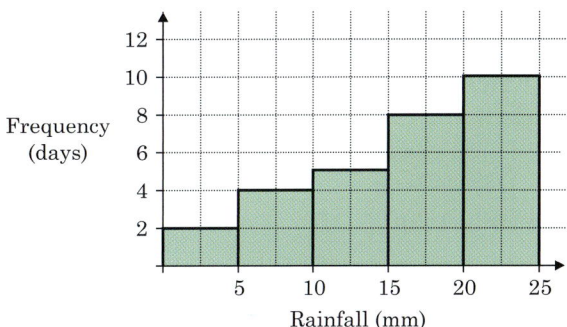

a) How many days were there in the month?

b) For how many days were there 10 mm or more of rain?

c) Chew Ling said, 'It rained more at the end of the month'. Explain whether Chew Ling is right or wrong.

8. The median age of five people was 11 and the range of their ages was 3. Copy each sentence below and state whether it is true, possible or false.

   a) Every person was either 10 or 11 years old.

b) The oldest person was 14 years old.

c) The mean age of the people was less than 11 years.

9. During 2022, Dilly kept a record of the highest temperature in London on the first day of each month. She recorded the temperatures to the nearest 1°C. Here are the results:

9 12 12 9 14 12 20 17 23 14 13 7

a) Complete the following stem-and-leaf diagram to display this data.

Stem	Leaf
0	
1	
2	

Key

1 | 2 means 12 °C

b) For the first day of the month in London during 2022, what was:

i) the median temperature

ii) the modal temperature

iii) the range of the temperatures?

10. Erica records the scores she gets in the written and practical elements of 10 physics tests.

Written	7	12	8	15	17	14	9	11	19	5
Practical	10	13	7	16	15	12	9	12	17	4

a) Draw a scatter diagram to show this information.

b) Describe the type of correlation shown by the scatter diagram.

c) Draw a line of best fit by eye.

d) Use your line of best fit to estimate the score that Erica would get in the practical element if she scored 10 in the written element.

11. Malak collects data on the length of some worms.

Length, x mm	**Frequency**
$10 \leqslant x < 20$	5
$20 \leqslant x < 30$	8
$30 \leqslant x < 40$	12
$40 \leqslant x < 50$	4
$50 \leqslant x < 60$	7

Draw a frequency polygon to show this information.

CALCULATOR

1. A sports magazine contains 24 articles on different national teams. The table shows how many articles there are of each team.

National team	Number of articles	Angle in a pie chart
Argentina	6	90°
Brazil	10	150°
England	3	
Egypt		

a) Complete the table. [3]

b) Complete the pie chart accurately and label the sectors for Brazil, England and Egypt. [2]

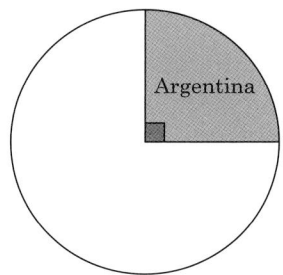

2. Manuela keeps a record of all her marks for vocabulary tests, as shown in the table below.

Mark	5	6	7	8	9	10
Frequency	1	5	10	9	7	3

a) i) How many vocabulary tests did Manuela do? [1]

ii) Write down the modal mark. [1]

iii) Find the median mark. [1]

iv) Calculate the mean mark. [3]

b) Manuela draws a pie chart to show this information. The sectors for her marks of 5, 6, 7 and 8 have already been drawn.

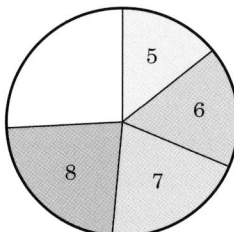

i) Calculate the angle of the sector for her mark of 9. [2]

ii) Complete the pie chart accurately. [1]

3. Which word describes the correlation in the scatter graph below?

Positive   Negative   None   [1]

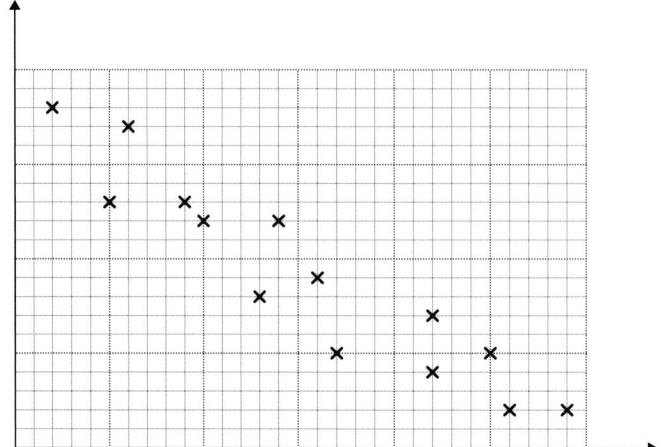

4. Amie plots a scatter diagram of midday temperature against the number of bottles of water she sells.

As the temperature increases, the number of bottles of water sold also increases. Which one of the following types of correlation will her scatter diagram show?

Positive   Negative   Zero   [1]

5. The table shows the daily profit, in dollars, of an ice cream seller in one week.

Monday	Tuesday	Wednesday	Thursday	Friday
756	738	760	759	767

For these values,

a) calculate the mean profit   [2]

b) find the median profit   [1]

c) find the range in profits.   [1]

6. Amnah selected a sample of 10 students from her school and measured their foot lengths and heights. The results are shown in the table below.

Foot length (cm)	15	18.5	22.5	26	19	23	17.5	25	20.5	22
Height (cm)	154	156	164	178	162	170	154	168	168	160

She calculated the mean foot length to be 20.9 cm and the range of the foot lengths to be 11 cm.

a) Calculate

   i) the mean height in cm   [2]

   ii) the range of the heights in cm.   [2]

b) In order to compare the two measures, she used a scatter diagram. The first three points are plotted on the grid.

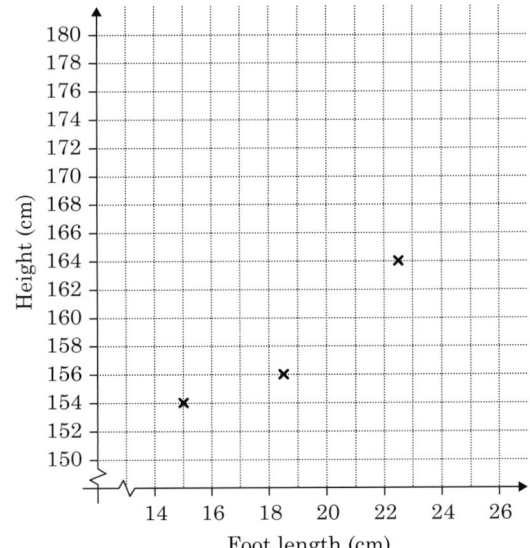

Foot length (cm)

i) Complete the scatter diagram by plotting the remaining 7 points. [2]

ii) Draw the line of best fit on the grid. [1]

iii) Use the line of best fit to estimate the height of a student with a foot length of 21 cm. [1]

iv) Which one of the following words describes the correlation?

Positive   Negative   Zero [1]

v) What does this suggest about the relationship between foot length and height? [1]

7. a) Neha records the scores of the 34 children in her class in a maths test.

4	10	5	6	4	8	6	4	7	3	9	7	4
7	3	5	4	6	5	10	7	5	5	6	4	7
7	6	6	5	5	3	5	6					

i) Using the list above copy and complete the frequency table. [3]

Score	3	4	5	6	7	8	9	10
Frequency								

ii) Calculate the mean of the scores. [3]

iii) Find the range of these scores. [1]

iv) Find the mode of these scores. [1]

v) Work out the median score. [2]

b) Faryal draws a bar chart to show the scores of the students in his class in a science test. [2]

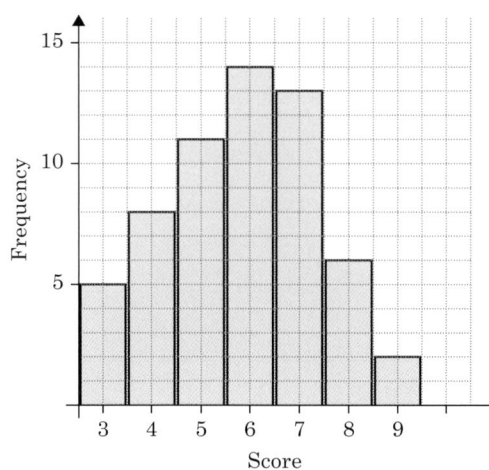

i) Copy the frequency table and use the information in the bar chart to complete it. [2]

Score	3 and 4	5 and 6	7 and 8	9 and 10
Frequency				

ii) Which is the modal class in the frequency table? [1]

8. The results of the men's 400-metre race from the 2016 and 2020 Olympic Games are displayed in the following stem-and-leaf diagram. The times are given to the nearest tenth of a second.

		2016					2020			
				9	43	0	8	9		
8	8	8	5	5	44	0	3	4	5	6
			1	0	45					

Key

5 | 44 | 3 means 44.5 seconds in 2016 and 44.3 seconds in 2020

a) Calculate

i) the median time for each year [2]

ii) the range of the times for each year. [2]

b) A newspaper report says, 'In 2020, the athletes were both faster and more consistent'.

Comment on the accuracy of this statement. [2]

6 Number 3

Bertrand Russell (1872–1970) tried to reduce all mathematics to formal logic. In particular, he realised that there were contradictions in the existing definitions used in set theory. He wrote to Gottlieb Frege, who was one of the developers of set theory, just as he was putting the finishing touches to a book that represented his life's work, pointing out that some of Frege's work was invalid. This led to further research in the field of logic and sets.

- Calculate with and convert between the following, including in contexts: proper fractions, improper fractions, mixed numbers, decimals and percentages.
- Simplify a ratio, divide a quantity by a ratio. Use ratios in context.
- Calculate a percentage of a quantity. Write one quantity as a percentage of another quantity. Solve various problems involving percentages including simple interest, compound interest, percentage increase and percentage decrease.
- Calculate with money including converting between currencies.
- Use and understand real-life graphs such as conversion graphs.

6.1 Fractions, decimals and percentages

Percentages are a convenient way of expressing fractions or decimals. '50% of $60' is the same as '$\frac{1}{2}$ of $60'. You need to be able to convert readily from one form to another.

Example

a) Change $\frac{7}{8}$ to a decimal.

$$\begin{array}{r} 0.\ 8\ 7\ 5 \\ 8{\overline{)7.^70^60^40}} \end{array}$$

Divide the numerator by the denominator, so divide 7 by 8.

$\frac{7}{8} = 0.875$

Add zeros after the decimal point.

b) Change 0.35 to a fraction.

$0.35 = \dfrac{35}{100}$

There are 2 decimal places, so this is thirty-five hundredths.

$\quad = \dfrac{7}{20}$

Simplify the fraction by dividing the numerator and denominator by 5.

c) Change $\frac{3}{8}$ to a percentage.

Multiply this fraction by 100%.

$\dfrac{3}{8} = \dfrac{3}{8} \times 100\%$

$\quad = \dfrac{300}{8}\%$

Multiply the numerator by 100 and then divide by 8.

$\quad = 37\dfrac{1}{2}\%$

Exercise 6.1A

1. Change the fractions to decimals.

a) $\dfrac{1}{4}$ b) $\dfrac{2}{5}$ c) $\dfrac{3}{8}$

d) $\dfrac{7}{20}$ e) $2\dfrac{5}{8}$ f) $9\dfrac{7}{10}$

2. Change the decimals to fractions and simplify.

   a) 0.2   **b)** 0.45   **c)** 0.36

   d) 0.125   **e)** 1.05   **f)** 0.007

3. Change to percentages.

   a) $\dfrac{1}{4}$ **b)** $\dfrac{1}{10}$ **c)** 0.72

   d) 0.075   **e)** 0.02   **f)** $1\dfrac{4}{5}$

4. Two shops had sale offers on an article which previously cost \$69. One shop had '$\dfrac{1}{5}$ off' and the other had '70% of old price'. Which price is lower?

5. Shareholders in a company can choose either '$\dfrac{1}{5}$ of \$5000' or '15% of \$5000'. Which is the greater amount?

6. Copy and complete the table.

	Fraction	Decimal	Percentage
a)	$\dfrac{1}{4}$		
b)		0.2	
c)			80%
d)	$\dfrac{1}{100}$		
e)			30%
f)	$4\dfrac{1}{20}$		

7. Arrange in order of size (smallest first).

   a) $\dfrac{1}{2}$; 45%; 0.6   **b)** 0.38; $\dfrac{6}{16}$; 4%   **c)** 0.111; 11%; $\dfrac{1}{10}$

For Questions **8** to **16**, use a calculator to evaluate. Round the answer to 2 decimal places.

   8. $\dfrac{1}{4} + \dfrac{1}{3}$ **9.** $\dfrac{2}{3} + 0.75$ **10.** $\dfrac{8}{9} - 0.24$

   11. $\dfrac{7}{8} + \dfrac{5}{9} + \dfrac{2}{11}$ **12.** $\dfrac{1}{3} \times 0.2$ **13.** $\dfrac{5}{8} \times \dfrac{1}{4}$

   14. $\dfrac{8}{11} \div 0.2$ **15.** $\left(\dfrac{4}{7} - \dfrac{1}{3}\right) \div 0.4$

   16. Pure gold is 24 carat gold. What percentage of 15 carat gold is pure gold?

6.2 Percentages

Example

a) Find 30% of 120.

    $10\% = 120 \div 10 = 12$ Find 10% by dividing by 10.

    $30\% = 12 \times 3 = 36$ Multiply this by 3 to find 30%.

b) Find 45% of 90.

    $10\% = 90 \div 10 = 9$ Find 10% by dividing by 10.

    $5\% = 9 \div 2 = 4.5$ Find 5% by halving 10%.

    $45\% = 9 \times 4 + 4.5 = 40.5$ Add four lots of 10% to 5% to find 45%.

c) Work out 22% of $40.

    $\dfrac{22}{100} \times \dfrac{40}{1} = \dfrac{880}{100}$ You can write your percentage as a fraction and multiply.

           $= 8.8$ Remember: 'of' means you multiply.

    Answer: $8.80

    [Alternatively, you can find 10% and 1%, and then find 11% and double it.]

d) Work out 16% of $85.

    Since $16\% = \dfrac{16}{100}$ you can replace 16% by 0.16.

    So 16% of $85 $= 0.16 \times 85 = \$13.60$

[Note that this method works best when you are allowed to use a calculator.]

e) Work out 120% of 500.

    $20\% = 100$ $20\% = \dfrac{20}{100} = \dfrac{1}{5}$, so you can find 20% by dividing by 5.

    $120\% = 500 + 100$ Add your 20% to the whole amount to find 120%.

           $= 600$

Exercise 6.2A

 Do not use a calculator for Questions **1** to **6**.

Work out:

1. 20% of $60   **2.** 10% of $80   **3.** 5% of $200

4. 6% of $50   **5.** 4% of $60   **6.** 130% of $80

7. 9% of $500   **8.** 18% of $400   **9.** 61% of $400

10. 112% of $80   **11.** 6% of $700   **12.** 11% of $800

13. 5% of 160 kg   **14.** 120% of 60 kg   **15.** 68% of 400 g

16. 15% of 300 m   **17.** 102% of 2000 km   **18.** 71% of $1000

19. 26% of 19 kg   **20.** 201% of 6000 g   **21.** 8.5% of $2400

22. Work out:

   a) $\dfrac{3}{4}$ of 64% of 0.3   **b)** 11% of $\dfrac{3}{5}$ of 0.25

23. Bharat has $450. He gives 35% of the money to his sister.
   How much money does Bharat give?

24. There are 20 000 people at a concert. 82% of the people are adults.
   Work out how many children went to the concert.

25. 35% of trains were late in a week.
   In that week there were 440 trains.
   Calculate how many trains were not late.

26. Jen wants to buy a computer that costs $600.
   She saves 30% of her wages each month.
   Each month Jen earns $400.
   Calculate how many months it will take Jen to save enough money to buy the computer.

27. Sofia earns $1200 a month. She spends 30% of this money on rent and 12% on bills.
   How much of the $1200 has she left?

6.3 Percentage increase and decrease

Example 1

A coat originally cost $24. Calculate the new price after a 5% reduction.

Price reduction = 5% of $24

$$= \frac{5}{100} \times \frac{24}{1} = \$1.20 \qquad \text{First find 5\% of \$24.}$$

New price of coat = $24 − $1.20   Then subtract from the
         = $22.80   original price.

Tip

A price reduction means that the price has decreased. This is also known as a **discount**.

Example 2

a) A CD originally cost $11.60. Calculate the new price after a 7% increase.

New price = 107% of $11.60	If you increase the price by 7%, the final price is 107% of the old price.
$= 1.07 \times 11.6$	Change the percentage to a decimal by dividing by 100.
$= \$12.41$ to the nearest cent.	

b) A coat originally cost $24. Calculate the new price after a 5% reduction.

New price = 0.95×24	$100\% - 5\% = 95\%$ so you multiply by 0.95.
$= \$22.80$	

> **Tip**
>
> This is an alternative method for answering Example 1.

> **Tip**
>
> You could work out 7% of $11.60 as in Example 1. However, here is a *quicker* way which you may prefer.

Exercise 6.3A

Find the new price after each percentage increase or decrease.

Do not use a calculator for Questions **1** to **6**.

Give your answer to the nearest cent where necessary.

	Original price	Increase or decrease
1.	$60	Increase by 5%
2.	$65	Decrease by 8%
3.	$440	Increase by 80%
4.	$66	Increase by 100%
5.	$63	Decrease by $33\frac{1}{3}\%$
6.	$91.50	Decrease by 50%
7.	$800	Decrease by 8%
8.	$82.50	Decrease by 6%
9.	$2000	Decrease by 2%
10.	$88.24	Increase by 25%
11.	$8.24	Increase by 46%
12.	$7.65	Increase by 24%
13.	$5.61	Increase by 31%

14.	$8.99	Decrease by 22%
15.	$11.12	Increase by 11%
16.	$17.62	Decrease by 4%

Exercise 6.3B

1. In a sale a shop reduces all its prices by 20%. Find the sale price of a sari which previously cost $44.

2. The price of a car was $5400 but it is increased by 6%. What is the new price?

3. The price of a small Persian rug was $245 but it has been reduced by 30%. What is the new price?

4. A music shop offers a 7% discount for cash. How much does a cash-paying customer pay for a CD advertised at $9.50?

5. A rabbit has a mass of 2.8 kg. After eating, its mass is increased by 1%. What is its mass now?

6. The insurance premium for a car is normally $90. With a 'no-claim bonus' the premium is reduced by 35%. What is the reduced premium?

7. The population of a town increased by 32% between 1975 and 2015. If there were 45 000 people in 1975, what was the 2015 population?

8. A restaurant adds a 12% 'service charge' on to the basic price of meals. How much do I pay for a meal with a basic price of $8.50?

9. A new-born baby has a mass of 3.1 kg. Her mass increases by 8% over the next two weeks. What is her new mass?

10. At the beginning of the year, a car is valued at $3250. During the year, its value falls by 15%. How much is it worth at the end of the year?

11. In a sale, a soft toy priced at $35 is reduced by 20%. At the end of the week, the *sale price* is reduced by a further 25%. Calculate:

    a) the price in the original sale

    b) the final price.

12. **a)** In 2021, a club had 40 members who each paid a fee of $120 per year. What was the total income from the members?

   b) In 2022, the fee is increased by 35% and the membership increases to 65.

     i) What is the 2022 fee?

     ii) What is the total income from the members in 2022?

6.4 Financial mathematics

Simple interest

When a sum of money $P is invested for T years at R% interest per (each) year then the interest gained, I, is given by:

$$I = \frac{P \times R \times T}{100}$$

This is known as **simple interest**.

Example

Joel invests $400 for 6 months at 5% simple interest per year. Work out the interest gained.

$P = \$400 \qquad R = 5 \qquad T = 0.5 \qquad$ (6 months is half a year)

so $I = \dfrac{400 \times 5 \times 0.5}{100}$

$\qquad I = \$10$

Exercise 6.4A

1. Calculate:

   a) the simple interest on $1200 for 3 years at 6% per year

   b) the simple interest on $700 at 8.25% per year for 2 years

   c) the length of time for $5000 to earn $1000 if invested at 10% simple interest per year

   d) the length of time for $400 to earn $160 if invested at 8% simple interest per year.

2. Khalid invests $6750 for 4 years in an account that pays 8.5% simple interest per year.

   a) How much interest does he earn?

   b) What is the total amount in Khalid's account after 4 years?

3. Petra invests $10 800. After 4 years she has earned $3240 in simple interest. At what annual rate of interest did she invest her money?

4. Ali invests $P at 6% simple interest for 7 years. He earns $504 in interest.

Work out the value of P.

5. Jenna invests $8000 at R% simple interest for 11 years. She earns $3080 in interest.

Work out the value of R.

6. Bilal invests $12 000 at 4.1% simple interest for T years. He earns $2214 in interest.

Work out the value of T.

Repeated percentage change

Suppose a bank pays a fixed interest rate of 10% on money in a savings account. A man puts $500 in the savings account.

After one year he has

$500 + 10% of $500 = $550

After two years he has

$550 + 10% of $550 = $605   [Check that this is $1.10^2 \times 500$]

After three years he has

$605 + 10% of $605 = $665.50   [Check that this is $1.10^3 \times 500$]

This type of interest is called **compound interest**. Compound interest is an example of repeated percentage change.

The formula for compound interest (and repeated percentage change in general) is

$$A = P \left(1 + \frac{r}{100} \right)^n$$

where A is the amount after the repeated change, P is the initial amount, r is the rate of change (for example, the rate of interest) and n is the number of units of time.

> **Tip**
>
> With compound interest, the interest rate is applied to interest earned in previous years, as well as to the initial amount paid in.

Exercise 6.4B

1. A bank pays compound interest of 9% per year.

Sabira puts $2000 in the bank. How much does she have after:

a) one year   **b)** two years   **c)** three years?

2. A bank pays compound interest of 11% per year. Jorge puts $5000 in the bank. How much does he have after:

a) one year   **b)** three years   **c)** five years?

3. A student gets a grant of $10000 a year. Assuming that her grant is increased by 7% each year, what will her grant be in four years' time?

4. Isoke's salary in 2019 was $30000 per year. Every year her salary is increased by 5%.

In 2020 her salary was 30000×1.05 $= \$31500$

In 2021 her salary was $30000 \times 1.05 \times 1.05$ $= \$33075$

In 2022 her salary was $30000 \times 1.05 \times 1.05 \times 1.05 = \$34728.75$

and so on.

What will her salary be in:

a) 2023   **b)** 2024?

5. The rental price of a dacha was $9000 in 2022. At the end of each month the price is increased by 6%.

Find the price of the dacha after:

a) 1 month   **b)** 3 months   **c)** 10 months.

6. Assuming an average inflation rate of 8%, work out the probable cost of the following items in 10 years:

a) motor bike $6500   **b)** phone $340   **c)** car $50000.

7. A new scooter is valued at $15000. At the end of each year its value is reduced by 15% of its value at the start of the year. What will it be worth after 3 years?

Earnings and tax

> ### Example
>
> Workers generally pay tax on their earnings. Sometimes they are entitled to a *tax-free* allowance before paying a percentage tax on the rest of their earnings.
>
> Vivien earns $42000 per year and she gets a tax-free allowance of $9000. If she pays 20% tax on the next $30000 and 40% on the rest, how much tax does she pay in total?
>
> | | |
> |---|---|
> | $42000 − $9000 = $33000 | Subtract the tax-free allowance. |
> | 20% of $30000 = $30000 ÷ 5 = $6000 | To find 20%, you can divide the number by 5. |
> | $33000 − $30000 = $3000 | Subtract the income she pays the lower tax rate on. |
> | 40% of $3000 = $3000 × 0.4 = $1200 | To find 40% you can multiply by 0.4. |
> | Total tax = $6000 + $1200 = $7200 | Add the two amounts of tax. |

Exercise 6.4C

1. Tomas earns $37 000 per year. He gets a tax-free allowance of $8000 and pays 25% tax on the rest. How much tax will he pay in a year?

2. Juliette earns $4500 per month. She gets a tax-free allowance of $10 000 per year and pays tax at 20% on the rest. How much tax will she pay in a year?

3. Elise gets a tax-free allowance of $6000 and pays tax at 25% on the next $20 000. She pays tax at a rate of 30% on the rest. If she earns $72 000 per year, how much tax must she pay?

4. Johan earns $650 per week and works for 48 weeks a year. If he gets a tax-free allowance of $8000, pays tax at a rate of 10% on the next $10 000 and 20% on the rest, how much tax will he pay in a year?

Making a profit

Example

A shopkeeper buys potatoes at a wholesale price of $420 per tonne and sells them at a retail price of 45c per kg.

How much profit does he make on one kilogram of potatoes?

He pays $420 for 1000 kg of potatoes.	Remember: 1 tonne = 1000 kg.
He pays $[420 ÷ 1000] for 1 kg of potatoes.	
So he pays $0.42 = 42c for 1 kg.	
He sells the potatoes for 45c per kg.	He sells the potatoes at a higher price than he paid for them, so he makes a profit.
Profit = 45c – 42c = 3c per kg	Profit = selling price – price paid.

Exercise 6.4D

Find the profit in each case.

Item	Retail price	Wholesale price	Profit
1. cans of drink	15c each	$11 per 100	profit per can?
2. rulers	24c each	$130 per 1000	profit per ruler?
3. pencils	22c each	$13 per 100	profit per pencil?
4. cans of soup	$0.27 each	$8.50 for 50 cans	profit per can?
5. newspapers	22c each	$36 for 200	profit per newspaper?
6. bars of chocolate	$0.37 each	$15.20 for 80	profit per bar?
7. rice	22c per kg	$160 per tonne	profit per kg?

8. carrots	38c per kg	$250 per tonne	profit per kg?
9. T-shirts	$4.95 each	$38.40 per dozen	profit per T-shirt?
10. eggs	96c per dozen	$60 per 1200	profit per dozen?
11. lychees	5 for 30c	$14 for 400	profit per lychee?
12. calculators	$19.50 each	$2450 for 200	profit per calculator?
13. fruit juice	55c for 100 ml	$40 for 10 litres	profit per 100 ml?
14. couscous	$16 per kg	$11 000 per tonne	profit per kg?
15. wire	23c per m	$700 for 10 km	profit per m?
16. cheese	$2.64 per kg	$87.50 for 50 kg	profit per kg?
17. string	46c per m	$160 for 500 m	profit per m?
18. apples	9c each	$10.08 for 144	profit per apple?
19. grass	$6.80 per m²	$1600 for 500 m²	profit per m²?
20. tins of paint	33c per tin	$72 for 400 tins	profit per tin?

6.5 Percentage change

Price changes are sometimes more significant when expressed as a percentage of the original price. For example, if the price of a car goes up from $7000 to $7070, this is only a 1% increase.

But if the price of a jacket went up from $100 to $170 this would be a 70% increase! In both cases, the actual increase is the same: $70.

$$\text{Percentage increase} = \frac{(\text{actual increase})}{(\text{original value})} \times \frac{100}{1}$$

Example

The price of a car is increased from $6400 to $6800.
Find the percentage increase.

$$\text{Percentage increase} = \frac{400}{6400} \times \frac{100}{1} = 6\frac{1}{4}\%$$

actual increase ↗ original value ↙

Remember to write the original value as the denominator.

For a decrease:

$$\text{Percentage decrease} = \frac{(\text{actual decrease})}{(\text{original value})} \times \frac{100}{1}$$

Exercise 6.5A

Calculate the percentage change and state whether it is an increase or a decrease.

	Original price	Final price
1.	$50	$54
2.	$80	$88
3.	$180	$225
4.	$100	$102
5.	$75	$78
6.	$800	$600
7.	$50	$40
8.	$120	$105
9.	$420	$280
10.	$6000	$1200

Exercise 6.5B

Find the percentage profit or percentage loss using one of these formulae:

$$\text{percentage profit} = \frac{(\text{actual profit})}{(\text{cost price})} \times \frac{100}{1} \quad \text{or} \quad \text{percentage loss} = \frac{(\text{actual loss})}{(\text{cost price})} \times \frac{100}{1}$$

Give the answers correct to one decimal place.

	Cost price	Selling price
1.	$11	$15
2.	$21	$25
3.	$36	$43
4.	$41	$50
5.	$411	$461
6.	$20	$18.47
7.	$17	$11
8.	$13	$9
9.	$211	$200
10.	$8.15	$7

Exercise 6.5C

1. The number of people employed by a company increased from 250 to 280. Calculate the percentage increase.

2. During the first four weeks of her life a baby's mass increases from 3000 g to 3870 g. Calculate the percentage increase in mass.

3. Before cooking, a joint of meat has a mass of 2.5 kg. After cooking, it has a mass of 2.1 kg. Calculate the percentage decrease in the mass.

4. When cold, an iron rod is 200 cm long. After being heated, the length increases to 200.5 cm. Calculate the percentage increase in length.

5. Juan buys a car for $4000 and sells it for $4600. Calculate his percentage profit.

6. Melanie buys berets for $6.20 and sells them for $9.99. Calculate her percentage profit correct to one decimal place.

7. Wei Wei buys rice at 20c per kg but has to sell it at only 17c per kg. Calculate her percentage loss per kilogram.

8. Before a service, the fuel consumption of a car was 5.1 km per litre. After the service, the fuel consumption improved to 5.8 km per litre. Calculate the percentage improvement in fuel consumption, correct to one decimal place.

9. The population of a town went down from 22 315 to 21 987. Calculate the percentage reduction, correct to one decimal place.

10. In 2021, a tennis player earned $2 530 700. In 2022 the same player earned $3 133 010. Calculate the percentage increase in her income, correct to one decimal place.

Exercise 6.5D

1. Gabir bought 40 articles for $10 and sold them at 32c each.

Calculate:

a) the cost price of each article

b) the total selling price of the 40 articles

c) the total profit

d) the percentage profit.

2. Ivan bought a box of 40 packets of sweets at 25c per packet.

a) Find the total cost of the box of sweets.

b) He sold 10 packets at 37c per packet, and the rest of the box at 35c per packet.

   i) How much profit did he make?

   ii) Express this profit as a percentage of his total cost price.

3. ABCD is a square of side 100 cm. Side AB is increased by 20% and side AD is reduced by 25% to form rectangle APQR.

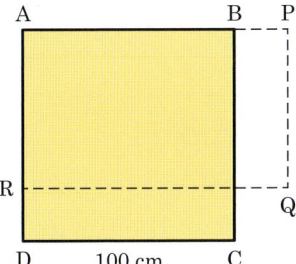

a) Calculate: **i)** the length of AP

   ii) the length of AR

   iii) the area of square ABCD

   iv) the area of rectangle APQR.

b) By what percentage has the area of the square been reduced?

4. When a house was built in 2021 the total cost was made up of the following:

wages $35 000

materials $18 000

overheads $4500

a) Find the total cost of the house in 2021.

b) In 2022 the cost of wages increased by 10%, the cost of materials increased by 5% and the overheads remained at their previous cost.

   i) Find the total cost of the house in 2022, correct to one decimal place.

   ii) Calculate the percentage increase from 2021 to 2022.

5. Four maths teachers calculate the area of the shape given and they all get different answers. Beatriz is wrong by 20% and Saghir is wrong by 5%.

Here are the four answers:

237.5 m², 250 m², 260 m², 300 m²

Which is the correct answer?

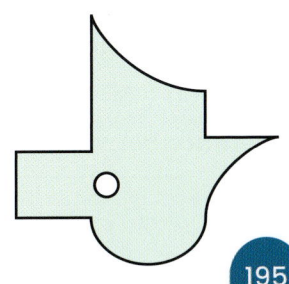

6.6 Map scales and ratio

The map below is drawn to a scale of 1:50 000. In other words, 1 cm on the map represents 50 000 cm on the land.

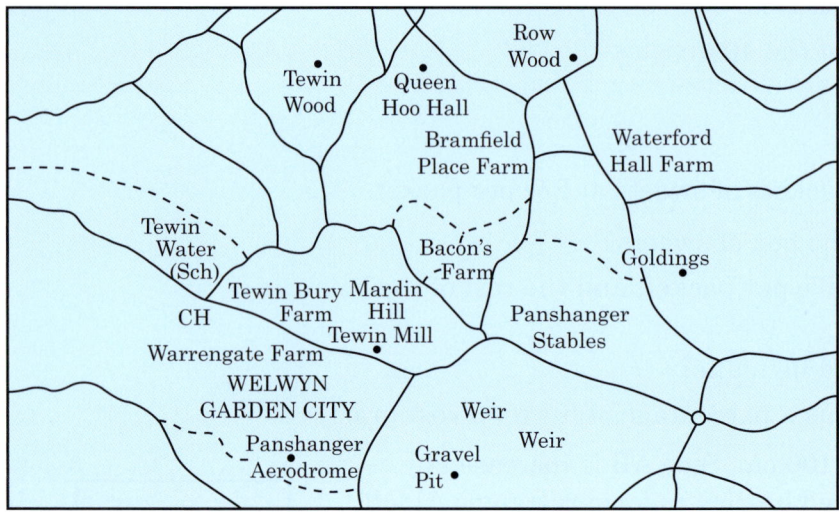

Example 1

On a map of scale 1:25 000 two towns appear 10 cm apart. What is the actual distance between the towns in km?

1 cm on map = 25 000 cm on land	
10 cm on map = 250 000 cm on land	Multiply 25 000 by 10.
250 000 cm = 2500 m	There are 100 cm in 1 m so divide by 100.
= 2.5 km	There are 1000 m in 1 km so divide by 1000.
The towns are 2.5 km apart.	

Example 2

The distance between two towns is 18 km.

How far apart will they be on a map of scale 1:50 000?

18 km = 18 000 m	Multiply by 1000 to find the number of metres.
18 000 m = 1 800 000 cm	Multiply by 100 to find the number of centimetres.
1 800 000 ÷ 50 000 = 36 cm	Divide by 50 000.
Distance between towns on map = 36 cm	

Exercise 6.6A

1. The scale of a map is $1:1000$. Find the actual length in metres represented on the map by 20 cm.

2. The scale of a map is $1:10\,000$. Find the actual length in metres represented on the map by 5 cm.

3. Copy and complete the table.

Map scale	Length on map	Actual length on land
a) $1:10\,000$	10 cm	1 km
b) $1:2000$	10 cm	... m
c) $1:25\,000$	4 cm	... km
d) $1:10\,000$	6 cm	... km

4. Find the actual distance in metres between two points which are 6.3 cm apart on a map whose scale is $1:1000$.

5. On a map of scale $1:300\,000$ the distance between Paris and Bonnieres is 8 cm. What is the actual distance in km?

6. A builder's plan is drawn to a scale of 1 cm to 10 m. How long is a road which is 12 cm on the plan?

7. The distance between two towns is 15 km. How far apart will they be on a map of scale $1:10\,000$?

8. The distance between two points is 25 km. How far apart will they be on a map of scale $1:20\,000$?

9. The length of a road is 2.8 km. How long will the road be on a map of scale $1:10\,000$?

10. The length of a reservoir is 5.9 km. How long will it be on a map of scale $1:100\,000$?

11. Copy and complete the table.

Map scale	Actual length on land	Length on map
a) $1:20\,000$	12 km	... cm
b) $1:10\,000$	8.4 km	... cm
c) $1:50\,000$	28 km	... cm
d) $1:40\,000$	56 km	... cm
e) $1:5000$	5 km	... cm

12. The scale of a drawing is 1 cm to 10 m. The length of a wall is 25 m. What will be the length of the wall on the drawing?

13. The length of a road on a map of scale $1:50\,000$ is 18 cm.

Bashir drives at a constant speed of 120 kilometres per hour.

Work out how long it takes Bashir to drive the length of the road.

Ratio

A map scale is an example of a **ratio**. A ratio shows the relative size of two or more quantities or values.

Simplifying a ratio is like simplifying a fraction. Find the highest common factor of the two (or more) numbers and divide.

You must ensure that the parts of the ratio are in the same units before simplifying.

Example 1

Simplify $6:8$ The highest common factor of 6 and 8 is 2.

$\div 2 \overset{6:8}{\underset{3:4}{}} \div 2$ Divide both parts of the ratio by 2.
$=$

Simplify $10:15:30$

$\div 5 \overset{10:15:30}{\underset{2:\ 3\ :6}{}} \div 5$ Divide all three parts of the ratio by 5.
$=$

Simplify $110\ \text{cm}:1\ \text{m}$

$1\ \text{m} = 100\ \text{cm}$ Write the quantities using the same units.

So ratio becomes $110:100$

$\div 10 \overset{110:100}{\underset{11:10}{}} \div 10$ Divide both parts by 10.
$=$

You can divide quantities into a given ratio.

Example 2

Share $\$60$ in the ratio $2:3$.

Total number of shares $= 2 + 3 = 5$

So one share $= \$60 \div 5 = \$12$

The two amounts are $2 \times \$12 = \$24$

and $3 \times \$12 = \$36$.

Exercise 6.6B

1. Simplify each of these ratios.

   a) 10:12　　　　**b)** 8:24　　　　**c)** 20:50　　　　**d)** 120:80

   e) 2:4:8　　　　**f)** 3:6:30　　　　**g)** 50:100:400

   h) 300 cm:1.2 m　　　　**i)** 900 g:1.4 kg

   j) 84 minutes:0.7 hours　　　　**k)** \$85:340 cents

2. Share \$30 in the ratio 1:2.

3. Share \$60 in the ratio 3:1.

4. Divide 880 g of rice between Millie and Mahnoor in the ratio 3:5.

5. Divide \$1080 between Isobel and Carlota in the ratio 4:5.

6. Share 126 litres of petrol between Maya and Tashu in the ratio 2:5.

7. Share \$60 in the ratio 1:2:3.

8. Anwar, Belusa and Nabila divided \$560 between them in the ratio 2:1:5. How much did Belusa receive?

9. A sum of \$120 is divided in the ratio 3:4:5. What is the largest share?

10. In an election, 7800 people voted Democrat, Socialist or Republican in the ratio 4:3:5. How many people voted Republican?

Example

In a basket, the ratio of melons to apples is 3:4.

If there are 9 melons, how many apples are there?

melons:apples = 3:4

Multiply both parts by 3.

melons:apples = 9:12

So there are 9 melons and 12 apples.

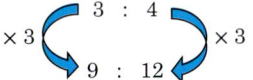

Exercise 6.6C

1. In a cafe, the ratio of coffees sold to cakes sold is $3:2$. If there are 12 coffees sold in an hour, how many cakes are sold?

2. In a pack of nuts, the ratio of pecans to walnuts is $4:1$. If there are 20 pecans, how many walnuts are there?

3. In a box, the ratio of nails to screws is $5:3$. If there are 15 nails, how many screws are there?

4. An alloy consists of copper, zinc and tin in the ratio $1:3:4$. If there is 10 g of copper in the alloy, find the mass of zinc and tin.

5. In a shop the ratio of oranges to apples is $2:5$. If there are 60 apples, how many oranges are there?

6. A recipe for 5 people calls for 1.5 kg of meat. How much meat is required if the recipe is adapted to feed 8 people?

7. A cake for 6 people requires 4 eggs. How many eggs are needed to make a cake big enough for 9 people?

8. A photocopier enlarges the original in the ratio $2:3$. The height of a tree is 12 cm on the original. How tall is the tree on the enlarged copy?

original   enlarged copy

9. A photocopier enlarges copies in the ratio $4:5$. The length of the headline 'BRIDGE COLLAPSES' is 18 cm on the original. How long is the headline on the enlarged copy?

10. A photocopier *reduces* copies in the ratio $5:3$. The height of a tower is 12 cm on the original. How tall is the tower on the reduced copy?

11. A cake with a mass of 550 g has three ingredients: flour, sugar and butter. There is twice as much flour as sugar and one and a half times as much sugar as butter. How much flour is there?

12. If $\frac{5}{8}$ of the children in a school are boys, what is the ratio of boys to girls?

13. A man and a woman share $1000 between them in the ratio $1:4$. The woman shares her part between herself, her mother and her daughter in the ratio $2:1:1$. How much does her daughter receive?

14. The number of pages in a newspaper is increased from 36 to 54. The price is increased in the same ratio. If the old price was 28c, what will the new price be?

15. Two friends bought a house for $220 000. Sam paid $140 000 and Joe paid the rest. Three years later they sold the house for $275 000. How much should Sam receive from the sale?

16. Concrete is made from 1 part cement, 2 parts sand and 5 parts aggregate. How much cement is needed to make 2 m³ of concrete?

17. A photocopier increases the sides of a square in the ratio $4:5$. By what percentage are the sides increased?

18. In an alloy the ratio of copper to iron to lead is $5:7:3$. What percentage of the alloy is lead?

19. Here is a recipe for making 8 mini sponge cakes.

    225 g butter

    200 g sugar

    4 eggs

    Half a lemon

    1 teaspoon vanilla

    250 g flour

    Calculate the ingredients required to make 20 mini sponge cakes.

20. Two different size cartons of orange juice are sold in a shop.

    | Carton A: | 500 ml | $0.79 |
    | Carton B: | 750 ml | $1.12 |

    Which of the two cartons is better value for money?

    Tip

    Work out the price per 100 ml. Which one is lower?

21. Three different size boxes of tea are sold by an online retailer.

    | Box 1: | 80 tea bags | $3.20 |
    | Box 2: | 100 tea bags | $3.90 |
    | Box 3: | 250 tea bags | $10.25 |

    Which box is the best value for money?

22. Tablo travels to work by train.

    There are two options for buying tickets. She can buy a ticket each day for $2.40, or a monthly travel pass for $45.50.

    On how many days per month does Tablo need to travel to work in order for the monthly pass to be better value for money?

6.7 Proportion

Example 1

If 11 gallons of petrol cost $20.46, find the cost of 27 gallons.

11 gallons cost $20.46

1 gallon costs $20.46 ÷ 11 = $1.86   Divide the total cost by
                                      the number of gallons.

27 gallons cost $1.86 × 27 = $50.22   Multiply the cost of
                                      1 gallon by 27.

Tip

The cost of petrol is **directly proportional** to the quantity bought.

Example 2

A farmer has enough hay to feed 5 horses for 6 days. How long would the hay last for 3 horses?

Tip

The length of time for which the horses can be fed is **inversely proportional** to the number of horses to be fed.

5 horses can be fed for 6 days

1 horse can be fed for 6 × 5 = 30 days   Multiply the number of days by the original
                                          number of horses.

3 horses can be fed for 30 ÷ 3 = 10 days. Divide by the new number of horses.

In the first example above it was helpful to work out the cost of *one* litre of petrol.
In the second example, we found the time for which *one* horse could be fed.
This is called the **unitary method** as it involves finding the amount for one unit.

To do these questions you need to think logically and decide whether the question involves direct or inverse proportion.

- If five men can paint a tower in 10 days, how long would it take one man?
  Questions like this are **inverse proportion**.
- If 33 books cost $280.50, how much will one book cost?
  Questions like this are **direct proportion**.

Exercise 6.7A

Questions 1 to 7 involve **direct** proportion.

1. If 5 pizzas cost $20, find the cost of 7 of these pizzas.

2. Magazines cost $16 for 8. Find the cost of 3 of these magazines.

3. Find the cost of 2 cakes if 7 cakes cost $10.50.

4. A machine fills 1000 bottles in 5 minutes. How many bottles will it fill in 2 minutes?

5. A train travels 100 km in 20 minutes. How long will it take to travel 50 km if travelling at the same rate?

6. Eleven pens cost $13.20. Find the cost of 4 of these pens.

7. Fishing line costs $1.40 for 50 m. Find the cost of 300 m of this fishing line.

Questions 8 to 14 involve **inverse** proportion.

8. If 12 men can build a house in 6 days, how long will it take 6 men to build the same house?

9. Six women can dig a hole in 4 hours. How long would it take 2 women to dig a hole of the same size?

10. A farmer has enough hay to feed 20 horses for 3 days. How long would the hay last for 60 horses?

11. Twelve people can clean an office building in 3 hours. How long would it take 4 people to clean this building?

12. Usually, it takes 12 hours for 8 men to do a job. How many men are needed to do the same job in 4 hours?

13. Five teachers can mark 60 exam papers in 4 hours. How long would it take one teacher to mark all 60 papers?

14. 10 ladybirds eat 400 greenflies in 3 hours.

Copy and complete the following:

a) 20 ladybirds eat ☐ greenflies in 3 hours.

b) 20 ladybirds eat ☐ greenflies in 9 hours.

c) 10 ladybirds eat 200 greenflies in ☐ hours.

d) ☐ ladybirds eat 4000 greenflies in 3 hours.

Exercise 6.7B

This is a mixed exercise. You need to decide if the question involves direct or inverse proportion.

1. If 7 packets of coffee cost $8.54, find the cost of 3 packets.

2. Find the cost of 8 bottles of cola, given that 5 bottles cost $11.90.

3. If 7 cartons of milk hold 14 litres, find how much milk there is in 6 cartons.

4. 5 builders took 6 hours to dig a trench. How long would it have taken 3 builders to dig an identical trench?

5. 10 people exploring a desert took enough water to last 5 days. How long would the water have lasted if there had been only 5 people?

6. A worker takes 8 minutes to make 2 circuit boards. How long would it take to make 7 circuit boards?

7. On a rose bush there are enough greenflies to last 9 ladybirds 4 hours. How long would the greenflies last if there were only 6 ladybirds?

8. The total mass of 8 tiles is 1720 g. Work out the mass of 17 tiles.

9. A machine can fill 3000 bottles in 15 minutes. How many of these bottles will it fill in 2 minutes?

10. A train travels 40 km in 120 minutes. How long will it take to travel 55 km at the same speed?

11. If 4 grapefruits can be bought for $2.96, how many can be bought for $8.14?

12. $15 can be exchanged for £9.74. How many British pounds can be exchanged for $37.50?

13. Usually, it takes 10 hours for 4 men to build a wall. How many men are needed to build the same wall in 8 hours?

14. A car travels 280 km on 35 litres of petrol. How much petrol is needed for a journey of 440 km?

15. Ten bags of corn will feed 60 hens for 3 days. Copy and complete the following:

   a) 30 bags of corn will feed ☐ hens for 3 days.

   b) 10 bags of corn will feed 20 hens for ☐ days.

   c) 10 bags of corn will feed ☐ hens for 18 days.

   d) 30 bags of corn will feed 90 hens for ☐ days.

16. Four machines produce 5000 batteries in 10 hours. How many batteries would 6 machines produce in 10 hours?

17. Newtonian spiders can spin webs in straight lines. If 15 spiders can spin a web of length 1 metre in 30 minutes, how long will it take 6 spiders to spin a web of the same length?

18. It takes b beavers n hours to build a dam. How long will it take half of the beavers to build the same size dam?

Currency exchange

Money is changed from one currency into another using the method of proportion.

Exchange rate for US dollars ($) in October 2022:

Country	Rate of exchange
Argentina (pesos)	$1 = 153.8 ARS
Australia (dollar)	$1 = 1.58 AUD
Eurozone (euros)	$1 = €1.02 EUR
India (rupees)	$1 = 82.68 INR
Japan (yen)	$1 = 149 JPY
Kuwait (dinars)	$1 = 0.31 KWD
UK (pounds)	$1 = £0.88 GBP

Example

Convert: **a)** $22.50 to dinars   **b)** €300 to dollars.

a) $1 = 0.31 dinars (KWD)

so $22.50 = 0.31 × 22.50 KWD

= 6.98 KWD

b)   €1.02 = $1

so $€1 = \$\dfrac{1}{1.02}$

so $€300 = \$\dfrac{1}{1.02} × 300$

= $294.12

Exercise 6.7C

Give your answers correct to two decimal places. Use the exchange rates given above.

1. Change the amount of dollars into the currency stated.
   - **a)** $20 [euros]
   - **b)** $70 [pounds]
   - **c)** $200 [pesos]
   - **d)** $1.50 [rupees]
   - **e)** $2.30 [yen]
   - **f)** 90c [dinars]

2. Change the amount of currency into dollars.
   - **a)** €500
   - **b)** £2500
   - **c)** 7.50 rupees
   - **d)** 900 dinars
   - **e)** 125.24 pesos
   - **f)** 750 AUD

3. A game costs £9.50 in London and $9.70 in Chicago. How much cheaper, in British money, is the game when bought in the US?

4. A wireless speaker costs €20.46 in Spain and £12.60 in the UK. Which is cheaper in dollars, and by how much?

5. The monthly rent of a flat in New Delhi is 32 860 rupees. How much is this in euros?

6. A Persian kitten is sold in several countries at the prices given below.

   Kuwait   150 dinars

   France   550 euros

   Japan   92 000 yen

   Write out in order a list of the prices converted into GBP.

7. An Australian in Germany has 700 AUD. If he changes the money, how many euros will he receive?

Conversion graphs

You can use a conversion graph to change from one unit to another.

Example

The graph shows the conversion from miles into kilometres.

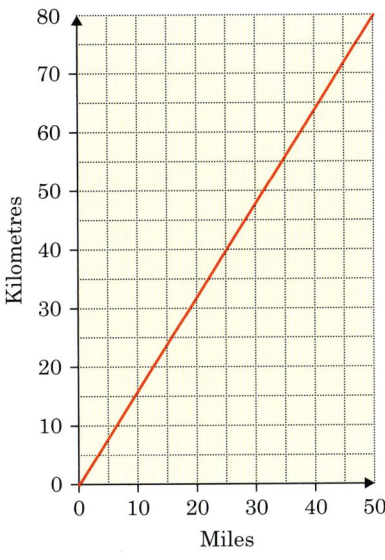

Use the graph to find:

a) the number of kilometres in 20 miles

b) the number of miles in 54 kilometres.

a) Draw a line from 20 on the *x*-axis up to the line and then across to the *y*-axis:

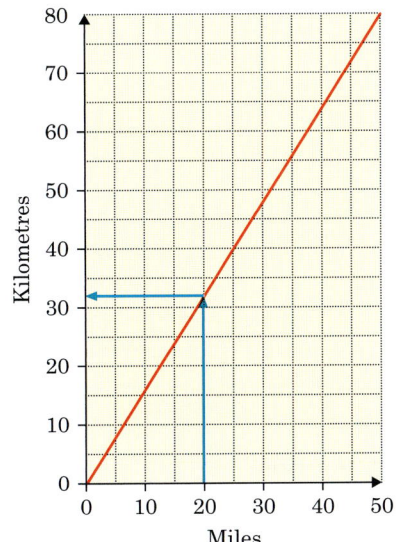

There are 32 km in 20 miles.

207

b) Draw a line from 54 on the *y*-axis across to the line and then down to the *x*-axis:

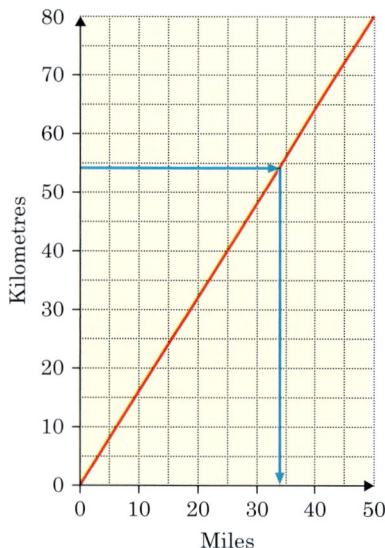

There are 34 miles in 54 km.

Exercise 6.7D

1. Give your answers as accurately as you can. [e.g. 3 lb = 1.4 kg approximately]

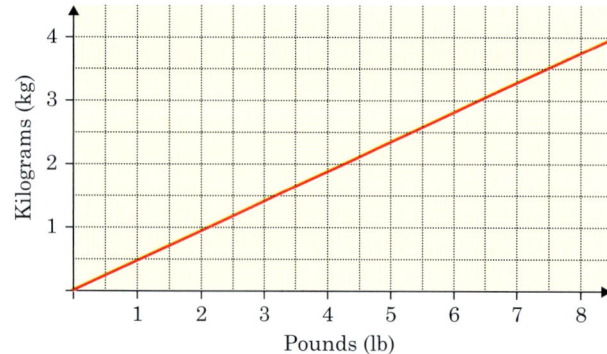

a) Convert into kilograms:

　i) 5.5 lb　　**ii)** 8 lb　　**iii)** 2 lb.

b) Convert into pounds:

　i) 2 kg　　**ii)** 3 kg　　**iii)** 1.5 kg.

c) A bag of sugar has a mass of 1 kg. What is its mass in pounds?

d) A washing machine has a load limit of 7 lb.

　What is the load limit in kilograms?

2. Between 1984 and 1994 the value of the British pound against the German mark (DM) changed. How much less in DM did you receive for £1 in 1994 compared with 1984?

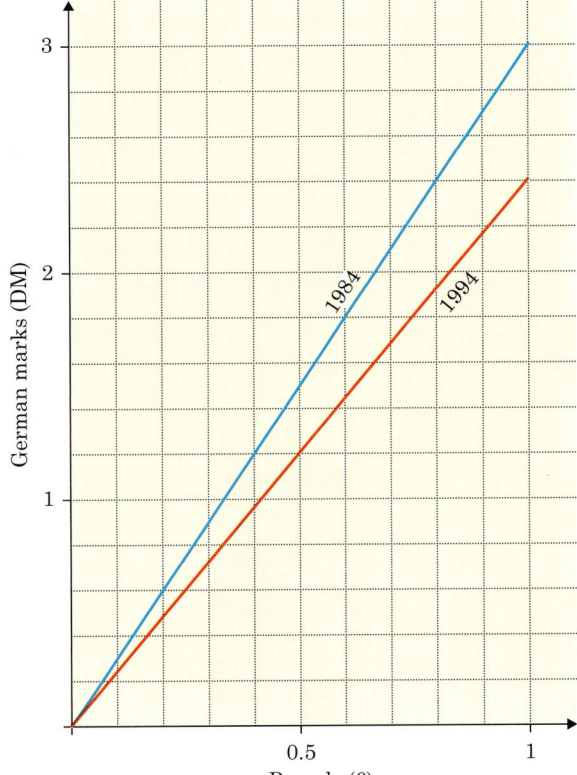

3. Temperature can be measured in °C or in °F. A conversion graph can be constructed using two points as follows:

Draw axes with a scale of 1 cm to 5° as shown.

$32\,°F = 0\,°C$ and $95\,°F = 35\,°C$.

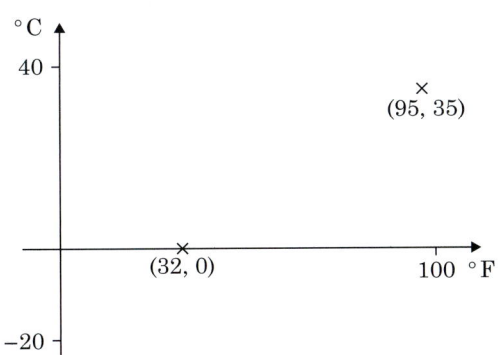

Draw a line through these two points. Use your graph to convert:

a) 50 °F into °C   **b)** 20 °C into °F   **c)** 0 °F into °C.

Revision exercise 6

1. A supermarket sells jam in two sizes.

 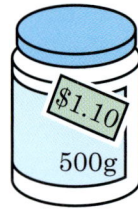

Which jar represents better value for money? (1 kg = 2.2 lb)

2. The pump shows the price of petrol in a garage.

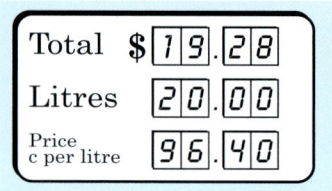

One day I buy $20 worth of petrol. How many litres do I buy?

3. a) On a map, the distance between two towns is 16 cm. Calculate the scale of the map if the actual distance between the towns is 8 km.

b) On another map, two towns are 1.5 cm apart and are actually 60 km apart. Calculate the scale of the map.

4. In December 2021, a factory employed 220 people, each person being paid $650 per week.

a) Calculate the total weekly wage bill for the factory.

b) In January 2022, the workforce of 220 was reduced by 10 per cent. Find the number of people employed at the factory after the reduction.

c) Also in January 2022, the weekly wage of $650 was increased by 10 per cent. Find the new weekly wage.

d) Calculate the total weekly wage bill for the factory in January 2022.

e) Calculate the difference between the total weekly wage bills in December 2021 and January 2022.

5. A model of a clock tower is made using a scale of 1 to 20.

a) The minute hand on the clock tower is 40 cm long. What is the length of the minute hand on the model?

b) The height of the model is 40 cm. What is the height, h, in metres, of the clock tower?

6. Without using a calculator, work out:

a) 20% of $65

b) 37% of $400

c) 85% of $2000.

7. A motorist travelled 800 km during May, when the cost of petrol was $1.80 per litre. In June, the cost of petrol increased by 10% and he travelled 5% less distance.

a) What was the cost, in dollars per litre, of petrol in June?

b) What distance did he travel in June?

8. Copy and complete the table. Write the missing fractions in their simplest form.

	Fraction	Decimal	Percentage
a)	$\dfrac{3}{5}$		
b)		0.75	
c)			5%
d)	$\dfrac{1}{8}$		

CALCULATOR

1. A club has 275 members.
   a) 99 members are women. What percentage of the members are women? [2]
   b) If, at the next meeting, more than 12.5% of the members were absent, what is the smallest number of members that could have been absent? [2]

2. In 2022, Lucy paid 589.95 US dollars ($) for a flight from Rome to New York.
   The return flight from New York to Rome cost her 402.5 euros (€).
   The exchange rate at the time of the return flight was €1 = $1.14.
   Calculate the difference, in US dollars, between the costs of the two flights.
   Give your answer correct to the nearest dollar. [2]

3. a) Abdul invests $300 for 2 years at 6.15% per year **simple** interest.
      Calculate how much **interest** Abdul receives. [2]
   b) Samia invests $500 for 2 years at 7% per year **compound** interest.
      Calculate how much **interest** Samia receives. [2]

4. Nicolas needs to borrow $4000 for 5 years. The bank offers him this choice.

Offer A	**Offer B**
Interest rate 7.4% per year	Interest rate 7% per year
Pay the interest at the end of	Pay the interest at the end of
each year	the 5 years

   Nicolas recognises that offer A is simple interest and offer B is compound interest.
   a) If he takes offer A, what is the total amount of interest he will pay? [2]
   b) If he takes offer B, how much interest will he pay? Give your answer correct to 2 decimal places. [3]

5. A bag of balloons costs $3.15. Henri and his friend share the cost in the ratio 4 : 3.
   How much does Henri pay? [2]

6. The scale on a map is 1 : 250 000. A road is 6.7 centimetres long on the map.
   Calculate the actual length of the road in kilometres. [2]

7. Aminata bought 25 metres of cloth at a cost of $90. She sold 12 metres of the cloth at $5.60 per metre and 13 metres at $3.20 per metre.
   a) Calculate the profit she made. [2]
   b) Calculate this profit as a percentage of the original cost. [1]

8. Write, in its simplest form, the ratio 4.5 kilograms : 900 grams. [2]

9. Write the following in order, with the smallest first.

    $$\frac{2}{5} \qquad 0.38 \qquad 42\%$$ [1]

10. $0.062 \qquad 62\% \qquad 0.602 \qquad \frac{6}{10} \qquad \frac{6}{100} \qquad 6.2\%$

    From the values listed above, write down:
    a) the smallest [1]
    b) the largest [1]
    c) the two which are equal. [1]

11. $0.08 \quad 80\% \quad \frac{8}{1000} \quad 8\% \quad 0.8 \quad \frac{8}{100} \quad 800\%$

    Write down the three numbers from the list above which have the same value. [1]

12. Scott changed 300 Australian dollars (AUD) into euros (€) when the rate was
    €1 = 1.56 AUD. How many euros did Scott receive? [2]

13. Joseph, Maria and Rebecca share $60 between them. Joseph gets $\frac{5}{12}$ of the $60.

    Maria gets 35% of the $60. Rebecca gets the rest of the $60.
    Calculate the amount each receives. [5]

14. Aida, Bernado and Cristiano need $40 000 to start a business.

    a) i) They borrow $\frac{3}{5}$ of this amount. Show that they still need $16 000. [1]

        ii) They provide the $16 000 themselves in the ratio
            Aida : Bernado : Cristiano = 4 : 3 : 1
            Calculate the amount each of them provides. [3]

    b) i) Office equipment costs 45% of the $40 000. Calculate the cost of the
            equipment. [2]

        ii) Office expenses cost another $7500. Write this as a fraction of $40 000.
            Give your answer in its lowest terms. [2]

        iii) How much remains of the $40 000 now? [1]

    c) They invest $12 000. After one year this has increased to $14 500. Calculate
        this percentage increase. [3]

7 Shape and space 2

Archimedes of Samos (287–212 BCE) studied at Alexandria as a young man. He was a practical man of common sense and was one of the first to apply scientific thinking to everyday problems. He developed methods for finding the area, the volume and the centre of gravity of many 2D and 3D shapes including circles, spheres, conics and spirals. By drawing polygons with many sides, he arrived at a value of π between $3\frac{10}{71}$ and $3\frac{10}{70}$. He was killed in the siege of Syracuse at the age of 75.

- Understand and use measures of rate such as pressure and density and including solving problems involving average speed.
- Calculate with time including in the 24-hour and 12-hour clock. Understand timetables and be able to read clocks.
- Use and interpret geometric terms and vocabulary.
- Use and convert between units of mass, length, area, volume and capacity including in solving problems.
- Solve problems involving triangles, rectangles, parallelograms, trapeziums and circles. Problems involving area and perimeter including the circumference and area of a circle, arc length and sector areas, and compound shapes.
- Solve problems involving cuboids, prisms, cylinders, spheres, pyramids and cones. Problems involving surface area and volume including compound solids.

7.1 Metric units

You need to know the following metric conversions:

Length	Mass	Volume
10 mm = 1 cm	1000 g = 1 kg	1000 ml = 1 litre
100 cm = 1 m	1000 kg = 1 t (t for tonne)	1000 l = 1 m^3
1000 m = 1 km		Also 1 ml = 1 cm^3

Exercise 7.1A

Copy and complete.

1. 85 cm =   m
2. 2.4 km =   m
3. 0.63 m =   cm
4. 25 cm =   m
5. 7 mm =   cm
6. 2 cm =   mm
7. 1.2 km =   m
8. 7 m =   cm
9. 0.58 km =   m
10. 815 mm =   m
11. 650 m =   km
12. 25 mm =   cm
13. 5 kg =   g
14. 4.2 kg =   g
15. 6.4 kg =   g
16. 3 kg =   g
17. 0.8 kg =   g
18. 400 g =   kg
19. 2 t =   kg
20. 250 g =   kg
21. 0.5 t =   kg
22. 0.62 t =   kg
23. 7 kg =   t
24. 1500 g =   kg
25. 800 ml =   l
26. 2 l =   ml
27. 1000 ml =   l
28. 4.5 l =   ml
29. 6 l =   ml
30. 3 l = cm^3
31. 2 m^3 =   l
32. 5.5 m^3 =   l
33. 0.9 l = cm^3
34. 600 cm^3 =   l
35. 15 m^3 =   l
36. 240 ml =   l
37. 28 cm =   m
38. 5.5 m =   cm
39. 305 g =   kg
40. 0.046 km =   m
41. 16 ml =   l
42. 208 mm =   m
43. 28 mm =   cm
44. 27 cm =   m
45. 788 m =   km
46. 14 t =   kg
47. 1.3 kg =   g
48. 90 l = m^3
49. 2.9 t =   kg
50. 19 ml =   l

51. Write down the most appropriate metric unit for measuring:
    a) the distance between Madrid and Barcelona
    b) the capacity of a bottle of water
    c) the mass of raisins needed for a cake
    d) the diameter of a small drill
    e) the mass of a car
    f) the area of a football pitch.

A 1 cm by 1 cm square measures 10 mm by 10 mm. The **area** of the square in mm² is therefore $10 \times 10 = 100$ mm².

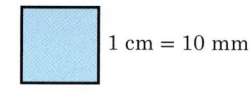

There are similar area conversions for m² into cm² and km² into m²:

$1 \text{ m}^2 = 100 \times 100 = 10\,000 \text{ cm}^2$

$1 \text{ km}^2 = 1000 \times 1000 = 1\,000\,000 \text{ m}^2$

Example 1

a) Convert 65 cm² to mm².

b) Convert 3.2 m² to cm².

a) $65 \times 100 = 6500$ mm²   There are 100 mm² in 1 cm² so you multiply by 100.

b) $3.2 \times 10\,000 = 32\,000$ cm²   There are 10 000 cm² in 1 m² so you multiply by 10 000.

Example 2

a) Convert 790 mm² to cm².

b) Convert 68 000 cm² to m².

a) $790 \div 100 = 7.9$ cm².   Each cm² is 100 mm² so you divide by 100.

b) $68\,000 \div 10\,000 = 6.8$ m².   Each m² is 10 000 cm² so you divide by 10 000.

A 1 cm by 1 cm by 1 cm cube measures 10 mm by 10 mm by 10 mm.
The **volume** of the cube is therefore $10 \times 10 \times 10 = 1000$ mm³.

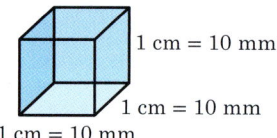

Similarly, $1 \text{ m}^3 = 100 \times 100 \times 100 = 1\,000\,000 \text{ cm}^3$

Example 3

a) Convert 4 cm³ to mm³.

b) Convert 25 m³ to cm³.

a) $4 \times 1000 = 4000$ mm³   There are 1000 mm³ in 1 cm³ so you multiply by 1000.

b) $25 \times 1\,000\,000 = 25\,000\,000$ cm³   There are 1 000 000 cm³ in 1 m³ so you multiply by 1 000 000.

Example 4

a) Convert $59\,000$ mm^3 to cm^3.

b) Convert $68\,000\,000$ cm^3 to m^3.

a) $59\,000 \div 1000 = 59$ cm^3 Each cm^3 is 1000 mm^3 so you divide by 1000.

b) $68\,000\,000 \div 1\,000\,000 = 68$ m^3 Each m^3 is $1\,000\,000$ cm^3 so you divide by $1\,000\,000$.

Exercise 7.1B

Copy and complete.

1. 2 cm$^2 = $ mm^2

2. 45 cm$^2 = $ mm^2

3. 1600 mm$^2 = $ cm^2

4. 48 mm$^2 = $ cm^2

5. 3 m$^2 = $ cm^2

6. 26 m$^2 = $ cm^2

7. 8600 cm$^2 = $ m^2

8. 760 cm$^2 = $ m^2

9. 5 km$^2 = $ m^2

10. $4\,500\,000$ m$^2 = $ km^2

11. 8 cm$^3 = $ mm^3

12. 21 cm$^3 = $ mm^3

13. $48\,000$ mm$^3 = $ cm^3

14. 6 m$^3 = $ cm^3

15. $28\,000\,000$ cm$^3 = $ m^3

Exercise 7.1C

A school has a machine that can produce centimetre cubes.

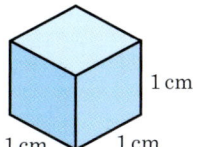

1 cm

1 cm   1 cm

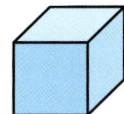

At the end of the afternoon the teacher, Mrs Evans, has one million cubes.

1. The million cubes could be stuck together with super glue to make a tower.

   Would the tower be as tall as:

   a) Nelson's Column

   b) the Empire State Building

   c) Mount Everest?

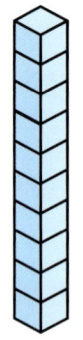

> **Tip**
>
> Use the internet to find the relevant heights.

2. If the cubes were placed in a single layer, would there be enough to cover:

   a) the floor of your classroom

   b) a 120 m by 55 m football pitch?

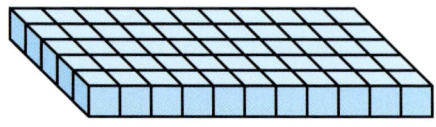

3. If the cubes were in a solid mass, would there be enough:

   a) to fill your classroom

   b) to fill a 1.2 m by 0.8 m by 0.9 m fridge-freezer?

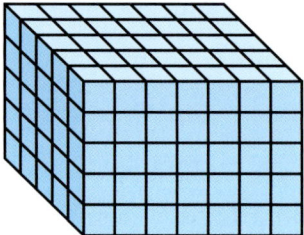

4. Mrs Evans sets out to make a *billion* of the cubes [1 000 000 000].

   a) Would that be enough to fill your classroom?

   b) Placed side by side, roughly how many times would they cover the distance from Cairo to Dubai (1500 km)?

7.2 Time

Analogue – 12-hour watch

Digital – 24-hour watch

The 24-hour clock

The times which most people use in their everyday lives are times measured from midnight or from mid-day (noon). In the morning

9 o'clock is 9 hours after midnight and is written 9.00 a.m. In the afternoon 4 o'clock is 4 hours after mid-day (noon) and is written 4.00 p.m.

Using the 24-hour clock, all times are measured from midnight.

This means 9.00 a.m. is written as 09 00 and 4.00 p.m. is written as 16 00.

Example 1

Convert each of these times to the 24-hour clock.

a) 8.00 a.m.

b) 9.30 p.m.

c) 3.15 p.m.

---

a) 8.00 a.m. = 08 00

b) 9.30 p.m. = 21 30

c) 3.15 p.m. = 15 15

> **Tip**
>
> When the time is p.m., add 12 hours to convert to the 24-hour clock.

Example 2

Convert each of these times to the 12-hour clock.

a) 06 30

b) 14 05

c) 22 35

---

a) 06 30 = 6.30 a.m.

b) 14 05 = 2.05 p.m.

c) 22 35 = 10.35 p.m.

> **Tip**
>
> If the time on the 24-hour clock is greater than 12, subtract 12 hours from the time and give your answer as p.m.

> **Tip**
>
> a.m. is an abbreviation of ante meridiem and means before mid-day.
>
> p.m. is an abbreviation of post meridiem and means after mid-day.

Exercise 7.2A

Write down the following in the 24-hour system.

1. 8.00 a.m.
2. 9.30 p.m.
3. 6.00 p.m.

4. 5.30 a.m.
5. 7.40 p.m.
6. 10.00 p.m.

7. 7.15 p.m.
8. 10.45 p.m.
9. 8.30 a.m.

10. 4.15 a.m.
11. 2.25 a.m.
12. 1.30 p.m.

13. 7.20 p.m.
14. 6.50 a.m.
15. 7.10 a.m.

16. Two minutes before midnight.
17. Two and a half hours before midnight.

18. Five minutes before noon.
19. Three and a half hours after noon.

20. One hour after midnight.
21. One and a half hours before noon.

22. Twenty minutes after midnight.
23. Five hours before midnight.

24. Six minutes after noon.
25. Fifty minutes after midnight.

Write the following in the 12-hour system.

26. 07 00
27. 19 30
28. 11 20
29. 04 45
30. 20 30

31. 21 15
32. 09 10
33. 11 45
34. 23 10
35. 20 00

36. 12 00
37. 01 40
38. 04 00
39. 07 07
40. 13 13

41. 12 15
42. 12 30
43. 15 45
44. 16 20
45. 05 16

Time intervals

Example

Find the time interval between 15 40 and 18 05.

From 15 40 to 16 00 is 20 minutes — Count on to the next hour.
From 16 00 to 18 05 is 2 hours 5 minutes — Count on from 16 00.
Altogether the time interval is 2 hours 25 minutes.

Tip

You could use a number line to help you to count on the time.

Exercise 7.2B

Find the number of hours and minutes between the following times.

1. 20 10 and 21 20
2. 21 40 and 23 50

3. 22 15 and 23 10
4. 19 30 and 20 05

5. 20 16 and 23 36

6. 11 25 and 13 10

7. 09 40 and 12 00

8. 21 17 and 23 10

9. 23 04 and 23 57

10. 17 45 and 23 10

11. 05 15 and 07 05

12. 11 26 and 14 40

13. 9.50 a.m. and 11.05 a.m.

14. 9.30 a.m. and 2.05 p.m.

15. 11.10 a.m. and 1.30 p.m.

16. 7.30 a.m. and 7.30 p.m.

17. 10.40 a.m. and 12.40 p.m.

18. 5.40 a.m. and 1.00 p.m.

19. 11.55 a.m. and 3.10 p.m.

20. 1.35 a.m. and 8.40 a.m.

21. 22 30 on Monday to 03 30 on Tuesday

22. 21 00 on Thursday to 01 40 on Friday

23. 17 30 on Monday to 02 00 on Tuesday

24. 23 45 on Saturday to 02 10 on Sunday

25. 22 50 on Thursday to 07 00 on Friday

26. 07 00 on Friday to 02 00 on Saturday

27. 09 30 on Monday to 04 30 on Tuesday

28. 09 15 on Wednesday to 02 45 on Thursday

29. 22 10 on Friday to 07 35 on Saturday

30. 06 30 on Friday to 16 30 on Saturday

Exercise 7.2C

Here is a timetable for trains from Florence to Rome.

Florence	07 10	08 35	11 35	13 50
Pontassieve	07 35	–	12 00	14 15
Figline Valdarno	07 55	09 21	12 21	–
Bucine	08 16	–	12 42	–
Arezzo	08 36	09 54	13 17	15 08
Orvieto	09 24	10 42	14 07	15 48
Orte	10 01	–	14 53	16 34
Rome	10 27	11 42	15 18	16 59

1. How long does it take the 07 10 from Florence to travel to:

   a) Pontassieve   **b)** Figline Valdarno   **c)** Bucine?

2. At how many stations does the 11 35 from Florence stop?

3. At what time does the 08 35 from Florence reach Arezzo?

4. If you had to be in Rome by 15 30 which train would you catch from Florence?

5. You arrive at Florence at 08 20. How long do you have to wait for the next train to Rome?

6. The 11 35 from Florence runs 10 minutes late. At what time will it reach Bucine?

7. How long does it take the 11 35 from Florence to travel to:

  a) Arezzo   **b)** Orvieto   **c)** Rome?

8. At how many stations does the 08 16 from Bucine stop before it reaches Orte?

9. At what time does the 12 21 from Figline Valdarno reach Orvieto?

10. If you had to be in Rome by 12 00 which train would you catch from Figline Valdarno?

11. You arrive at Florence at 11 08. How long do you have to wait for the next train to Rome?

12. The 08 35 from Florence runs 19 minutes late. At what time will it reach Rome?

13. A flight from London to New York takes exactly 8 hours. The time in New York is 5 hours earlier than the time in London. If a plane takes off from London at 15 00, what time will it be in New York when it lands?

14. A flight from Paris to Istanbul takes 3 hours and 25 minutes. The time in Istanbul is one hour later than the time in Paris. If a plane takes off from Paris at 08 37, what time will it be in Istanbul when the plane lands?

15. A flight from Tokyo to Melbourne takes 10 hours and 25 minutes. The time in Tokyo is 2 hours earlier than the time in Melbourne. If a plane leaves Tokyo at 09 15, what time will it be in Melbourne when the plane lands?

There are 7 days in a week.

There are 365 days in a year (except for a leap year when there are 366 days).

Of the 12 months of the year, April, June, September and November have 30 days.

The rest of the months have 31 days, except February.

February has 28 days in a year, except in a leap year when it has 29 days.

Use this information to answer these questions.

16. How many complete weeks are there in a year?

17. How many days are there between 1 January and 31 March (inclusive) in a non-leap year?

18. How many complete weeks are there between 1 April and 30 September (inclusive)?

19. 2024 is a leap year. How many days will there be between 1 February and 30 November (inclusive)?

20. On 1 March, Joshva says his birthday is in 102 days' time. When is Joshva's birthday?

7.3 Rates

In mathematics, a **rate** is the ratio of one quantity measured against another quantity. These two quantities will usually have different units.

Speed is a common measure of rate. A speed given in kilometres per hour tells us how many kilometres an object travels in one hour.
Other common measures of rate include:

- litres per minute, when filling a bath with water, for example
- kilowatt hours per day, when measuring energy consumption
- density, which is the amount of mass in a given volume
- pressure, which is the amount of force applied to a given area
- population density, which is the number of people who live in a given area
- hourly rates of pay.

Speed, distance and time

Calculations involving these three quantities are simpler when the speed is **constant**. The formulae connecting the quantities are as follows:

- $\text{speed} = \dfrac{\text{distance}}{\text{time}}$
- $\text{distance} = \text{speed} \times \text{time}$
- $\text{time} = \dfrac{\text{distance}}{\text{speed}}$

> **Tip**
>
> The formulae for rates can be remembered by thinking about the units that are used to measure them. You just need to remember that 'per' means 'divided by'.
>
> This means that because speed can be measured in km per hour, and because km is a measure of distance and hours are a measure of time, then the formula for speed is distance divided by time.

A helpful way of remembering these formulae is to write the letters D, S and T in a triangle,

thus:

to find D, cover D and you have $S \times T$

to find S, cover S and you have $\dfrac{D}{T}$

to find T, cover T and you have $\dfrac{D}{S}$

Be careful with the units when you are answering questions involving these formulae.

Example 1

A man runs at a speed of 8 km/h for a distance of 5200 metres. Find the time taken in minutes.

5200 metres = 5.2 km   1 km = 1000 m

Time taken in hours $= \dfrac{D}{S} = \dfrac{5.2}{8}$

$= 0.65$ hours

Time taken in minutes $= 0.65 \times 60$

$= 39$ minutes

Tip

You learned how to use your calculator for converting between decimal fractions and hours and minutes in Chapter 4.

Example 2

A cheetah runs 87.5 metres in 3.5 seconds.
Find its speed in m/s.

Speed $= \dfrac{D}{T} = \dfrac{87.5}{3.5}$

$= 25$ m/s

Example 3

A motorcycle travels at a constant speed of 42 km/h for 1 hour 15 minutes.

Find the distance travelled in kilometres.

Distance = speed × time

1 hour 15 minutes is 1.25 hours

Distance = $42 \times 1.25 = 52.5$ km

Tip

There are 60 minutes in an hour so 15 minutes is $\dfrac{15}{60} = 0.25$ hours.

Example 4

Change a speed of 54 km/h into metres per second.

54 km/hour = 54 000 metres per hour	1 km = 1000 m
$= \dfrac{54\,000}{60}$ metres per minute	There are 60 minutes in an hour so divide by 60.
$= \dfrac{54\,000}{60 \times 60}$ metres per second	There are 60 seconds in a minute so divide by 60.
$= 15$ m/s	

Exercise 7.3A

1. Find the time taken for the following journeys:

   a) 100 km at a speed of 40 km/h

   b) 250 miles at a speed of 80 miles per hour

   c) 15 metres at a speed of 20 cm/s (answer in seconds)

   d) 10^4 metres at a speed of 2.5 km/h.

2. Change the units of the following speeds as indicated:

   a) 72 km/h into m/s

   b) 108 km/h into m/s

   c) 300 km/h into m/s

   d) 30 m/s into km/h

   e) 22 m/s into km/h

   f) 0.012 m/s into cm/s

   g) 9000 cm/s into m/s

   h) 600 miles/day into miles per hour.

3. Find the speeds of the bodies which move as follows:

   a) a distance of 600 km in 8 hours

   b) a distance of 31.64 km in 7 hours

   c) a distance of 136.8 m in 18 seconds

   d) a distance of 4×10^4 m in 100 seconds

   e) a distance of 5×10^5 cm in 5 seconds

   f) a distance of 10^8 mm in 30 minutes (in km/h)

   g) a distance of 500 m in 10 minutes (in km/h).

4. Find the distance travelled (in metres) in the following:

   a) at a speed of 55 km/h for 2 hours

   b) at a speed of 40 km/h for $\dfrac{1}{4}$ hour

   c) at a speed of 338.4 km/h for 10 minutes

   d) at a speed of 15 m/s for 5 minutes

   e) at a speed of 14 m/s for 1 hour

f) at a speed of 4×10^3 m/s for 400 seconds

g) at a speed of 8×10^5 cm/s for 2 minutes.

5. A cyclist travels 39 km in 3 hours and 15 minutes.
   What is their average speed?

6. A motorcyclist travels at an average speed of 48 km/h for 2 hours and 45 minutes.
   How far do they travel?

7. A train covers a distance of 120 km at an average speed of 75 km/h.
   How long does the journey take? Give your answer in hours and minutes.

8. A car travels 60 km at 30 km/h and then a further 180 km at 160 km/h. Find:
   a) the total time taken
   b) the average speed for the whole journey.

> **Tip**
>
> The average speed is the total distance travelled divided by the total time taken.

9. A cyclist travels 25 kilometres at 20 km/h and then a further 80 kilometres at 25 km/h. Find:
   a) the total time taken
   b) the average speed for the whole journey.

10. A swallow flies at a speed of 50 km/h for 3 hours and then at a speed of 40 km/h for a further 2 hours. Find the average speed for the whole journey.

Other rates

Exercise 7.3B

1. Find the following rates in the units given.
   a) 4 litres in 5 minutes (litres per minute)
   b) 12 litres in 45 minutes (litres per hour)
   c) 78 litres in 12 minutes (litres per hour)
   d) 800 kilowatt hours in 2 months (kilowatt hours per year)
   e) 12 kilowatt hours in 3 days (kilowatt hours per year)

> **Tip**
>
> There are 365 days in a year.

2. Find the time taken (in minutes) to fill the following containers:

   a) a 3-litre bowl at a rate of 2 litres per minute

   b) a 30-litre bucket at a rate of 0.2 litres per second

   c) a 120-litre hot water tank at a rate of 80 litres per hour

   d) a 300-ml beaker at a rate of 0.5 litres per hour.

3. A bath is filled with 80 litres of water in 6 minutes. Find the rate at which it is being filled.

4. A typical household uses 4600 kilowatt hours of energy in a year. Find the rate at which the household uses energy in kilowatt hours per day.

5. Water is dripping from a tap at a rate of 5 millilitres per second. How long will it take, in minutes, to fill a bowl with a capacity of 2.5 litres?

6. A rain butt with a capacity of 60 litres fills completely with water each day.

   a) Find the rate of fill in millilitres per hour.

   b) A gardener can use all of the water to hose his garden in 15 minutes. Find the rate of flow of the water from the hose in millilitres per second.

7. Find the pressure exerted by a force of 10 000 N on an area of 40 m^2.

8. The pressure exerted on a table is 500 N/m^2. If the force is 750 N, what is the area of the table?

9. In 2020, the population density of Tokyo was 6402 people per square kilometre. The size of Tokyo is 2194 km^2. How many people lived in Tokyo in 2020? Give your answer accurate to 2 significant figures.

> **Tip**
>
> Use the formula
> $$\text{pressure} = \frac{\text{force}}{\text{area}}$$
> to give pressure in N/m^2.

> **Tip**
>
> Use the formula
> $$\text{population density} = \frac{\text{population}}{\text{land area}}$$
> to give population density in people per km^2.

10. In 2020, there were 28.64 million people living in Texas, which has an area of 695 662 km^2. What was the population density in people per km^2? Give your answer correct to 3 significant figures.

11. A metal block has a mass of 1200 g and a volume of 900 cm³.
    What is the density of the block?

12. A crystal has a density of 1.3 g/cm³ and a volume of 2.5 cm³.
    What is the mass of the crystal?

13. A piece of wood has mass 650 grams. The density of the wood
    is 0.78 g/cm³. What is the volume of the wood? Give your answer
    to the nearest cm³.

14. Bilal earns $450 for doing a 9-hour shift in a hospital.
    Calculate Bilal's hourly rate of pay.

15. Maryssa's hourly rate of pay is $23.50. She works a 7.5 hour shift.
    Calculate how much Maryssa gets paid for her shift.

16. Frank's hourly rate of pay is $9.30. He earns $348.75 per week.
    Calculate the number of hours that Frank works in a week.

> **Tip**
>
> Use the formula
> $$\textbf{density} = \frac{\text{mass}}{\text{volume}}$$
> to give density in
> g/cm³.

7.4 Area and perimeter

Perimeter is the total distance around the edge of a shape.

- The perimeter of a square, side length x, is $4x$.

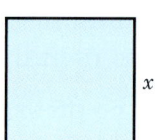

- The perimeter of a rectangle, side lengths l and w, is $2l + 2w$.

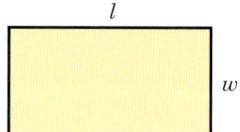

- The perimeter of a triangle, side lengths a, b and c, is $a + b + c$.

You also need to be able to work out the area of geometrical shapes.
Area is the space covered by a shape.

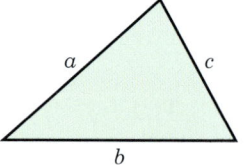

Square, rectangle and triangle

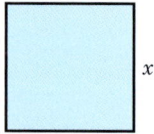

Square: area $= x^2$

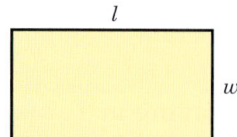

Rectangle: area $= l \times w$

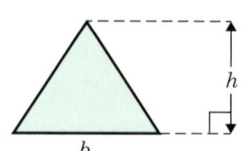

Triangle: area $= \dfrac{b \times h}{2}$

Example 1

Work out the area of this shape. All lengths are in cm.

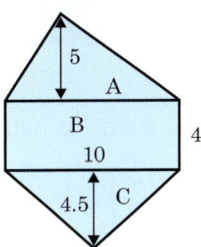

Shape A is a triangle: area $= \dfrac{10 \times 5}{2} = 25$ cm^2

Shape B is a rectangle: area $= 10 \times 4 = 40$ cm^2

Shape C is a triangle: area $= \dfrac{10 \times 4.5}{2} = 22.5$ cm^2

Area of whole shape $= 25 + 40 + 22.5 = 87.5$ cm^2

Example 2

Work out the perimeter and area of this shape. All lengths are in cm.

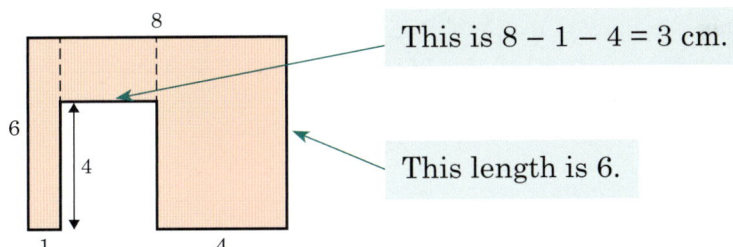

This is $8 - 1 - 4 = 3$ cm.

This length is 6.

First, find the missing lengths in the shape, as shown.

The perimeter is the total distance round the edge of the shape:

$$8 + 6 + 4 + 4 + 3 + 4 + 1 + 6 = 36 \text{ cm}$$

To find the area, there are two methods.

Method 1

Find the area of the 'complete' rectangle and subtract the area of the smaller 'removed' rectangle:

$$8 \times 6 - 3 \times 4 = 48 - 12 = 36 \text{ cm}^2$$

Method 2

Split the shape into rectangles as shown by the dotted lines:

$6 \times 1 + 3 \times 2 + 4 \times 6 = 6 + 6 + 24 = 36$ cm^2

> **Tip**
>
> You could split the shape differently, but you should always get the same answer!

Exercise 7.4A

In this exercise, all dimensions are in cm.

In Questions **1** and **2**, work out the area and perimeter.

1.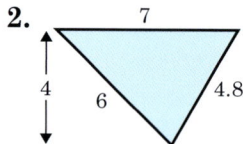

2.

In Questions **3** to **8**, work out the area.

3.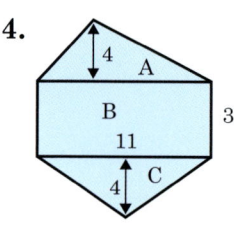

4.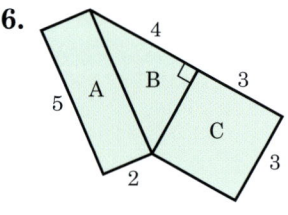

5.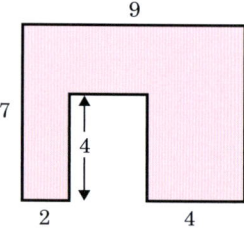

6.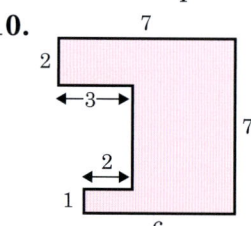

7.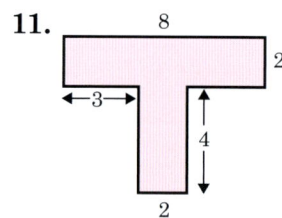

8.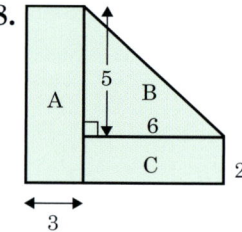

In Questions **9** to **16**, work out the area and perimeter.

9.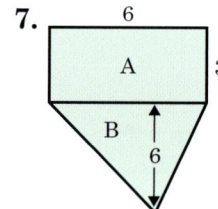

10.

11.

12.

13.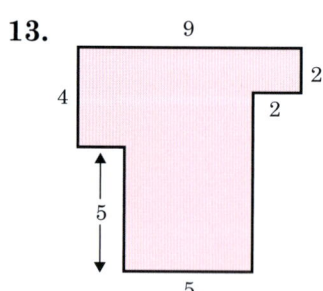

14.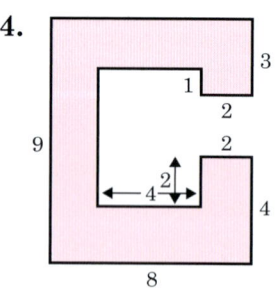

15.

16.

Exercise 7.4B

1. a) Work out the areas of triangles A, B and C.

b) Work out the area of the square enclosed by the broken lines.

c) Hence work out the area of the shaded triangle.

Give your answers in square units.

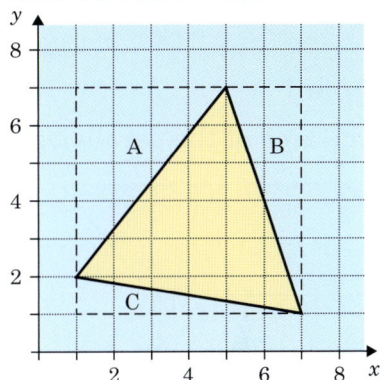

2. a) Work out the areas of triangles A, B and C.

b) Work out the area of the rectangle enclosed by the broken lines.

c) Hence work out the area of the shaded triangle.

Give your answers in square units.

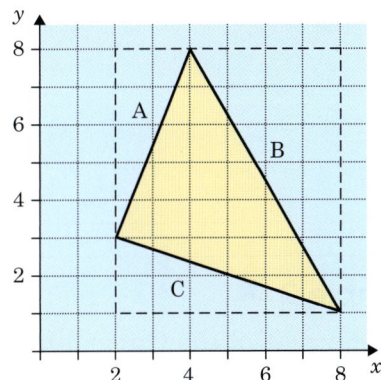

For Questions **3** to **7**, draw a pair of axes similar to those in Questions **1** and **2**. Plot the points in the order given and find the area of the shape enclosed.

3. (1, 4), (6, 8), (4, 1)

4. (1, 7), (8, 5), (4, 2)

5. (2, 4), (6, 1), (8, 7), (4, 8), (2, 4)

6. (1, 4), (5, 1), (7, 6), (4, 8), (1, 4)

7. (1, 6), (2, 2), (8, 6), (6, 8), (1, 6)

Trapezium and parallelogram

Trapezium (two parallel sides)

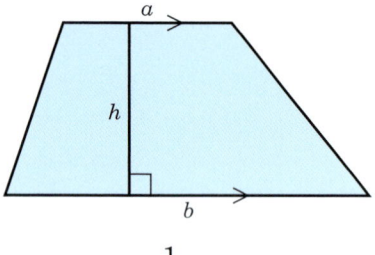

$$\text{area} = \frac{1}{2}(a + b) \times h$$

Parallelogram

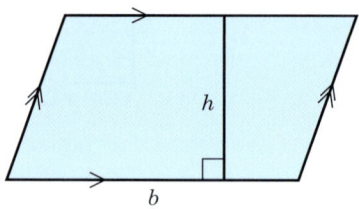

$$\text{area} = b \times h$$

Example 1

Find the area of the parallelogram.

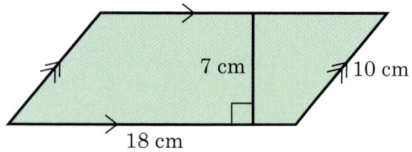

Area of paralellogram = 18 × 7

$$= 126 \text{ cm}^2$$

Tip

Make sure that you use the perpendicular height in the formula for the area of a parallelogram, not the length of the sloping side.

Example 2

Find the area of the trapezium.

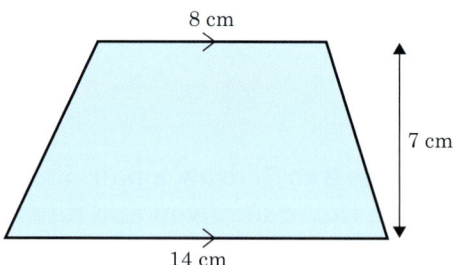

Area of trapezium $= \dfrac{1}{2}(8 + 14) \times 7$

$$= \frac{1}{2} \times 22 \times 7$$

$$= 77 \text{ cm}^2$$

Exercise 7.4C

In Questions **1** to **5** find the area of each shape. All lengths are in cm.

1.
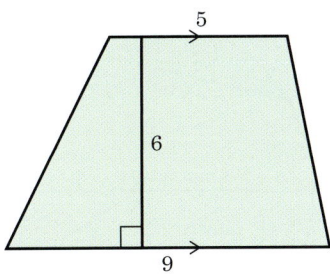

2.
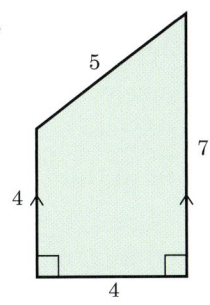

3.
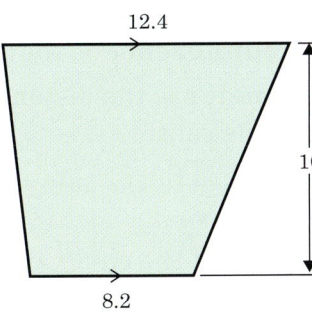

4.
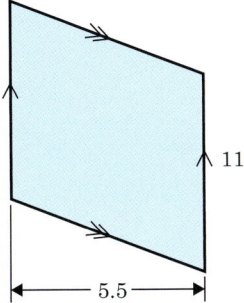

5.
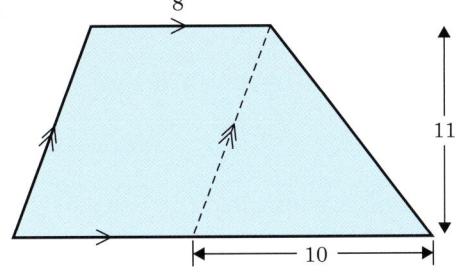

6. Large areas of land are measured in hectares. 1 hectare = 10 000 m². Copy and complete the statements below:

area of square = _____ m²

area of square = _____ hectares

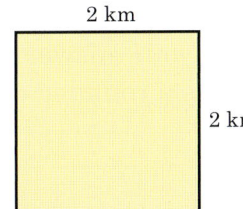

7. The field shown is sprayed with fertiliser at the rate of 2 litres per hectare. The cost of the spray is $25 for 100 litres.

How much will it cost to spray this field, to the nearest dollar?

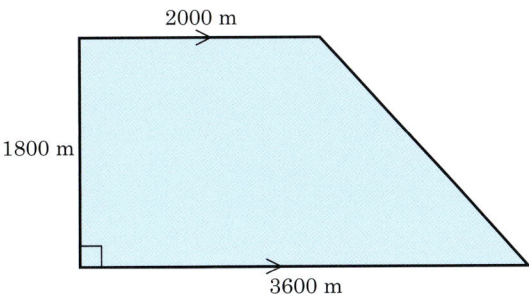

233

7.5 Circle calculations

You need to be able to identify parts of the circle.

The **circumference** is the perimeter of the circle.

The **diameter** is the distance across the circle, passing through the centre.

The **radius** is the distance from the centre to the circumference.

Other parts of the circle are shown in the diagram.

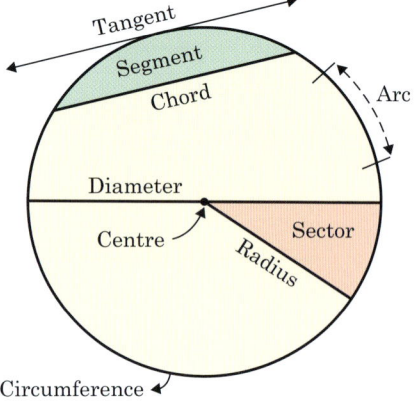

Circumference of a circle

The circumference of a circle of diameter d is given by $C = \pi d$.

Example 1

Find the circumference of this circle.

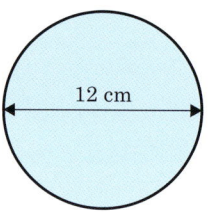

12 cm

$C = \pi \times 12$ cm

$C = 12\pi = 37.7$ cm (to 3 s.f.)

You could be asked to give your answer as an exact multiple of π, or as a rounded decimal.

Example 2

Find the circumference of this circle.

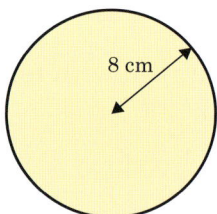

8 cm

$d = 2 \times 8 = 16$ cm

$C = \pi \times 16 = 16\pi = 50.3$ cm (to 3 s.f.)

Tip

You are given the radius, so work out the diameter first.

Example 2 above shows how to find the circumference when given the radius, r.

Since the diameter is two times the radius, you can use the alternative formula $C = 2\pi r$.

Exercise 7.5A

In Questions **1** to **4**, find the circumference without using a calculator. Give the answers in terms of π.

1.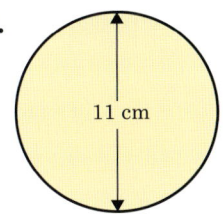
11 cm

2.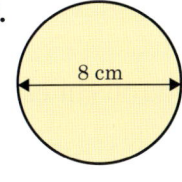
8 cm

3.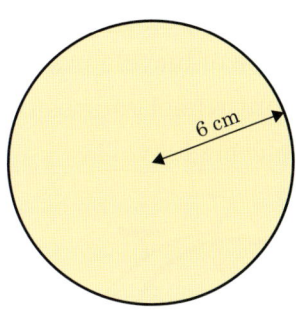
6 cm

4.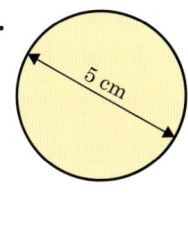
5 cm

In Questions **5** to **8**, find the circumference.

Use the π button on a calculator or take $\pi = 3.142$. Give the answers correct to 3 significant figures.

5.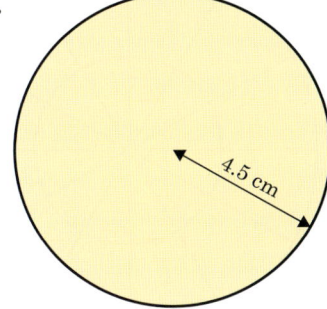
4.5 cm

6.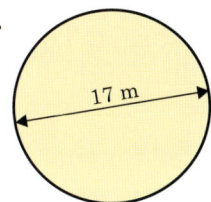
17 m

7.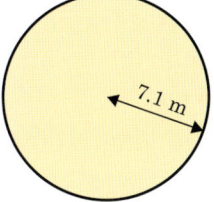
7.1 m

8.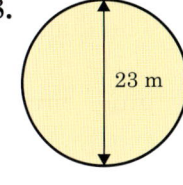
23 m

9. A circular pond has a diameter of 2.7 m.

Calculate the length of the perimeter of the pond.

10. How many complete revolutions does a cycle wheel of diameter 60 cm make in travelling 400 m?

11. A running track has two semi-circular ends of radius 34 m and two straight sections of length 93.2 m, as shown.

Calculate the total distance around the track to the nearest metre.

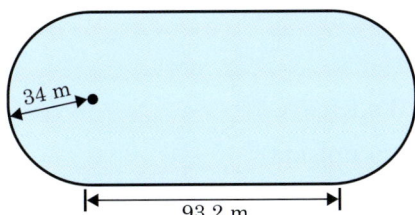
34 m
93.2 m

12. The minute hand of a clock is 14.4 cm long.

How far does the tip of the minute hand move between 12 00 and 12 15?

13. An old-fashioned type of bicycle is shown.

In a journey the front wheel rotates completely 156 times.

How far does the bicycle travel?

radius
0.84 m

radius
0.2 m

14. The diagram shows a framework for a target. The radius of the outer circle is 30 cm and the radius of the inner circle is 15 cm.

Calculate the total length of wire needed for the whole framework.

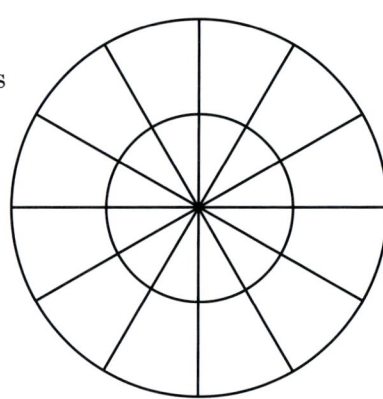

Area of a circle

The area of a circle of radius r is given by $A = \pi r^2$.

Example

Find the area of this circle.

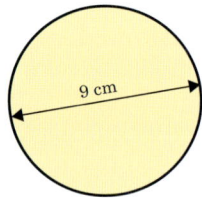

9 cm

In this circle $r = 9 \div 2 = 4.5$ cm

Area of circle $= \pi \times 4.5^2$

$\qquad\qquad = 63.6$ cm^2 (to 3 s.f.)

Remember: the formula is $\pi(r^2)$ not $(\pi r)^2$.

Tip

You need to work out the radius of the circle first before using the area formula.

Exercise 7.5B

📝 In Questions **1** to **4** find the area of the circle without using a calculator. Give the answers in terms of π.

1.

11 cm

2.

5 cm

3.

3 m

4.

7 m

In Questions **5** to **8** find the area of the circle.

Use the π button on a calculator or use π = 3.142.

Give the answers correct to three significant figures.

5.

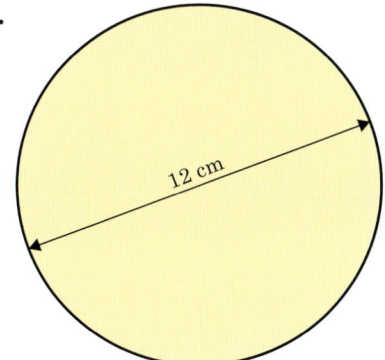

12 cm

6.

8 cm

7.

5 m

8.

11 cm

9. A spinner of radius 7.5 cm is divided into six equal sectors. Calculate the area of each sector.

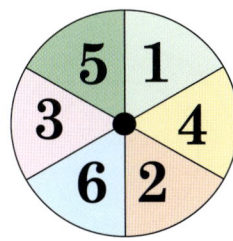

10. A circular swimming pool of diameter 12.6 m is to be covered by a plastic sheet.
Work out the surface area the sheet must cover.

11. A circle of radius 5 cm is inscribed inside a square as shown.
Find the area shaded.

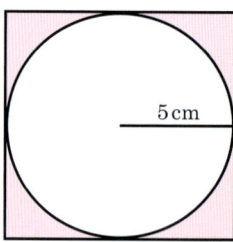

12. Each square metre of a lawn requires 2 g of weedkiller. How much weedkiller is needed for a circular lawn of radius 27 m?

13. Discs of radius 4 cm are cut from a rectangular plastic sheet of length 84 cm and width 24 cm.

a) How many complete discs can be cut out?

Find:

b) the total area of the discs cut

c) the area of the sheet wasted.

14. A circular pond of radius 6 m is surrounded by a path of width 1 m.
Find the area of the path.

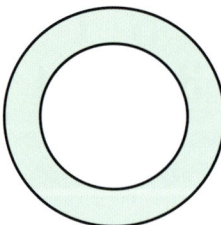

15. The diagram below shows a lawn (unshaded) surrounded by a path of uniform width (shaded). The curved end of the lawn is a semi-circle of diameter 10 m.

Calculate the total area of the path.

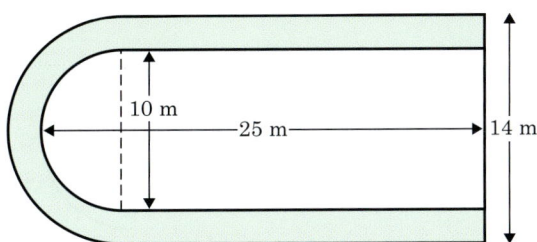

16. Sophia says: 'The area of a circle with radius 6 cm is $36\pi^2$ cm^2'.

Is Sophia correct? Explain your answer.

More complicated shapes

Example

The shape is made up of a semi-circle and a rectangle.
Find:

a) the perimeter   **b)** the area.

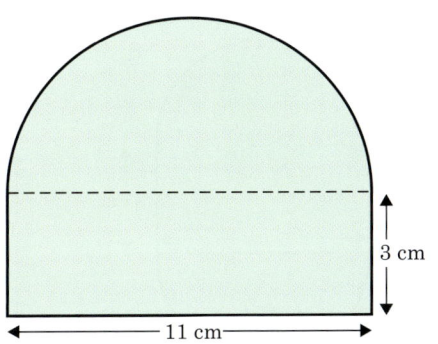

a) Perimeter $= \left(\dfrac{\pi \times 11}{2} \right) + 11 + 3 + 3$ Add the semi-circular arc to the three sides of the rectangle.

$= 34.3$ cm (3 s.f.)

A semi-circle is half a circle.

b) Area $= \left(\dfrac{\pi \times 5.5^2}{2} \right) + (11 \times 3)$ Add the area of the semi-circle to the area of the rectangle.

$= 80.5$ cm^2 (3 s.f.)

Divide the area of the whole circle by 2.

Exercise 7.5C

Use the π button on a calculator or take π = 3.142. Give the answers correct to 3 s.f.
For each shape, find the perimeter.

1.

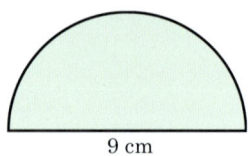

9 cm

2.

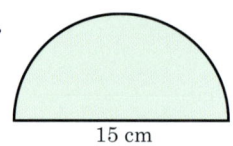

15 cm

3.

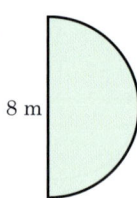

8 m

4.

3.2 cm

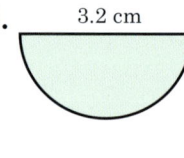

5.

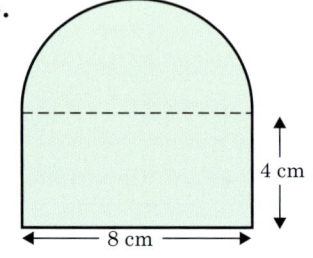

4 cm

8 cm

6.

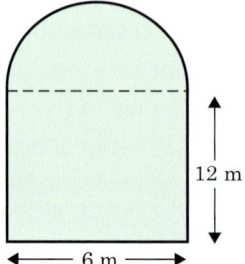

12 m

6 m

7.

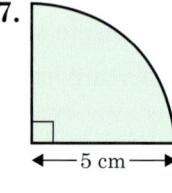

5 cm

8.

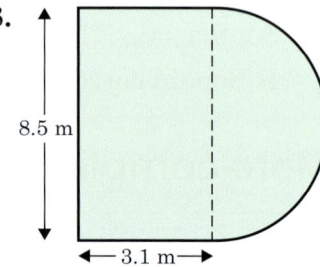

8.5 m

3.1 m

9.

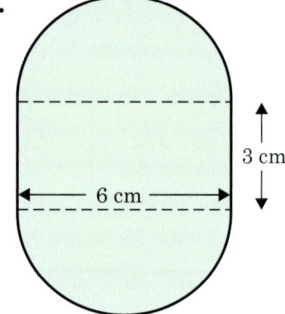

3 cm

6 cm

10.

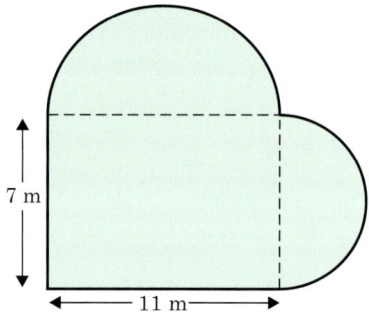

7 m

11 m

11.

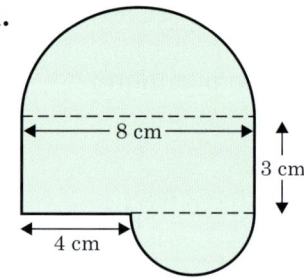

8 cm

3 cm

4 cm

Exercise 7.5D

Find the shaded area for each shape. All lengths are in cm.

1.

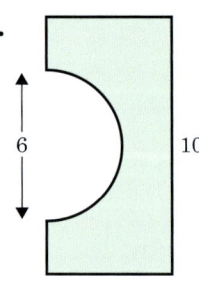

6

10

5

2.

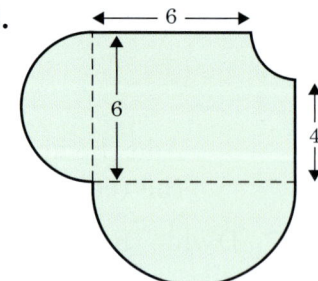

6

6

4

3.

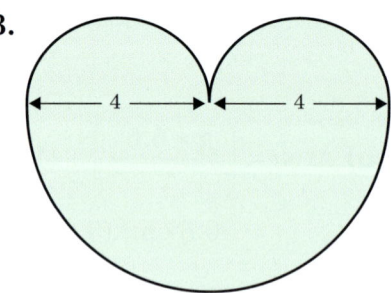

4

4

4.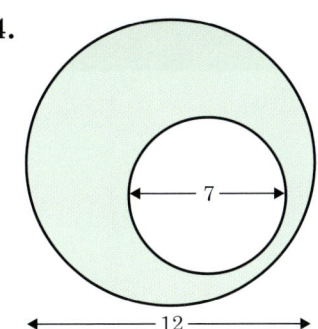

5.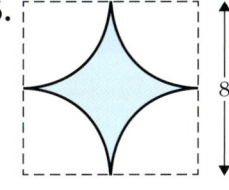

6.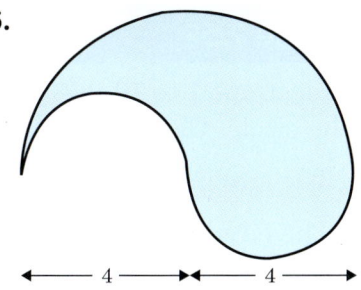

7. a) Find the area of triangle OAD.

    b) Hence find the area of the square ABCD.

    c) Find the area of the circle.

    d) Hence find the shaded area.

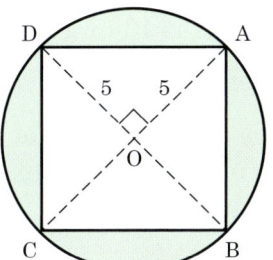

Finding the radius of a circle

Sometimes it is difficult to measure the diameter or radius of a circle accurately but the circumference can be measured using a tape measure.

Example 1

The circumference of a circle is 60 cm.

Find the radius of the circle.

$C = \pi d$	
$60 = \pi d$	Substitute the value of C into the formula.
$\dfrac{60}{\pi} = d$	Divide by π.
$r = \dfrac{(60\,/\,\pi)}{2} = 9.55$ cm (to 3 s.f.)	Divide this by 2 to find the radius.

Example 2

The area of a circle is 18 m².

Find the radius of the circle.

$\pi r^2 = 18$	Substitute the value of A into the formula.
$r^2 = \dfrac{18}{\pi}$	Divide by π.
$r = \sqrt{\left(\dfrac{18}{\pi}\right)} = 2.39$ m (to 3 s.f.)	Square root this to find the radius.

Exercise 7.5E

In Questions **1** to **10** use the information given to calculate the radius of the circle. Use the π button on a calculator or take π = 3.142.

1. The circumference is 15 cm.

2. The circumference is 28 m.

3. The circumference is 7 m.

4. The area is 54 cm².

5. The area is 38 cm².

6. The area is 49 m².

7. The circumference is 16 m.

8. The area is 60 cm².

9. The circumference is 29 cm.

10. The area is 104 cm².

11. An odometer is a wheel used for measuring long distances. The circumference of the wheel is exactly one metre. Find the radius of the wheel.

12. A sheet of paper is 32 cm by 20 cm. It is made into a hollow cylinder of height 20 cm with no overlap.

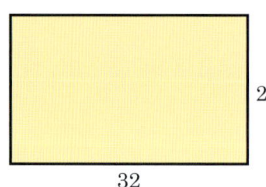

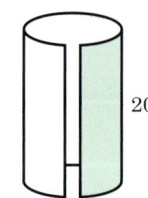

Find the radius of the cylinder.

13. The area of the centre circle on a football pitch is 265 m². Calculate the radius of the circle to the nearest 0.1 m.

14. Eight sections of curved railway track can be joined to make a circular track. Each section is 23 cm long. Calculate the diameter of the circle.

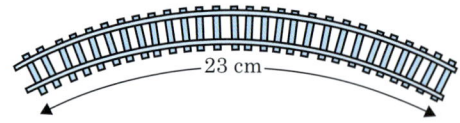

15. Calculate the radius of a circle whose area is equal to the sum of the areas of three circles of radii 2 cm, 3 cm and 4 cm.

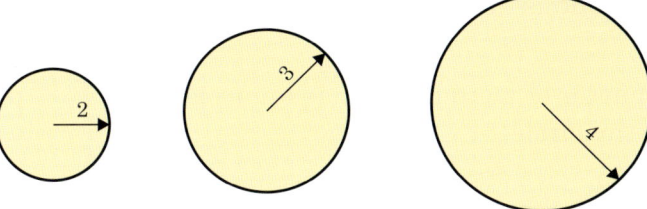

16. The handle of a paint tin is a semi-circle of wire which is 28 cm long. Calculate the diameter of the tin.

17. A television transmitter is designed so that people living inside a circle of area 120 000 km² can receive pictures. What is the radius of this circle? Give your answer to the nearest km.

18. The circle and the square have the same area. Find the radius of the circle.

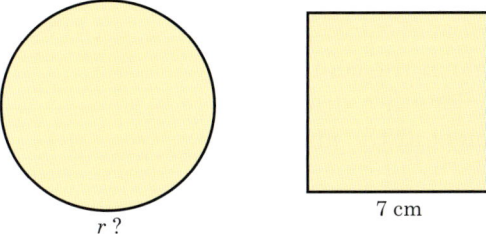

r ?

7 cm

19. The circumference of this circle is 52 m. Find its area.

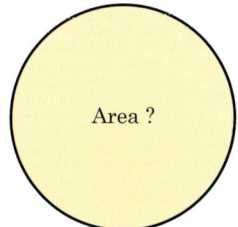

Area ?

20. The area of a circular target is 1.2 m². Find the circumference of the target.

21. The perimeter of a circular pond is 85 m long. Work out the area of the pond.

22. The sector shown is one quarter of a circle and has an area of 23 cm². Find the radius of the circle.

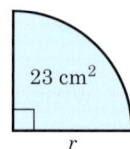

23. Grass seed is sown at a rate of 40 grams per square metre and one box of seed contains 2.5 kg. The seed is just enough to sow a circular lawn. Calculate the radius of this lawn to the nearest 0.1 metre.

7.6 Arc length and sector area

The **arc length** is a fraction of the whole circumference that depends on the angle at the centre of the circle.

Arc length, $l = \dfrac{\theta}{360} \times \pi d$ or $l = \dfrac{\theta}{360} \times 2\pi r$

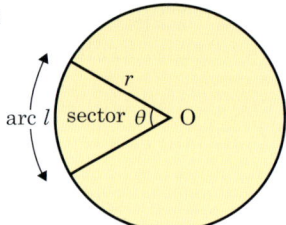

Tip

You say that the arc **subtends** the angle at the centre. This means that the two straight lines, length r, drawn from the ends of the arc to the centre meet at the angle θ.

Similarly, the area of a sector is a fraction of the area of the whole circle that depends on the angle at the centre of the circle.

Sector area, $A = \dfrac{\theta}{360} \times \pi r^2$

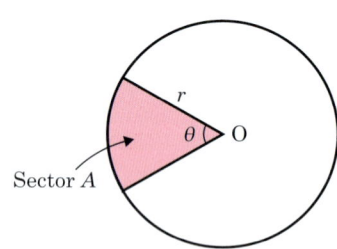

Example

a) Find the length of an arc which subtends an angle of 120° at the centre of a circle of radius 12 cm.

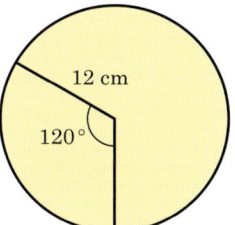

b) A sector of a circle of radius 10 cm has an area of 25π cm². Find the angle at the centre of the circle.

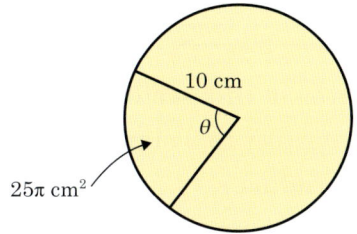

a) Arc length $= \dfrac{120}{360} \times 2 \times \pi \times 12$

$\qquad\qquad = 8\pi$

$\qquad\qquad = 25.1$ cm (3 s.f.)

b) Let the angle at the centre of the circle be θ.

$\dfrac{\theta}{360} \times \pi \times 10^2 = 25\pi$

$\therefore \quad \theta = \dfrac{25\pi \times 360}{\pi \times 100}$

$\theta = 90°$

The angle at the centre of the circle is $90°$.

Exercise 7.6A

In Questions **1** to **4** find:

a) the arc length of the sector

b) the perimeter of the sector

c) the area of the sector.

Give the answers as multiples of π.

1.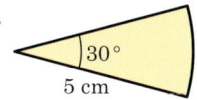

2.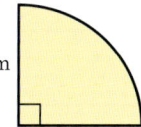

3.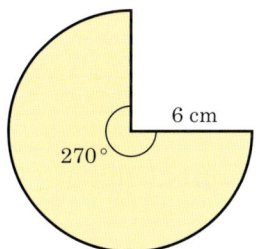

4.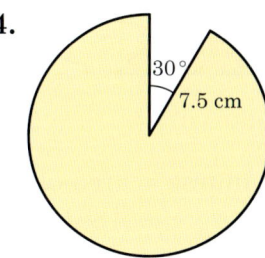

5. Find the shaded areas.

a)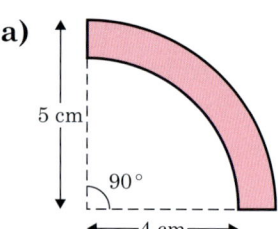

b)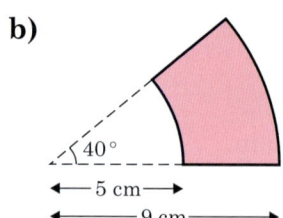

6. The area of this sector is 37.83 cm².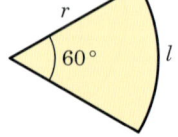

Find, correct to 3 significant figures:

a) the radius r

b) the arc length l.

7. The length of the minor arc AB of a circle, centre O, is 2π cm and the length of the major arc is 22π cm. Find: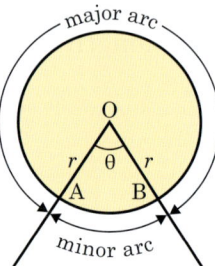

a) the radius of the circle

b) the acute angle AOB.

8. A wheel of radius 10 cm is turning at a rate of 5 revolutions per minute. Calculate:

a) the angle through which the wheel turns in 1 second

b) the distance moved by a point on the rim in 2 seconds.

7.7 Volume and surface area

Prisms and cuboids

A prism is an object with a uniform cross-section.

A cuboid is a prism whose cross-section is a rectangle.

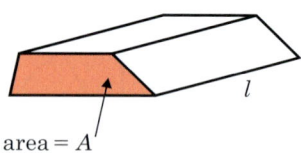

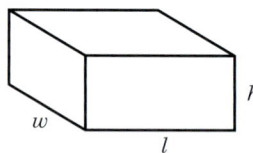

Volume = area of cross-section × length
$$= A \times l$$

Volume = length × width × height
$$= l \times w \times h$$
Surface area $= 2 \times (wl + hl + wh)$

To find the surface area of a solid, find the area of every face and add them all together.

Example 1

Here is a cuboid.

Find:

a) the volume

b) the total surface area.

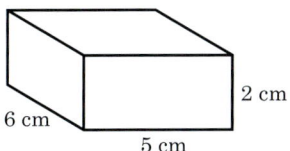

a) Volume = $lwh = 5 \times 6 \times 2 = 60$ cm³

b) To find the surface area, you can think of the cuboid being 'opened out' to make a net.

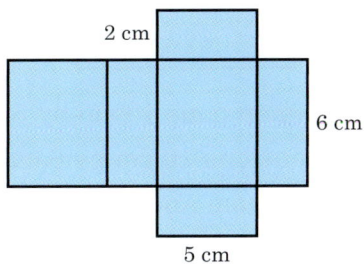

There are two rectangles with dimensions 6 cm by 5 cm.

There are two rectangles with dimensions 5 cm by 2 cm.

There are two rectangles with dimensions 6 cm by 2 cm.

Total surface area = $2 \times 6 \times 5 + 2 \times 5 \times 2 + 2 \times 6 \times 2 = 60 + 20 + 24 = 104$ cm²

Example 2

Here is a triangular prism.

Find:

a) the volume

b) the total surface area.

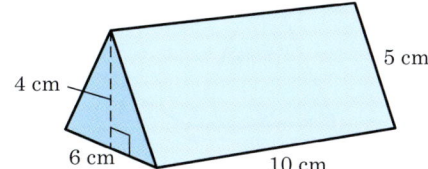

a) Volume = cross-section area × length

Area of triangular cross-section = $\dfrac{6 \times 4}{2} = 12$ cm²

Hence $V = 12 \times 10 = 120$ cm³

b) To find the surface area, again think of the shape as being 'opened out'.

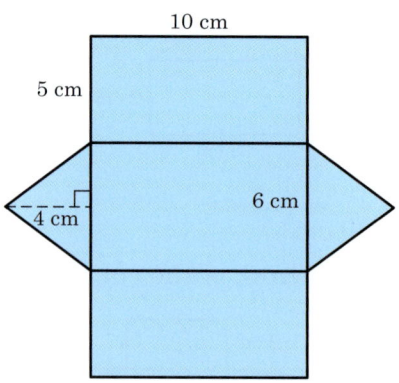

There are two rectangles with dimensions 10 cm by 5 cm.

There is one rectangle with dimensions 10 cm by 6 cm.

There are two triangles with base 6 cm and perpendicular height 4 cm.

Total surface area $= 2 \times 10 \times 5 + 10 \times 6 + 2 \times \dfrac{6 \times 4}{2} = 100 + 60 + 24 = 184$ cm^2

Exercise 7.7A

Find the volume of each prism in Questions **1** and **2**.

1. Area of end $= 15$ cm^2

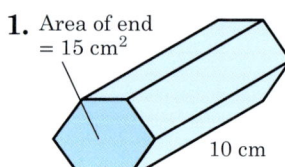

10 cm

2.

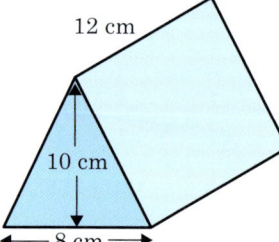

12 cm

10 cm

8 cm

Find the volume and surface area of each prism in Questions **3** to **5**.

3.

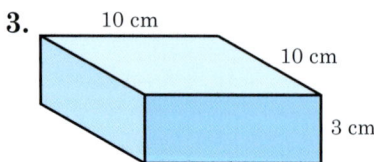

10 cm

10 cm

3 cm

4.

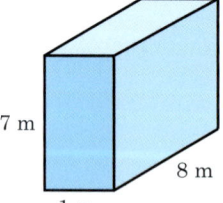

7 m

1 m

8 m

5.

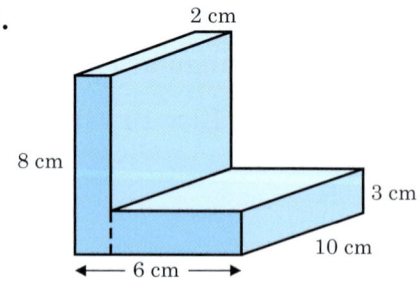

2 cm

8 cm

3 cm

6 cm

10 cm

Find the volume of each prism in Questions **6** to **8**.

6.

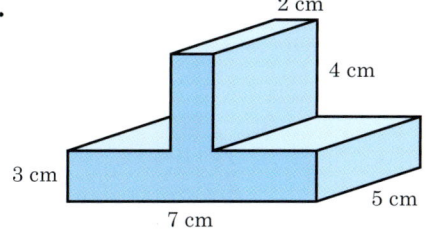

7.

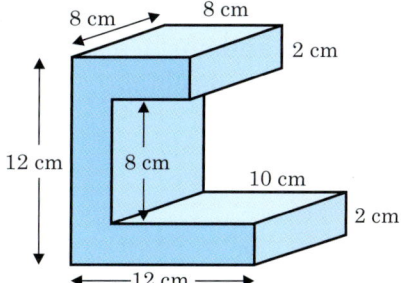

8.

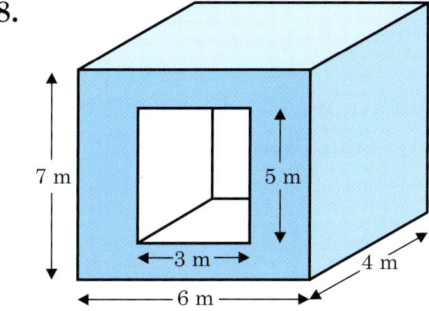

9. A wooden cuboid has the dimensions shown.

a) Calculate the total surface area.

b) One tin of paint can cover 3 m². How many cuboids can be painted using the paint in one tin?

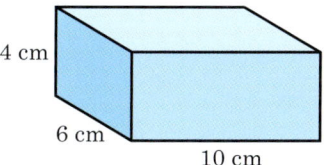

Cylinders

A cylinder is a 3D shape with uniform circular cross-section. You can find the volume and surface area in a similar way as for a prism.

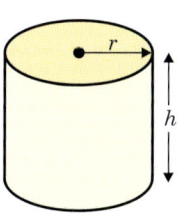

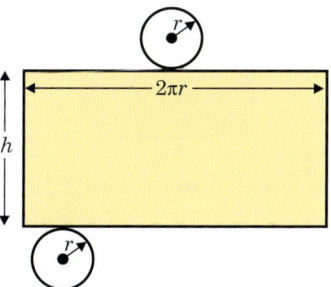

Tip

When you 'unwrap' a cylinder, it forms a rectangle. The width of the rectangle is equal to the circumference of the circular end faces.

Volume = area of circle × height
$$= \pi r^2 h$$

Surface area = area of circular ends
+ curved surface area
$$= 2\pi r^2 + 2\pi rh$$

Exercise 7.7B

 Find the volume and surface area of each cylinder in Questions **1** to **3**. Give your answers as exact multiples of π.

1.
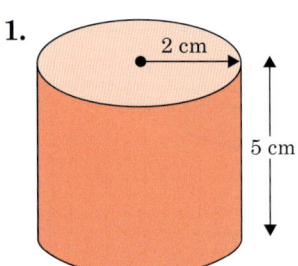
2 cm
5 cm

2.
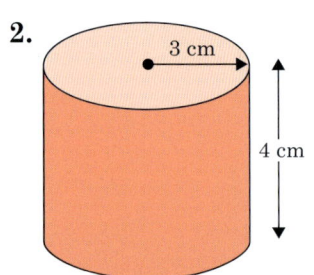
3 cm
4 cm

3.
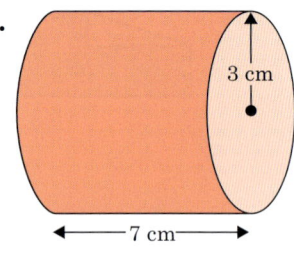
3 cm
7 cm

Find the volume and surface area of each cylinder in Questions **4** to **10**. Use the π button on a calculator and give the answers correct to 3 significant figures.

4.
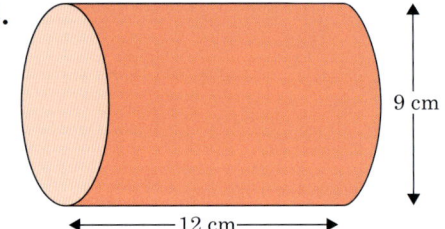
9 cm
12 cm

5.
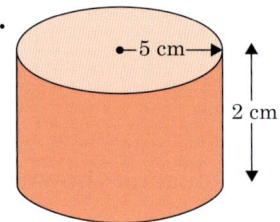
5 cm
2 cm

6.
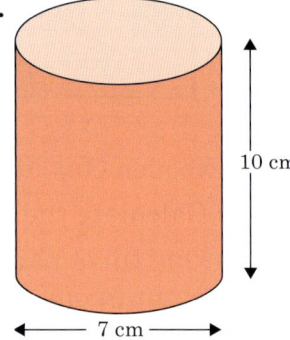
10 cm
7 cm

7. radius = 7 cm, height = 5 cm

8. diameter = 8 m, height = 3.5 m

9. diameter = 11 m, height = 2.4 m

10. radius = 3.2 cm, height = 15.1 cm

11. The diagram shows an oil drum. Find the capacity in litres. ($1000 \text{ cm}^3 = 1$ litre)

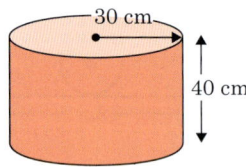
30 cm
40 cm

12. Cylinders are cut along the axis of symmetry to form the objects below. Find the volume of each object.

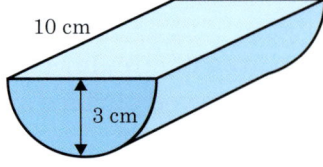
10 cm
3 cm

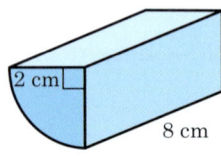
2 cm
8 cm

Spheres

Volume $= \dfrac{4}{3}\pi r^3$

Surface area $= 4\pi r^2$

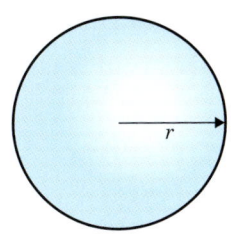

Cones

Volume $= \dfrac{1}{3}\pi r^2 h$

Surface area = area of circular base + area of curved surface

$\qquad\qquad = \pi r^2 + \pi r l$

where l is the slant height of the cone.

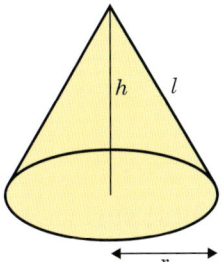

Pyramids

Volume $= \dfrac{1}{3}$ (base area) × height

(Note the similarity with the formula for a cone.)

There is no general formula given for the surface area of a pyramid.

Also, note that the base of a pyramid can be any shape.
It does not have to be square.

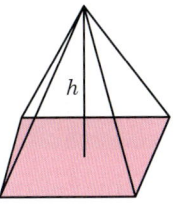

Example

a) A pyramid has a square base of side 5 m
and vertical height 4 m. Find its volume.

$\qquad$ Volume of pyramid $= \dfrac{1}{3}(5 \times 5) \times 4$ $\qquad\qquad$ The pyramid has a square base of area 5 × 5.

$\qquad\qquad\qquad\qquad = 33\dfrac{1}{3}\ \text{m}^3$

b) Calculate the radius of a sphere of
volume 500 cm³.

$\qquad$ Let the radius of the sphere be r cm.

$\qquad \dfrac{4}{3}\pi r^3 = 500$ $\qquad\qquad\qquad\qquad$ Write the formula with $V = 500$.

$\qquad\qquad r^3 = \dfrac{3 \times 500}{4\pi}$ $\qquad\qquad\qquad$ Multiply by 3 and divide by 4π.

$\qquad\qquad r = \sqrt[3]{\left(\dfrac{3 \times 500}{4\pi}\right)} = 4.92\ \ (3\ \text{s.f.})$ $\qquad$ Cube root to find r.

$\qquad$ The radius of the sphere is 4.92 cm.

c) Calculate the volume and total surface area of this cone.

Give your answers as exact multiples of π.

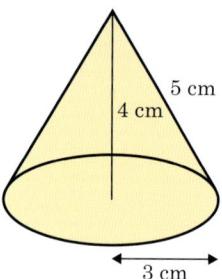

5 cm

4 cm

3 cm

Volume of cone $= \dfrac{1}{3}\pi r^2 h$

$= \dfrac{1}{3}\pi \times 3^2 \times 4$ Substitute into the formula.

$= 12\pi \text{ cm}^3$ Write your answer as an exact multiple of π.

Total surface area $= \pi r^2 + \pi r l$ Total surface area = area of circular base

$= \pi \times 3^2 + \pi \times 3 \times 5$ + curved surface area

$= 9\pi + 15\pi$ As before, write your answer as an exact

$= 24\pi \text{ cm}^2$ multiple of π.

Exercise 7.7C

Find the volume and total surface area of the shapes in Questions **1** to **4**. Give your answers as exact multiples of π.

1. Cone: height = 8 cm, radius = 6 cm, slant height = 10 cm

2. Sphere: radius = 5 cm

3. Sphere: radius = 6 cm

4. Cone: height = 12 cm, radius = 5 cm, slant height = 13 cm

Find the volume and surface area of the shapes in Questions **5** to **8**. Give your answers to 3 significant figures.

5. Sphere: diameter = 8 cm

6. Cone: height = 7 cm, radius = 24 cm, slant height = 25 cm

7. Sphere: radius = 0.1 m

8. Half sphere: radius = 3 cm

> **Tip**
>
> A half sphere has half the volume of a sphere. To calculate the surface area, add half the surface area of the sphere to the area of the circle formed when the sphere is cut into two.
>
>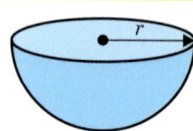
>
> r

Find the volume of the shapes in Questions **9** to **11**.

9. Pyramid: rectangular base 7 cm by 8 cm, height = 5 cm

10. Pyramid: square base of side = 4 m, height = 9 m

11. Pyramid: equilateral triangular base of side = 8 cm, height = 10 cm

12. Find the height of a pyramid of volume 20 m³ and base area 12 m².

13. Find the radius of a sphere of volume 60 cm³.

14. Find the height of a cone of volume 2.5 litres and radius 10 cm.

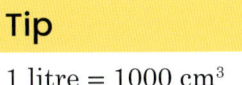

Tip

1 litre = 1000 cm³

15. A spherical ball is immersed in water contained in a vertical cylinder.

Assuming the water covers the ball, calculate the rise in the water level if:

a) sphere radius = 3 cm, cylinder radius = 10 cm

b) sphere radius = 2 cm, cylinder radius = 5 cm.

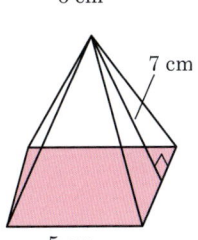

16. The diagram shows a square-based pyramid.

a) Calculate the area of one of the sloping faces.

b) Hence, find the total surface area of the pyramid.

17. Find the total surface area of this square-based pyramid.

Exercise 7.7D

This exercise contains a mixture of questions involving the volumes of a wide variety of different objects.

Where necessary give answers correct to 3 s.f.

1. A cylinder has a cross-sectional area of 12 cm² and a length of two metres. Calculate its volume in cm³.

2. The diagram represents a building.

    a) Calculate the area of the shaded end.

    b) Calculate the volume of the building.

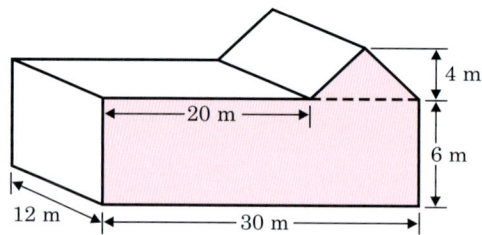

3. A rectangular block has dimensions 20 cm by 7 cm by 7 cm. Find the volume of the largest solid cylinder that can be cut from this block.

4. Metal discs are made with a circular cross-section, as shown.

    a) Find the area of the cross-section of the disc.

    b) The discs are 0.2 cm thick. Calculate the volume of the disc.

    c) Find, in cm³, the volume of brass needed to make 10 000 of these discs.

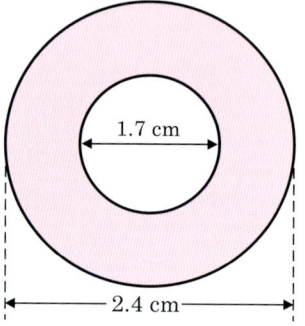

5. a) A cylindrical water tank has internal diameter 40 cm and height 50 cm. Calculate the volume of the tank.

    b) A cylindrical mug has internal diameter 8 cm and height 10 cm. Calculate the volume of the mug.

    c) If the tank is full, how many mugs can be filled from the tank?

6. In the diagram all the angles are right angles and the lengths are in cm. Find the volume of the shape.

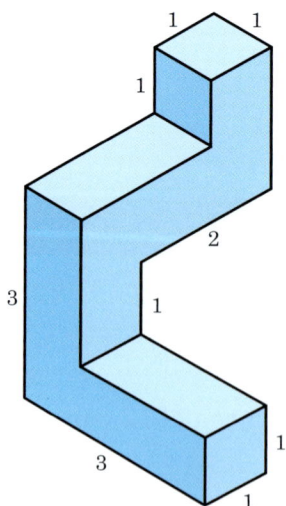

7. The diagram shows the cross-section of a steel bar which is 4 m long.

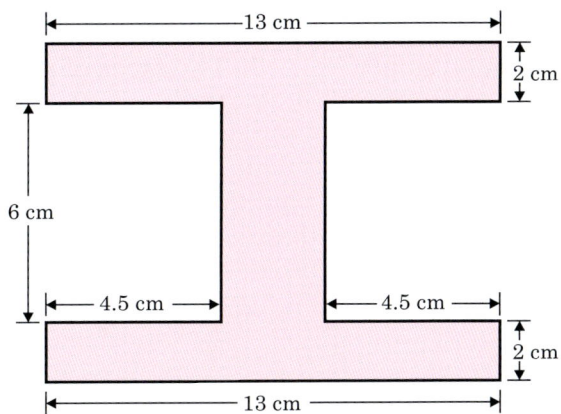

a) Calculate the cross-sectional area in cm².

b) Calculate the volume of the bar in cm³.

c) If 1 cm³ of steel has a mass of 7.8 g, find the mass of the bar in kg.

d) How many bars can be carried on a lorry if it cannot carry more than 8 tonnes? (1 tonne = 1000 kg)

8. Rahim decided to build a garage. The garage was to be 6 m by 4 m and 2.5 m in height. Each brick measures 22 cm by 10 cm by 7 cm. Rahim estimated that he would need about 40 000 bricks.

Is this a reasonable estimate? Explain your answer.

9. A cylindrical tin of height 15 cm and radius 4 cm is filled with sand from a rectangular box that is 50 cm by 40 cm by 20 cm. How many times bigger is the capacity of the box than the cylinder?

10. The diagram shows a trough. Water pours into the trough at a rate of 2 litres/min. How long, to the nearest minute, will it take to fill the trough? (1 litre = 1000 cm³)

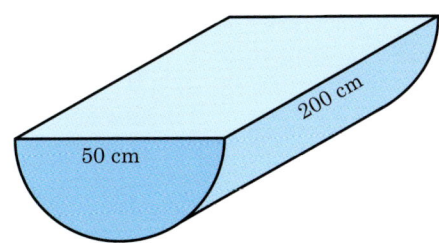

Revision exercise 7

1. **a)** Calculate the speed (in metres per second) of a slug which moves a distance of 30 cm in 1 minute.

   b) Calculate the time taken for a bullet to travel 8 km at a speed of 5000 m/s.

   c) Calculate the distance flown, in a time of four hours, by a bird that flies at a speed of 12 m/s.

2. A train travels between Milan and Pina, a distance of 108 km, in 45 minutes, at a steady speed. It passes through Rosta 40 minutes after leaving Milan. How far, in km, is it from Rosta to Pina?

3. An athlete runs 25 laps of a track in 30 minutes 10 seconds.

   a) How many seconds does he take to run 25 laps?

   b) How long does he take to run one lap, if he runs the 25 laps at a constant speed?

4. Work out the difference between one ton and one tonne.

   Give your answer to the nearest kg.

   > **Tip**
   >
   > 1 tonne = 1000 kg
   > 1 ton = 2240 lb
   > 1 lb = 454 g

5. A motorist travelled 200 km in five hours. Her average speed for the first 100 km was 50 km/h. What was her average speed for the second 100 km?

6. Find the area, correct to 3 s.f.

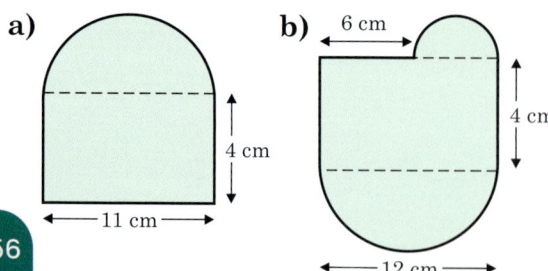

a) 4 cm, 11 cm

b) 6 cm, 4 cm, 12 cm

7. The faces of a round and square clock have exactly the same area. If the round clock has a radius of 10 cm, how wide is the square clock?

8. A large metal cylinder has a length 7 cm and radius 3 cm.

   a) Calculate the volume, in cm³, of the cylinder.

   The cylinder is to be melted down and used to make cylindrical coins of thickness 3 mm and radius 12 mm.

   b) Calculate the volume, in mm³, of each coin.

   c) Calculate the number of coins that can be made from the large cylinder.

9. Two girls walk at the same speed from A to B. Aruni takes the large semi-circle and Deepa takes the three small semi-circles. Who arrives at B first? Show working to explain your answer.

10. In Figure 1, a circle of radius 4 cm is drawn inside a square. In Figure 2, a square is drawn inside a circle of radius 4 cm.

Calculate the shaded area in each diagram.

Figure 1   Figure 2

11. A cylinder of radius 8 cm has a volume of 2 litres. Calculate the height of the cylinder.

12. Twenty-seven small cubes fit exactly inside a cube-shaped box without a lid. How many of the small cubes are touching the sides or the bottom of the box?

13. The square shown has sides of length 3 cm and the arcs have centres at the corners. Find the shaded area.

14. A circle with radius 10 cm has the same area as a sector of another circle with radius r cm and an angle of $40°$. Find the value of r.

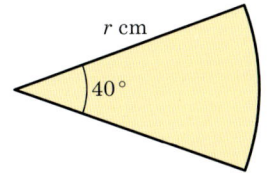
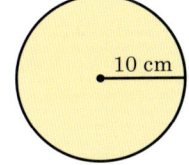

15. A metal ball has a volume of 64 cm³. The density of the metal is 1.32 g/cm³.

Work out the mass of the ball.

16. A force of 85 newtons is applied to a surface with area 1.7 m².

Work out the pressure that is applied.

17. In 2022 the population of Cairo was 9.54 million.

The area of Cairo is 3085 km².

Calculate, correct to 3 significant figures, the population density of Cairo.

18. 🚫 Work out the surface area and volume of a sphere with radius 4 cm.

Give your answers as exact multiples of π.

19. A cone has a curved surface area of 75 cm² and a slant height of 6 cm.

   a) Calculate the radius of the cone.

   Give your answer to 3 significant figures.

   b) Given that the height of the cone is 4.5 cm, calculate the volume of the cone.

   Give your answer to 3 significant figures.

CALCULATOR

1. A train leaves Paris at 9 51 and arrives in Marseille at 13 20. How long
    does the journey take? Give your answer in hours and minutes. [1]

2.

Wythall	10 50	11 20	11 50	12 30
Hollywood	11 00	11 30	12 00	12 40
Kings Heath	11 25	11 55	12 25	13 05

The table above is part of a bus timetable.

a) The 11 20 bus left Wythall a minute early and arrived at Kings Heath
    4 minutes late.

    How many minutes did it take to reach Kings Heath? [1]

b) Zac walked to the bus stop at Hollywood and arrived there at 11 51.

    The next bus arrived on time.

    How many minutes did Zac wait for the bus? [1]

3. a) Write down:

    i) the number of cm² in 2 m² [1]

    ii) the number of cm³ in 4000 mm³ [1]

    iii) how many litres there are in 0.1 m³. [1]

    b) A bath tub measures 1.5 metres by 0.8 metres by 0.7 metres. Work out
        how long it will take to fill the bath tub if water flows into it at a rate of
        50 litres per minute. [3]

4. Find the area of a circle of radius 6.4 cm. Write down your answer:

    a) exactly as it appears on your calculator [1]

    b) correct to the nearest square centimetre. [1]

5. The diagram shows a square tile of side 8 centimetres with 4 identical
    quarter circles shaded.

    Calculate the area of the *unshaded* region. [4]

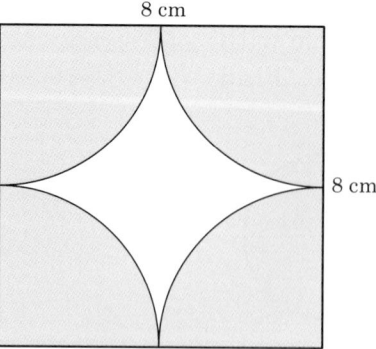

8 cm

8 cm

6. The area of a square is 17.64 cm².

   Work out the perimeter of the square. [1]

7. A 400-metre running track has two straight sections, each of length 110 metres, and two semi-circular ends.

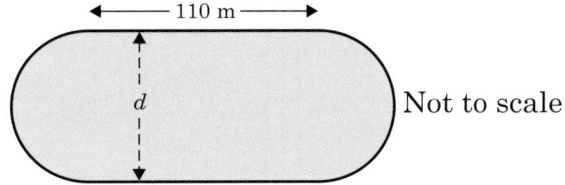

       Not to scale

   a) Calculate the *total* length of the *curved* sections of the track. [1]
   b) Calculate d, the distance between the parallel straight sections of the track. [2]

8.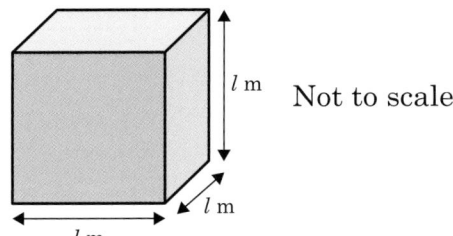

   Not to scale

   A cube of side l metres has a volume of 30 cubic metres. Calculate the length l.
   Give your answer to the nearest cm. [2]

9. a) Calculate the exact volume of a cylinder of radius 40 cm and height 128 cm. [2]
   b) Write your answer to part **(a)** in cubic metres, correct to 3 significant figures. [1]

10.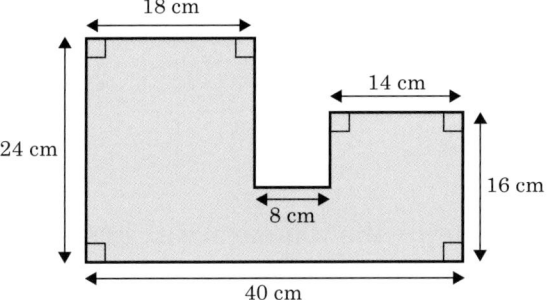

   For the shape above, work out:
   a) the perimeter [1]
   b) the area. [2]

11. A ship is flying two flags.

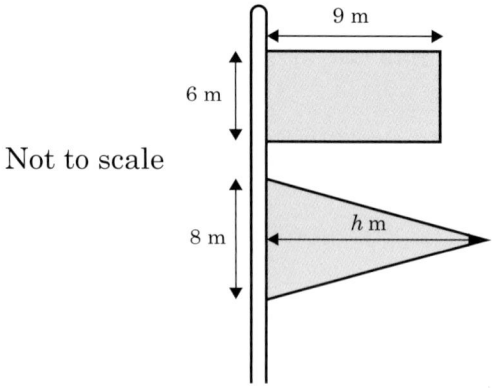

9 m

6 m

Not to scale

8 m

h m

The first is a rectangle 6 metres by 9 metres.

The second is an isosceles triangle with base 8 metres and height *h*.

The flags have the same area.

Find the value of *h*.   [2]

12. A candle, made from wax, is in the shape of a cylinder.

The radius is 1.4 centimetres and the height is 25 centimetres.

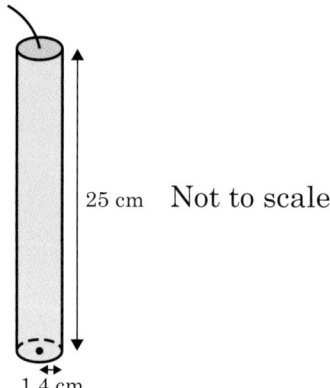

25 cm   Not to scale

1.4 cm

a) Calculate, correct to the nearest cubic centimetre, the volume of wax in the candle.

[The volume of a cylinder, radius *r*, height *h*, is $\pi r^2 h$.]   [2]

b) The candle burns 0.9 cm³ of wax every minute. How long, in hours and minutes,
will it last? Write your answer correct to the nearest minute.   [3]

c) The candles are stored in boxes which measure *x* cm by 28 cm by 25 cm. Each box
contains 120 candles. Calculate the minimum value of *x*.   [2]

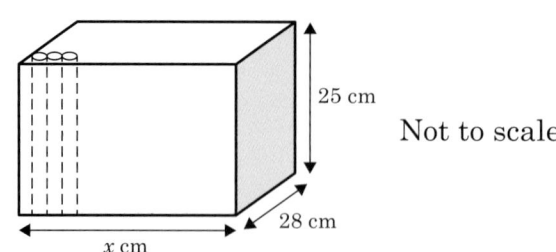

25 cm

Not to scale

28 cm

x cm

d) A shopkeeper pays \$35 for one box of 120 candles. He sells all the candles for 38 cents each.

    i) How much profit does he make? [2]

    ii) Calculate his profit as a percentage of his cost price. [3]

13. A circular pizza with a diameter of 30 cm is shared equally between 5 friends.

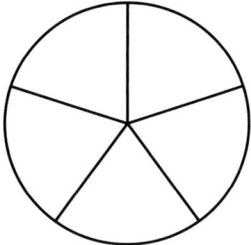

    a) What is the area of each person's slice of pizza? [2]

    b) If you had a circular pizza with the same area as each person's slice, what would its diameter be? [2]

14. Three spherical tennis balls, each with radius r cm, are packed tightly into a cylindrical box, also with radius r, so that the top and bottom balls are in contact with the top and bottom of the box.

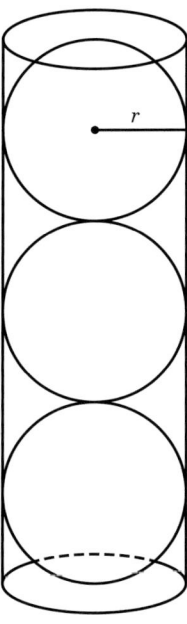

What fraction of the space inside the box is taken up by the balls?

You may use the formula

volume of a sphere $= \dfrac{4}{3}\pi r^3$ [4]

15. A metal cuboid has a mass of 680 grams and a volume of 400 cm³.

    a) Calculate the density of the metal. [2]

    The cuboid has a height of 8 cm. It exerts a force of 6.67 newtons on a surface.

    b) Calculate the pressure exerted by the cuboid on the surface.

       Give your answer in newtons per square metre. [3]

16. A cylindrical barrel has a height of 1.4 m and a radius of 0.5 m.

    a) Calculate the total surface area of the barrel. [2]

    The barrel is filled with oil at a rate of 0.7 litres per minute.

    b) Calculate the amount of time it takes for the barrel to completely fill with oil.

       Give your answer in hours and minutes, to the nearest minute. [4]

René Descartes (1596–1650) was one of the greatest philosophers of his time. Strangely his restless mind only found peace and quiet as a soldier and he apparently discovered the idea of 'Cartesian' geometry in a dream before the battle of Prague. The word 'Cartesian' is derived from his name and his work formed a link between geometry and algebra which led to the development of calculus. He finally settled in Holland, where he lived for ten years, but later moved to Sweden where he soon died of pneumonia.

Ingrid Daubechies (1954–) is a Belgian physicist and mathematician, best known for her work on image compression technology. Her work has also enabled scientists to extract information from samples of bones and teeth. The image processing methods she has helped to develop can be used to establish the age and authenticity of works of art and have been used on paintings by artists such as Vincent van Gogh and Rembrandt.

- Use, understand and draw real-life graphs such as travel graphs and conversion graphs.
- Use, understand, sketch and draw linear, quadratic and reciprocal graphs and solve equations graphically, e.g. finding the intersection of a line and a curve.
- Understand and use Cartesian coordinates.
- Draw graphs of linear equations.
- For straight-line graphs, find, use and interpret: the gradient from a graph, the equation of the line, and the gradient and equation of a parallel line.

8.1 Drawing graphs

Linear graphs

Vertical lines

Vertical lines all have equations of the form $x = a$.

For example, the graph shows the lines with equations $x = -3$, $x = 1$ and $x = 4$.

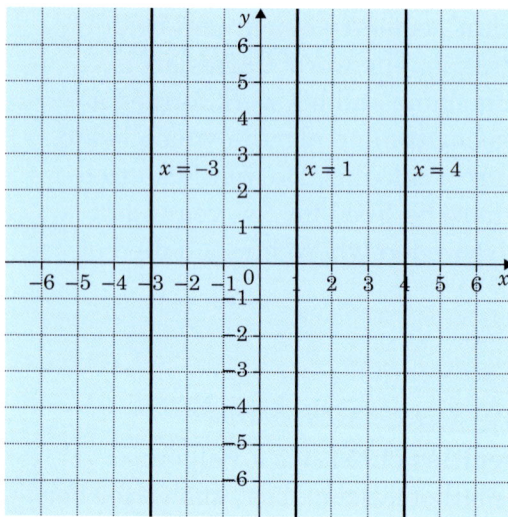

Horizontal lines

Horizontal lines all have equations of the form $y = b$.

For example, the graph shows the lines with equations $y = 3$, $y = 0$ and $y = -2$.

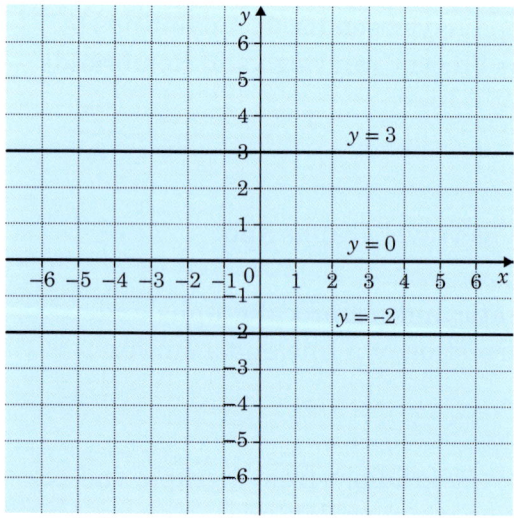

You can also draw the graphs of straight lines that are sloping.

Example

Draw the graph of $y = 4 - 2x$ for values of x from -2 to $+3$.

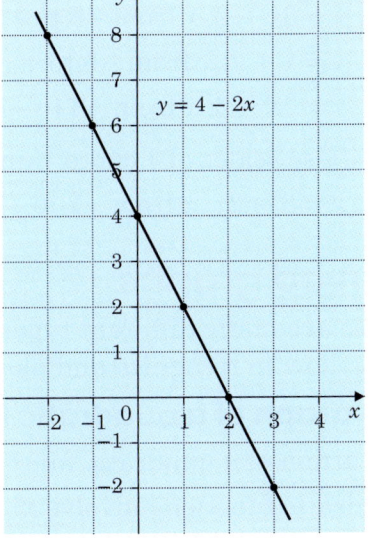

$y = 4 - 2x$

a) Substitute the x-values into the equation.

Some examples are shown below the table.

x	-2	-1	0	1	2	3
y	8	6	4	2	0	-2

$\boxed{4 - 2(-2) = 8}$ $\boxed{4 - 2(0) = 4}$ $\boxed{4 - 2(3) = -2}$

b) Plot the values of x and y from the table as coordinates.

Plot $(-2, 8)$, $(-1, 6)$, $(0, 4)$, $(1, 2)$, $(2, 0)$ and $(3, -2)$.

Join the points with a straight line.

Exercise 8.1A

1. Using the same set of axes with both x and y numbered from -6 to $+6$, draw these straight lines:

a) $x = 2$

b) $x = -4$

c) $y = 1$

d) $y = -5$

For Questions **2** to **12** make a table of values and then draw the graph. Suggested scales: 1 cm to 1 unit on both axes, unless otherwise stated.

2. $y = 2x + 1$; x from -3 to $+3$

x	-3	-2	-1	0	1	2	3
y	-5	-3					

3. $y = 3x - 5$; x from -2 to $+3$

4. $y = x + 2$; x from -4 to $+4$

5. $y = 2x - 7$; x from -2 to $+5$

6. $y = 4x + 1$; x from -3 to $+3$

(Use scales of 1 cm to 1 unit on the x-axis and 1 cm to 2 units on the y-axis.)

7. $y = x - 3$; x from -2 to $+5$

8. $y = 2x + 4$; x from -4 to $+2$

9. $y = 3x + 2$; x from -3 to $+3$

10. $y = x + 7$; x from -5 to $+3$

11. $y = 4x - 3$; x from -3 to $+3$

(Use scales of 1 cm to 1 unit on the x-axis and 1 cm to 2 units on the y-axis.)

12. $y = 8 - 2x$; x from -2 to $+4$

Curved graphs

There are many different types of curved graphs. You need to know how to recognise and plot two of them.

The graph of a quadratic function

The graph of a quadratic function is a curve called a **parabola**.

Quadratic functions involve x^2 and no higher powers of x.

Examples of quadratic functions are $y = x^2 + 3$ and $y = 2 - x - x^2$.

Quadratic graphs with a positive x^2 term look like this:

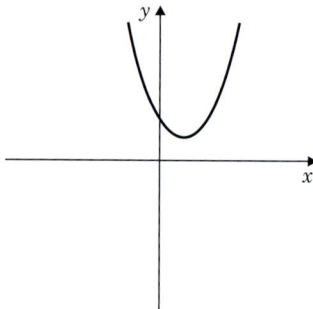

Quadratic graphs with a negative x^2 term look like this:

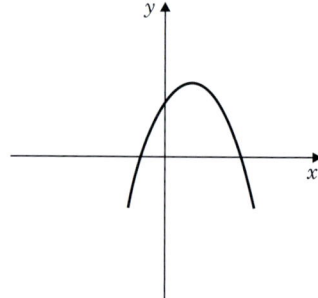

Note that quadratic graphs have a line of symmetry.

You can draw the graph of a quadratic function using a table of values.

Example 1

Draw the graph of $y = x^2 + x - 2$ for values of x from -4 to $+3$. Draw the line of symmetry and write its equation.

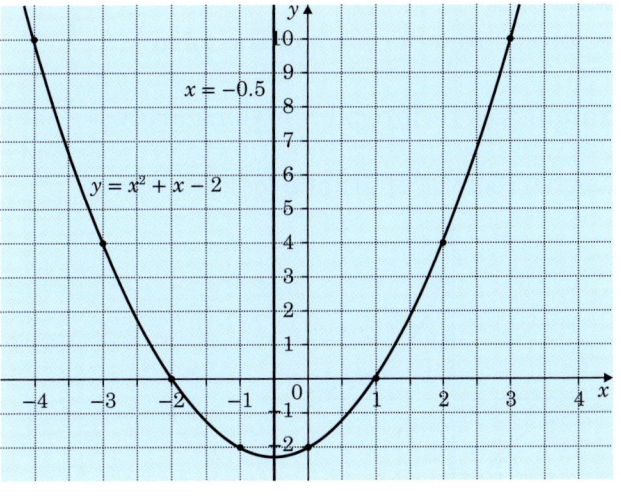

a) Substitute the x-values into the equation.

Some examples are shown below the table.

x	-4	-3	-2	-1	0	1	2	3
y	10	4	0	-2	-2	0	4	10

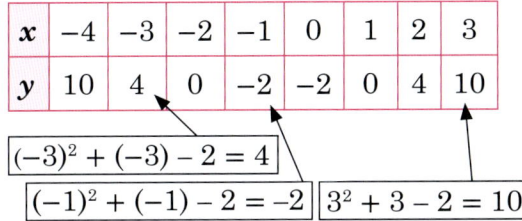

b) Plot the values of x and y from the table as coordinates.

Plot $(-4, 10)$, $(-3, 4)$, $(-2, 0)$, $(-1, -2)$, $(0, -2)$, $(1, 0)$, $(2, 4)$ and $(3, 10)$.

Join the points with a smooth curve.

c) The line of symmetry is drawn on the graph.

Since it passes through -0.5 on the x-axis, it has equation $x = -0.5$.

Example 2

Draw the graph of $y = 2x - x^2$ for values of x from -2 to $+4$. Draw the line of symmetry and write its equation.

Be careful! The '$-x^2$' term means $-(x^2)$.

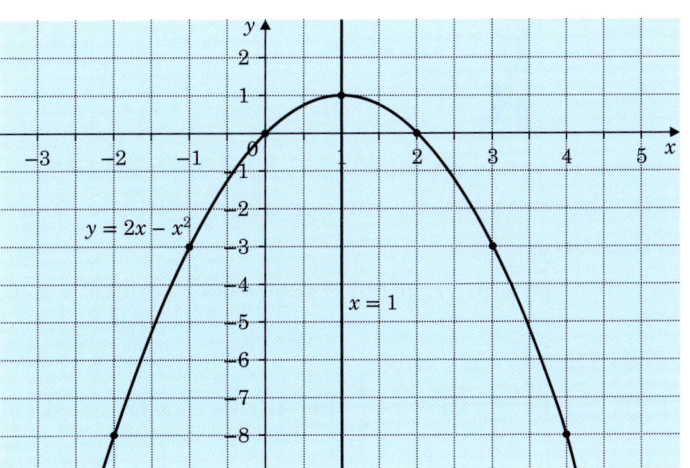

a)

x	-2	-1	0	1	2	3	4
y	-8	-3	0	1	0	-3	-8

b) Plot the x and y values from the table as coordinates.

Join the points with a smooth curve.

c) The line of symmetry is drawn on the graph.

Since it passes through 1 on the x-axis, it has equation $x = 1$.

The graph of a reciprocal function

The graph of a reciprocal function is a curve called a **rectangular hyperbola**.
Reciprocal functions are of the form $y = \dfrac{a}{x}$.

Examples of reciprocal functions are $y = \dfrac{1}{x}$ and $y = -\dfrac{3}{x}$.

Positive reciprocal graphs look like this:

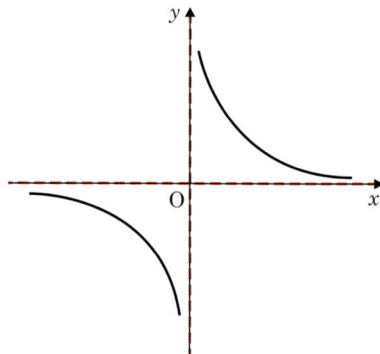

Negative reciprocal graphs look like this:

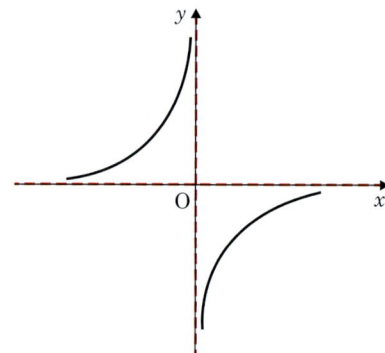

Since you cannot divide by zero, a reciprocal function is undefined when $x = 0$ or when $y = 0$. This is shown by the dotted lines in the diagrams above. These lines are called **asymptotes**.

You can draw the graph of a reciprocal function using a table of values.

Example 3

Draw the graph of $y = \dfrac{4}{x}$ for values of x from -4 to $+4$.

a) Substitute the x-values into the equation.

x	-4	-3	-2	-1	0	1	2	3	4
y	-1	-1.33	-2	-4	undefined	4	2	1.333	1

---

b) Plot the points and join them with two smooth curves.

Remember that the function is undefined for $x = 0$, so do not join the left and the right parts of the graph.

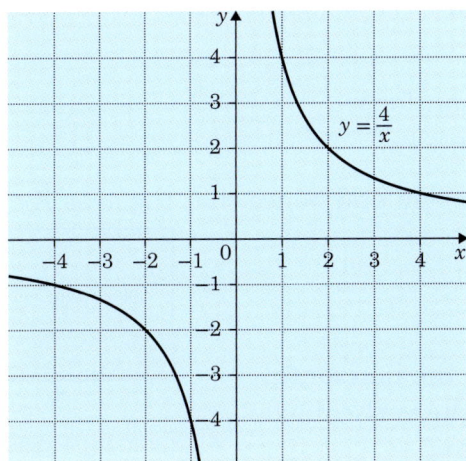

$y = \dfrac{4}{x}$

Exercise 8.1B

For Questions **1** to **12** make a table of values and then draw each graph on a separate grid. Draw the line of symmetry and write down its equation.

Suggested scales: 2 cm to 1 unit on the x-axis and 1 cm to 1 unit on the y-axis.

1. $y = x^2 + 2$; x from -3 to $+3$

x	-3	-2	-1	0	1	2	3
y	11	6	3				

2. $y = x^2 + 5$; x from -3 to $+3$

3. $y = x^2 - 4$; x from -3 to $+3$

4. $y = x^2 - 8$; x from -3 to $+3$

5. $y = x^2 + 2x$; x from -4 to $+2$

x	-4	-3	-2	-1	0	1	2
y	8	3					8

6. $y = x^2 + 4x$; x from -5 to $+1$

7. $y = x^2 + 4x - 1$; x from -5 to $+1$

8. $y = x^2 + 2x - 5$; x from -4 to $+2$

9. $y = x^2 + 3x + 1$; x from -4 to $+2$

10. $y = 6 - x^2$; x from -3 to $+3$

11. $y = 3x - x^2$; x from -1 to $+4$

12. $y = 2 + x - x^2$; x from -3 to $+4$

The graphs for Questions **13** and **14** are more difficult. Make a table of values and then draw each graph on a separate grid.

13. $y = \dfrac{12}{x}$; x from 1 to 12

Suggested scale: 2 cm to 1 unit on the x-axis and 1 cm to 1 unit on the y-axis.

14. $y = \dfrac{16}{x}$; x from 1 to 10

Suggested scale: 1 cm to 1 unit for x; 1 cm to 1 unit for y.

15. A rectangle has a perimeter of 14 cm and length x cm.

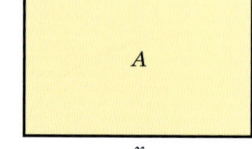

a) Show that the width of the rectangle is $(7 - x)$ cm and hence that the area A of the rectangle is given by the formula

$$A = x(7 - x).$$

b) Draw the graph of $A = x(7 - x)$, plotting x on the horizontal axis with a scale of 2 cm to 1 unit, and A on the vertical axis with a scale of 1 cm to 1 unit. Take x from 0 to 7.

From the graph find:

c) the area of the rectangle when $x = 2.25$ cm

d) the dimensions of the rectangle when its area is 9 cm²

e) the maximum area of the rectangle

f) the length and width of the rectangle corresponding to the maximum area.

16. Here are four sketch graphs.

Copy and complete the table below.

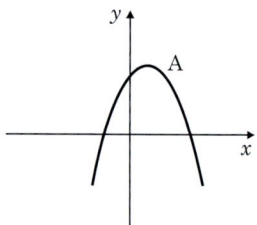 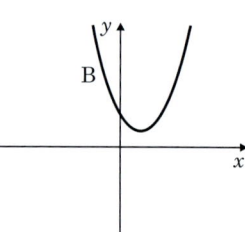 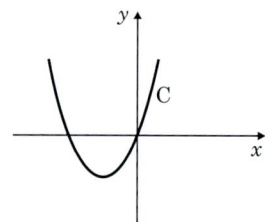 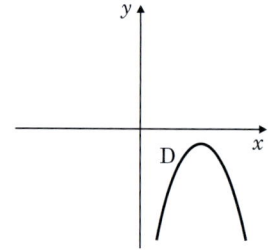

Positive x^2	Negative x^2
B	

17. Here is a sketch graph.

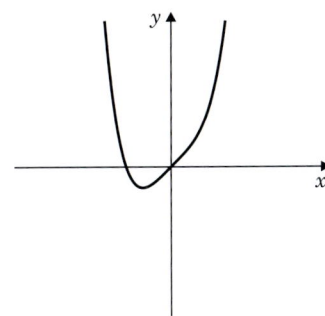

Rehan says: 'The graph is curved in a 'U' shape so it must be a quadratic graph'.

Is Rehan correct? Explain your answer.

8.2 Gradient and the form $y = mx + c$

The gradient of a line tells us how steep it is.

If you know the coordinates of two points on a line, you can find the gradient using the formula

$$\text{Gradient} = \frac{\text{vertical distance between the points}}{\text{horizontal distance between the points}}$$

Consider the line which joins (1, 2) and (3, 6).

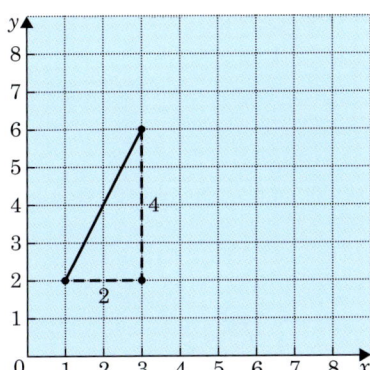

The vertical distance between the points is 4 and the horizontal distance is 2.

The gradient $= \dfrac{4}{2} = 2$

Notice that:

- a line sloping upwards to the right has a positive gradient
- a line sloping downwards to the right has a negative gradient.

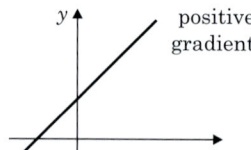
positive gradient

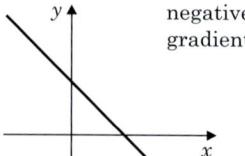
negative gradient

[Some people think of a capital 'N' for negative.]

Exercise 8.2A

1. The graph shows some line segments.

   Find the gradient of the line joining

   a) (1, 3) and (2, 6)   **b)** (2, 3) and (4, 7)

   c) (4, 3) and (8, 5)   **d)** (3, 9) and (9, 11).

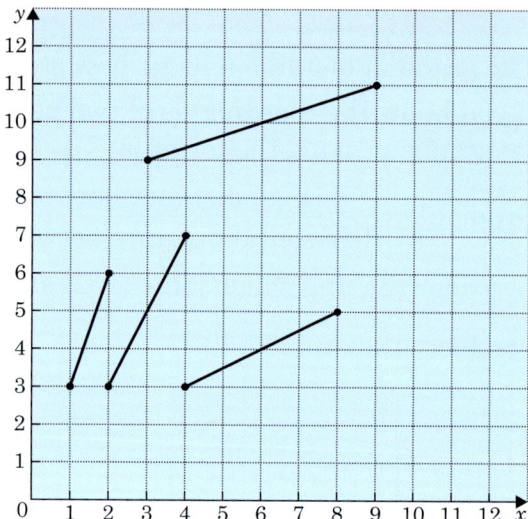

2. By plotting points on a graph, or otherwise, find the gradient of the line joining

   a) (1, 4) and (3, 2)   **b)** (2, 5) and (5, −1)

   c) (6, 2) and (2, 10)   **d)** (3, −2) and (−3, 2)

   e) (−2, −4) and (−1, 2)   **f)** (2, −3) and (−2, 6).

3. Find the gradient of the line joining

   a) A and B   **b)** B and C

   c) C and D   **d)** D and A.

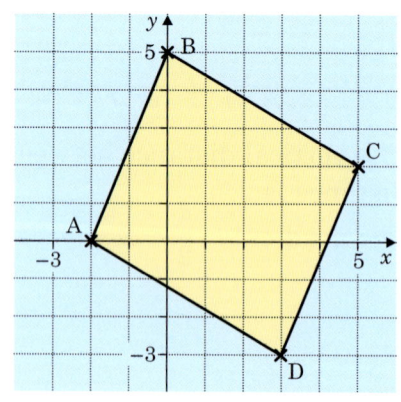

Gradient and intercept

A straight line can be described in terms of

- its gradient
- where it crosses the y-axis (the y-intercept).

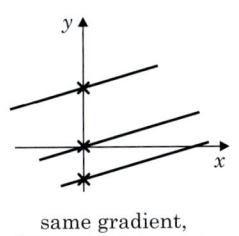

same gradient,
different y-intercepts

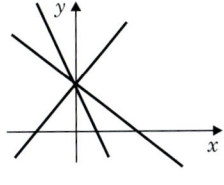

same y-intercept,
different gradients

You can draw a straight line when you know the gradient and y-intercept.

Example

Draw the lines with

a) gradient 2, y-intercept 1

b) gradient -3, y-intercept 5.

a) Plot the y-intercept and then create 'steps' going two up for every one across as shown. Once you have plotted three points, draw a straight line through the points.

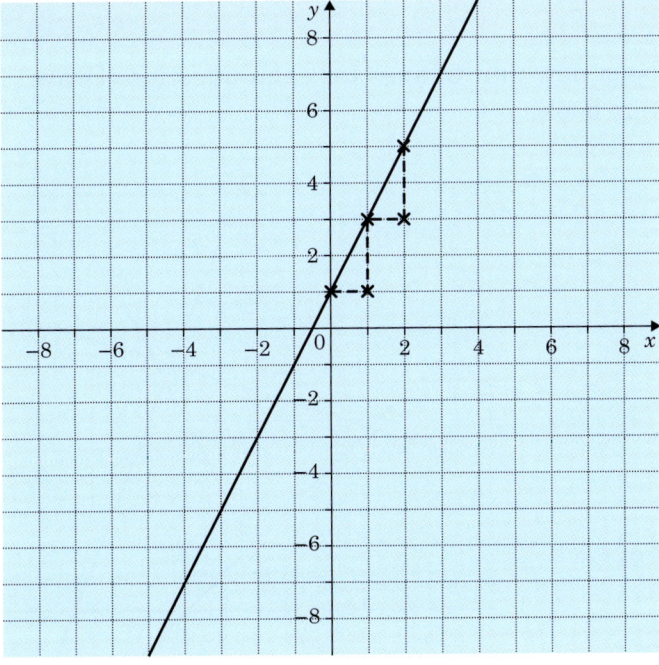

b) Since the gradient is negative, create steps that go three *down* for every one across. Again, once you have three points, draw a straight line passing through them.

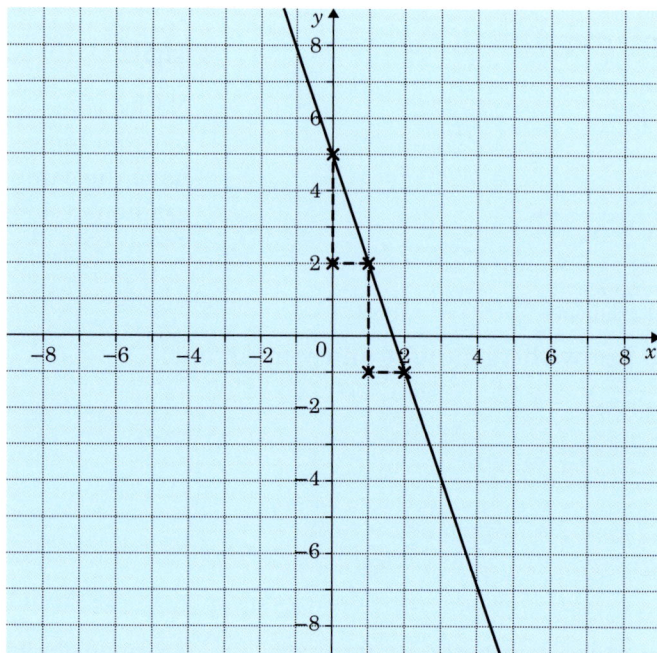

Exercise 8.2B

Draw the following straight lines. Use a new pair of axes for each question. Draw about six diagrams on one page of your book.

1. Gradient 2, y-intercept 3
2. Gradient 1, y-intercept -3
3. Gradient 2, y-intercept 0
4. Gradient -1, y-intercept 4
5. Gradient -3, y-intercept 0
6. Gradient -2, y-intercept -2

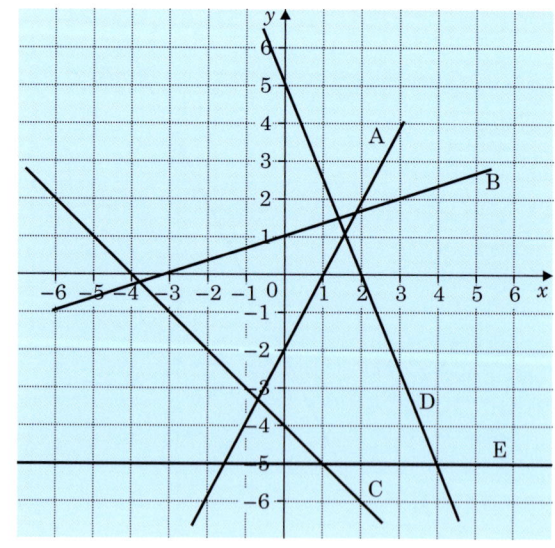

7. Give the gradient and y-intercept of each line in the graph on the right.

The line $y = mx + c$

$y = mx + c$ is the equation of a straight line with

- gradient m, and
- y-intercept c.

Example 1

Sketch the line with equation $y = 3x - 1$.

Plot the y-intercept $(0, -1)$.

Draw a line with positive gradient through the y-intercept.

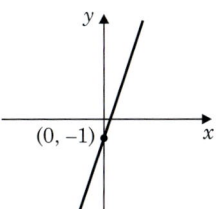

Example 2

a) Sketch the line $2x + y = 6$.

b) Write down the gradient of the line.

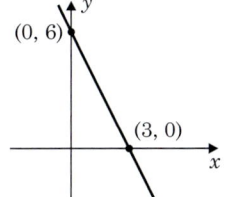

a) The y-intercept is when $x = 0$: $y = 6$

The x-intercept is when $y = 0$: $2x = 6$ so $x = 3$

Plot these two points and join them up.

b) The line has a gradient of $\dfrac{-6}{3} = -2$

Exercise 8.2C

Write down the gradient and y-intercept of each of the following lines.

1. $y = 2x - 3$ **2.** $y = 3x + 2$

3. $y = -x - 4$ **4.** $y = x + 3$

5. $y = -2x - 4$ **6.** $y = 2 - 3x$ ← Be careful!

7. $y = 4 - 7x$ **8.** $y = 2x - 1$

9. $y = 3 - x$ **10.** $y = 7 - 2x$

In Questions **11** to **14** sketch each graph and write down the gradient.

11. $3x + y = 6$ **12.** $3x - y = 9$

13. $y - 2x = 8$ **14.** $6x + 3y = 18$

Example

Find the equation of the line that is parallel to the line $y = 2x + 3$ and which passes through (2, 1).

The line has gradient 2 and will be of the form $y = mx + c$.

Since $m = 2$, $y = 2x + c$.

Substitute the point (2, 1) into this equation: $\qquad 1 = 2 \times 2 + c$

Hence: $\qquad c = 1 - 4 = -3$

The line has equation $y = 2x - 3$.

Exercise 8.2D

In Questions **1** to **6** match each diagram with the correct equation from the list below.

1.

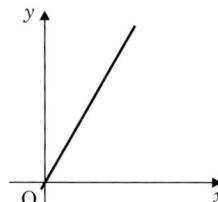

2.

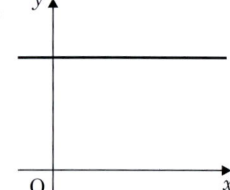

3.

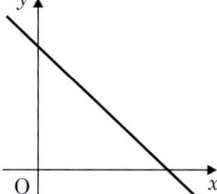

4.

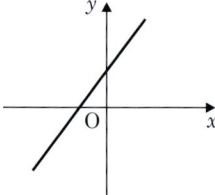

5.

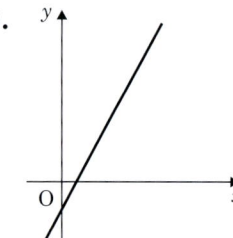

6.

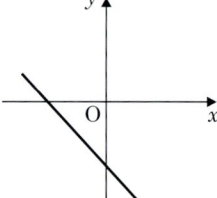

a) $y = -x - 4$ **b)** $y = 2x - 1$ **c)** $y = 2x + 3$

d) $y = 3x$ **e)** $y = 3 - x$ **f)** $y = 5$

7. Find the equations of the lines A and B.

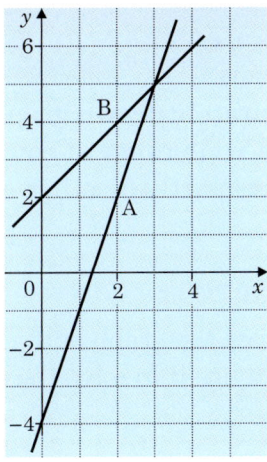

8. Find the equations of the lines C and D.

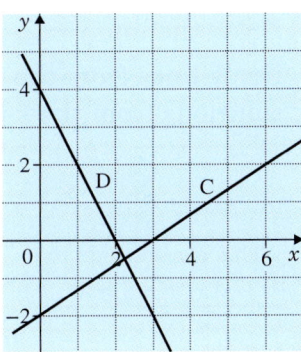

9. Look at the graph.

a) Find the equation of the line which is parallel to line A and which passes through the point (1, 8).

b) Find the equation of the line which is parallel to line B and which passes through the point (2, 1).

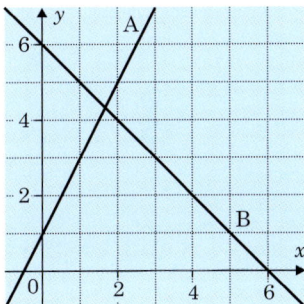

10. Look at the graphs in Question **7**.

a) Find the equation of the line which is parallel to line A and which passes through the point (4, 13).

b) Find the equation of the line which is parallel to line B and which passes through the point (1, −1).

8.3 Graphical solutions to equations

Accurately drawn graphs enable approximate solutions to be found for a wide range of equations, many of which are impossible to solve exactly by other methods.

Example

Draw the graph of the function $y = x^2 - x - 3$ for $-2 \leqslant x \leqslant 3$ and use it to solve the equations:

a) $x^2 - x - 3 = 2$

b) $x^2 - x - 3 = 0$

a) Draw the line $y = 2$.

Draw lines from the intersection points of the line and the curve – read the solutions from the x-axis.

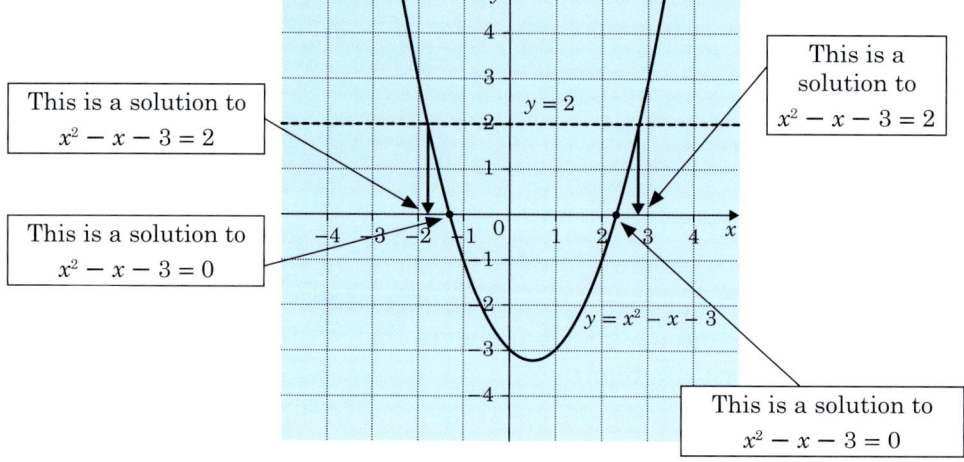

This is a solution to $x^2 - x - 3 = 2$

This is a solution to $x^2 - x - 3 = 2$

This is a solution to $x^2 - x - 3 = 0$

$y = 2$

$y = x^2 - x - 3$

This is a solution to $x^2 - x - 3 = 0$

$x = -1.8$ and $x = 2.8$ approximately

b) The line $y = 0$ is the x-axis.

$x = -1.3$ and $x = 2.3$

The solutions when the quadratic expression = 0, i.e. the points where the graph crosses the x-axis, are called **roots**.

Exercise 8.3A

1. In the diagram shown, the graphs of $y = x^2 - 2x - 3$, $y = 3$ and $y = -2$ have been drawn.

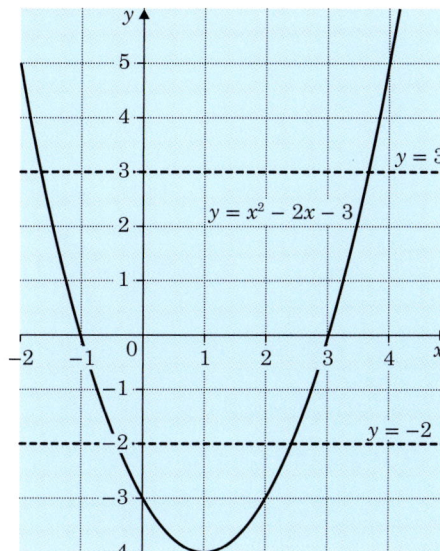

Use the graphs to find approximate solutions to the following equations:

a) $x^2 - 2x - 3 = 3$ **b)** $x^2 - 2x - 3 = -2$ **c)** $x^2 - 2x - 3 = 0$

In Questions **2** to **5** use a scale of 2 cm to 1 unit for x and 1 cm to 1 unit for y.

2. Draw the graph of the function $y = x^2 - 2x$ for $-1 \leqslant x \leqslant 4$. Hence find approximate solutions of the equations

a) $x^2 - 2x = 1$ **b)** $x^2 - 2x = 0$

> **Tip**
>
> $-1 \leqslant x \leqslant 4$ means use x-values from -1 to 4.

3. Draw the graph of the function $y = -x^2 + 3x - 5$ for $-1 \leqslant x \leqslant 5$. Hence find approximate solutions of the equations

a) $-x^2 + 3x - 5 = -5$ **b)** $-x^2 + 3x - 5 = -8$

4. Draw the graph of $y = -x^2 + 2x - 2$ for $-2 \leqslant x \leqslant 4$. By drawing other graphs, solve the equations

a) $-x^2 + 2x - 2 = -8$ **b)** $-x^2 + 2x - 2 = -3$

5. Draw the graph of $y = x^2 - 7x$ for $0 \leqslant x \leqslant 7$.

a) Use the graph to find approximate solutions of the equation $x^2 - 7x = -3$.

b) Explain why the equation $x^2 - 7x = -14$ does not have a solution.

8.4 Interpreting graphs

Travel graphs

Example

The graph shows a journey by a cyclist.

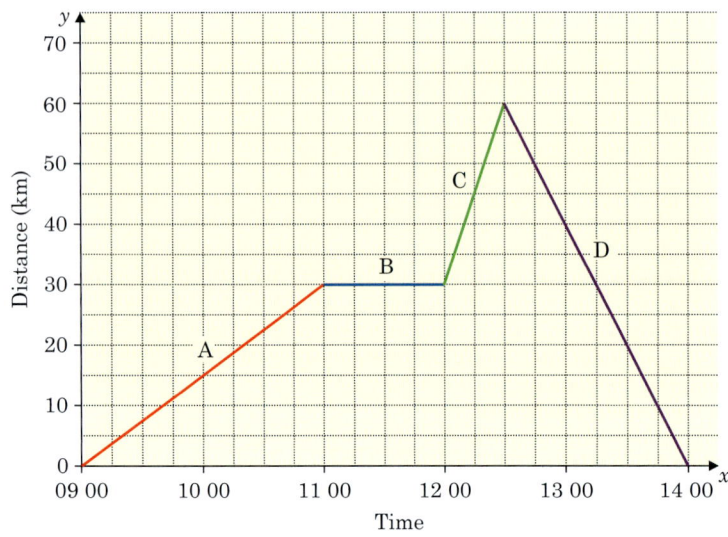

a) How far did the cyclist travel before stopping?

b) For how long did the cyclist rest?

c) What was the cyclist's speed between 09 00 and 11 00?

d) During which part of the journey did the cyclist travel the fastest?

a) The cyclist travelled 30 km before stopping.   This is represented by part A of the graph.

b) The cyclist rested for 1 hour.   This is represented by part B of the graph.

c) Distance = 30 km

Time = 2 hours

Speed = distance ÷ time = 30 ÷ 2 = 15 km/h   Speed is a rate of change and is represented by the gradient of the line segment.

d) The cyclist travelled fastest in part C.   The fastest speed is when the gradient is steepest.

Exercise 8.4A

1. The graph shows a return journey by car from Grenoble to Sisteron.

   a) How far is it from Grenoble to Gap?

   b) How far is it from Gap to Sisteron?

   c) At which two places does the car stop?

   d) How long does the car stop at Sisteron?

   e) When does the car
      i) arrive in Gap
      ii) arrive back in Grenoble?

   f) What is the speed of the car
      i) from Grenoble to Gap
      ii) from Gap to Sisteron
      iii) from Sisteron to Grenoble?

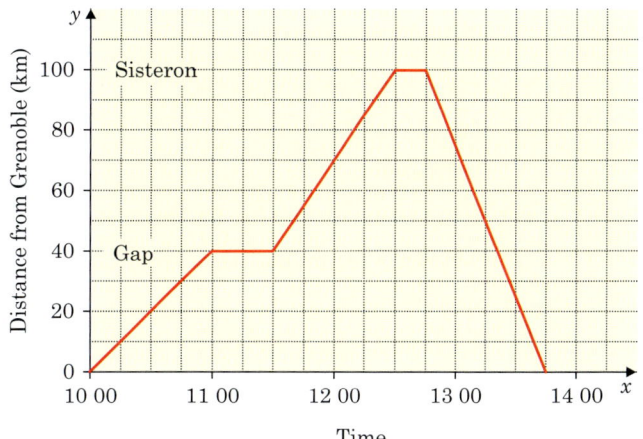

2. Yousef cycles to a friend's house but on the way his bike breaks, and he has to walk the remaining distance. At his friend's house, he repairs the bike and then returns home. On the way back, he stops at a shop.

   a) How far is it to his friend's house?

   b) How far is it from his friend's house to the shop?

   c) At what time did his bike break?

   d) How long did he stay at his friend's house?

   e) At what speed did he travel:
      i) from home until the bike broke
      ii) after the bike broke to his friend's house
      iii) from his friend's house to the shop
      iv) from the shop back to his own home?

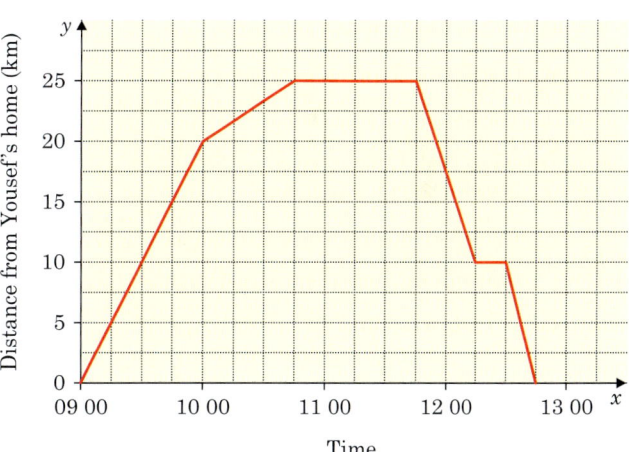

3. Eberto and Markus use the same road to travel between Aston and Borton.

   a) At what time did:

       i)   Eberto arrive in Borton

       ii) Markus leave Aston?

   b) **i)**   When did Eberto and Markus pass each other?

       ii) In which direction was Eberto travelling?

   c) Find the following speeds:

       i)   Markus from Aston to Stanley

       ii) Eberto from Aston to Borton

       iii) Markus from Stanley to Borton

       iv) Eberto from Borton back to Aston.

   d) When did Markus arrive in Borton?

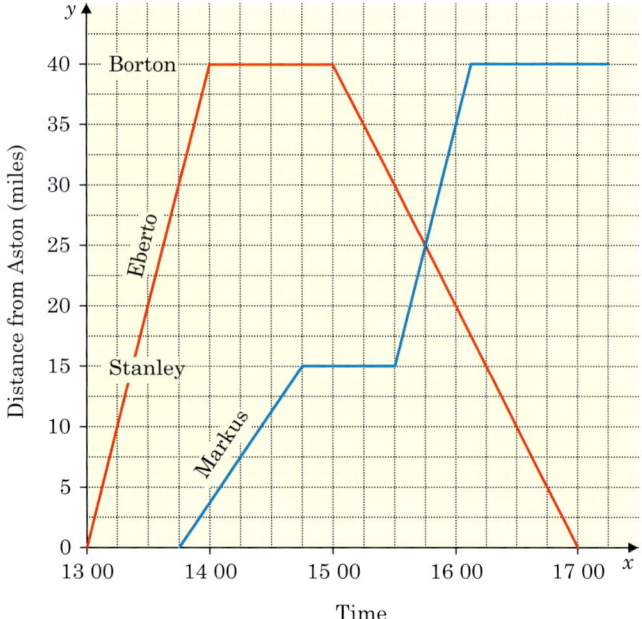

4. The graph shows the journeys made by a van and a car starting at Toledo, travelling to Madrid and returning to Toledo.

   a) For how long was the van stationary during the journey?

   b) At what time did the car first overtake the van?

   c) At what speed was the van travelling between 09 30 and 10 00?

   d) What was the greatest speed attained by the car during the entire journey?

   e) What was the average speed of the car over its entire journey?

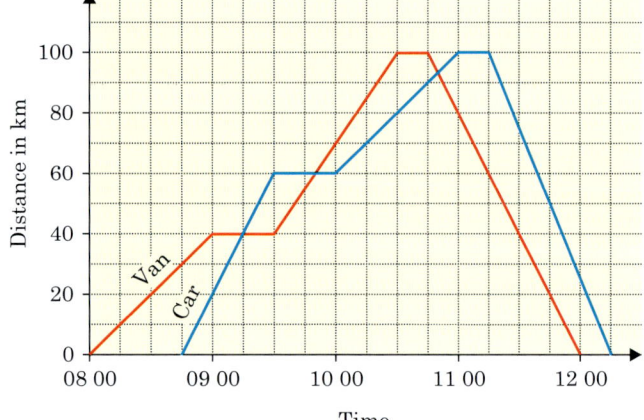

5. The graph shows the journeys of a bus and a car along the same road. The bus goes from Bangkok to Tainan and back to Bangkok. The car goes from Tainan to Bangkok and back to Tainan.

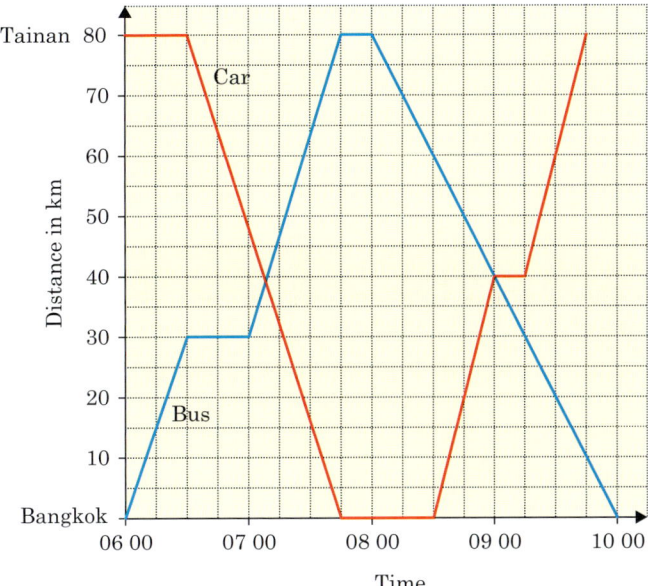

a) When did the bus and the car meet for the second time?

b) At what speed did the car travel from Tainan to Bangkok?

c) What was the average speed of the bus over its entire journey?

d) Approximately how far apart were the bus and the car at 09 45?

e) What was the greatest speed attained by the car during its entire journey?

In Questions **6**, **7** and **8**, draw a travel graph to illustrate the journey described. Draw axes with similar scales to Question **5**.

6. Mrs Chuong leaves home at 08 00 and drives at a speed of 50 km/h. After $\frac{1}{2}$ hour she reduces her speed to 40 km/h and continues at this speed until 09 30. She stops from 09 30 until 10 00 and then returns home at a speed of 60 km/h.

Use a graph to find the approximate time at which she arrives home.

7. Cillian leaves home at 09 00 and drives at a speed of 20 km/h. After $\frac{3}{4}$ hour he increases his speed to 45 km/h and continues at this speed until 10 45. He stops from 10 45 until 11 30 and then returns home at a speed of 50 km/h.

Use a graph to find the approximate time at which he arrives home.

8. At 10 00 Akram leaves home and cycles to his grandparents' house which is 70 km away. He cycles at a speed of 20 km/h until 11 15, at which time he stops for $\frac{1}{2}$ hour. He then completes the journey at a speed of 30 km/h. At 11 45 Akram's sister, Hameeda, leaves home and drives her car at 60 km/h. Hameeda also goes to her grandparents' house and uses the same road as Akram. At approximately what time does Hameeda overtake Akram?

Real-life graphs

Exercise 8.4B

1. The graph shows how the share price of a chemical firm varied over a period of weeks. The share price is the price in cents paid for one share in the company.

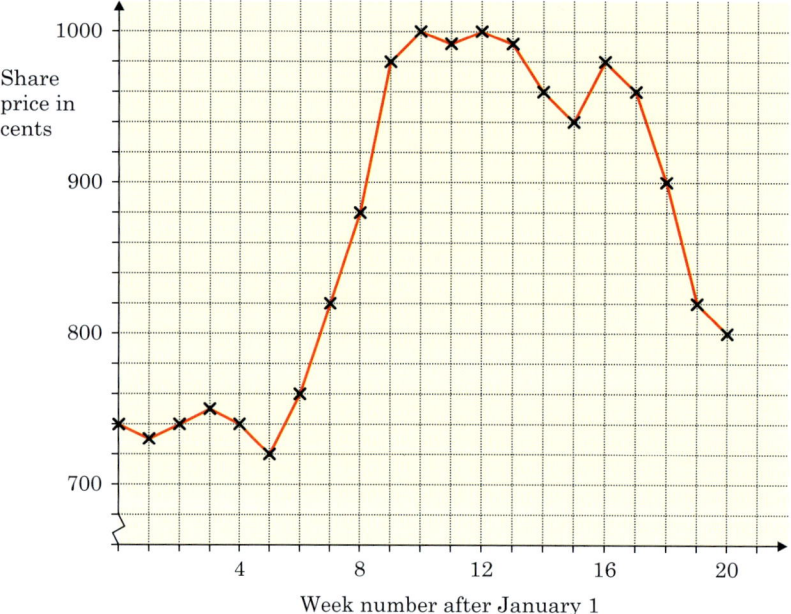

a) What was the share price in Week 4?

b) Kasia bought 200 shares in Week 6 and sold them all in Week 18. How much profit did she make?

c) Nhean buys 5000 shares when they are at their cheapest and sells them when they are at their most expensive. What is the maximum profit he could make?

2. The graph shows the number of students in a school at different times on one day.

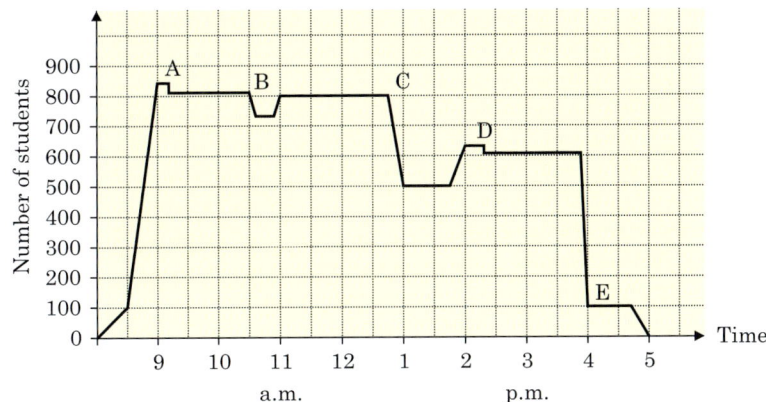

Referring to the points A, B, C, D, E, describe briefly what happened to the number of students during the day. Give an explanation of why you think this happened.

3. The graph below shows average television and radio audiences throughout a typical day.

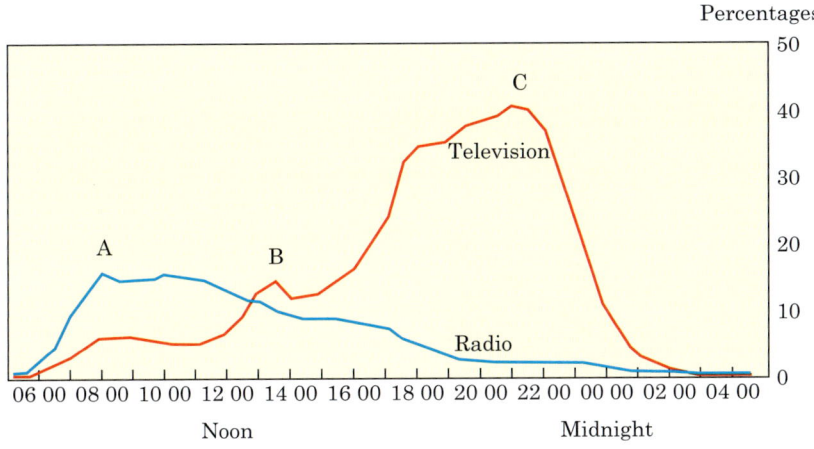

a) When are the 'peak times' for
   i) radio audiences
   ii) television audiences?
b) Give reasons to explain the shape of the graphs at times A, B and C.

4. The graph below shows how to convert between miles and kilometres.

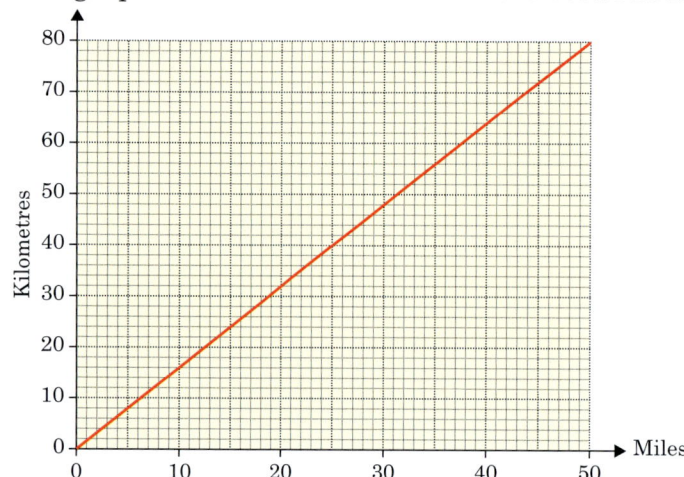

a) How many kilometres is one small square worth on the vertical axis?

b) Use the graph to find how many kilometres are the same as:

    i) 25 miles   **ii)** 15 miles   **iii)** 45 miles   **iv)** 5 miles

c) Use the graph to find how many miles are the same as:

    i) 64 km   **ii)** 56 km   **iii)** 16 km   **iv)** 32 km

5. The graph below shows how to convert between dollars and euros on a given day.

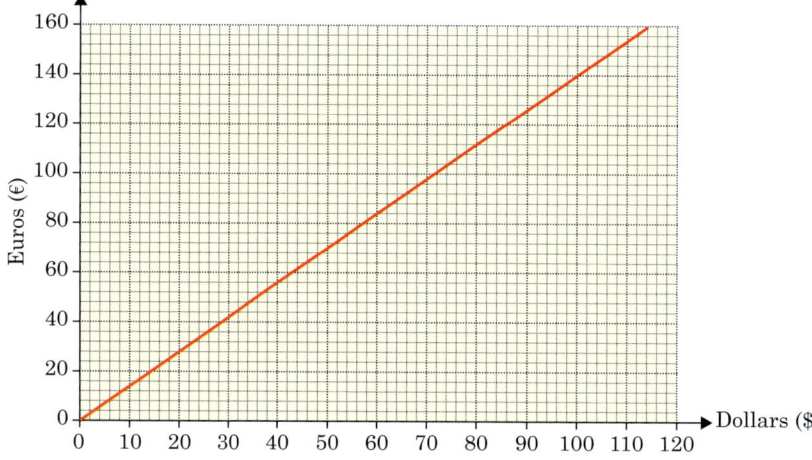

a) Use the graph to find how many euros are the same as:

    i) $20   **ii)** $80   **iii)** $50

b) Use the graph to find how many dollars are the same as:

    i) €56   **ii)** €84   **iii)** €140

c) Tim spends €154 on clothes in Paris. How many dollars has he spent?

6. A car travels along a motorway. The amount of petrol in its tank at different distances is shown on the graph.

   a) How much petrol was bought at the first stop?

   b) What was the petrol consumption in km per litre:

      i) before the first stop

      ii) between the two stops?

   c) What was the average petrol consumption over the 200 km?

   d) After it leaves the second service station, the car's petrol consumption is reduced to 4 km per litre for 20 km. After that, the car travels a further 80 km during which time the consumption is 8 km per litre. Copy the graph and extend it to show the next 100 km. How much petrol is in the tank at the end of the journey?

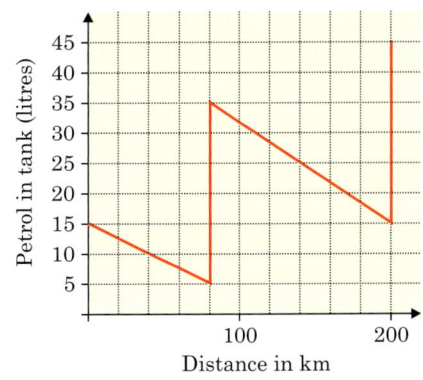

7. Kendal Motors hires out vans.

   KENDAL MOTORS $35
   plus 20 cents per km (including tax!)

   Copy and complete the table, where x is the number of km travelled and C is the total cost in dollars.

x	0	50	100	150	200	250	300
C	35			65			95

   Draw a graph of C against x, using scales of 2 cm for 50 km on the x-axis and 1 cm for $10 on the C-axis.

   Use the graph to find the number of km travelled when the total cost was $71.

8. Jeff is a plumber. This is what he charges.

   24 hr PLUMBING
   Call out $18   plus $15 per hour
   0707 874561   NO tax

   Copy and complete the table, where C stands for his total charge and h stands for the number of hours he works.

h	0	1	2	3
C		33		

Draw a graph with h across the page and C up the page. Use scales of 2 cm to 1 hour for h and 2 cm to \$10 for C.

Use your graph to find how long he worked if his charge was \$55.50.

Sketch graphs

Exercise 8.4C

1. Which of the graphs A to D below best fits the following statement?
   'Unemployment is still rising but by less each month.'

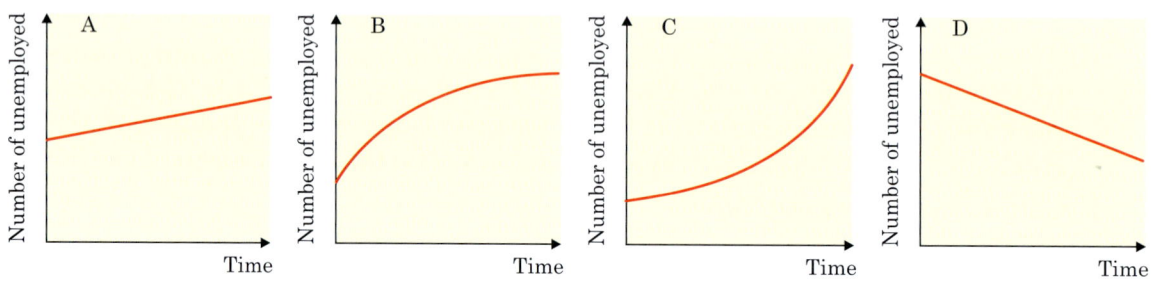

2. Which of the graphs A to D best fits the following statement? 'The price of oil was rising more rapidly in 2022 than at any time in the previous ten years.'

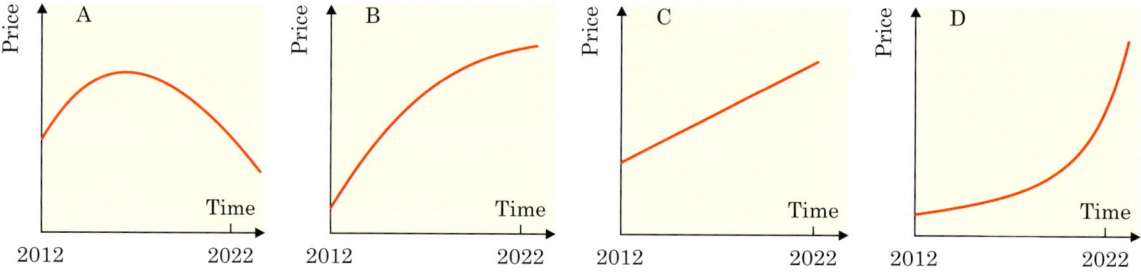

3. Which of the graphs A to D below best fits each of the following statements?

   a) The birth rate was falling but is now steady.
   b) Unemployment, which rose slowly until 2020, is now rising rapidly.
   c) Inflation, which has been rising steadily, is now beginning to fall.
   d) The price of gold has fallen steadily over the last year.

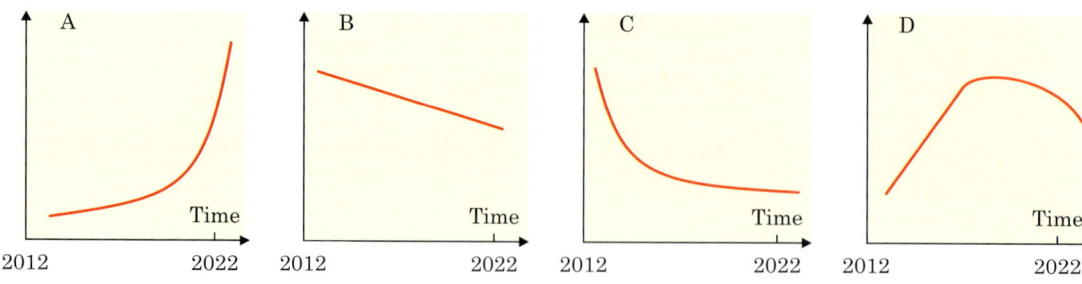

4. The graph shows the motion of three cars A, B and C, along the same road.

   Answer the following questions, giving estimates where necessary.

   a) Which car is in front after

      i) 10 s   **ii)** 20 s?

   b) When is B in front?

   c) When are B and C going at the same speed?

   d) When are A and C going at the same speed?

   e) Which car is going fastest after 5 s?

   f) Which car starts slowly and then goes faster and faster?

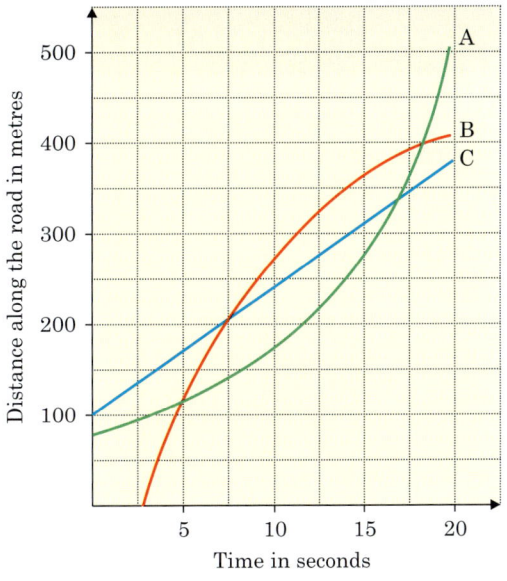

5. Three girls, Hanna, Fateema and Carine, took part in a race. Describe what happened during the race, giving as many details as possible.

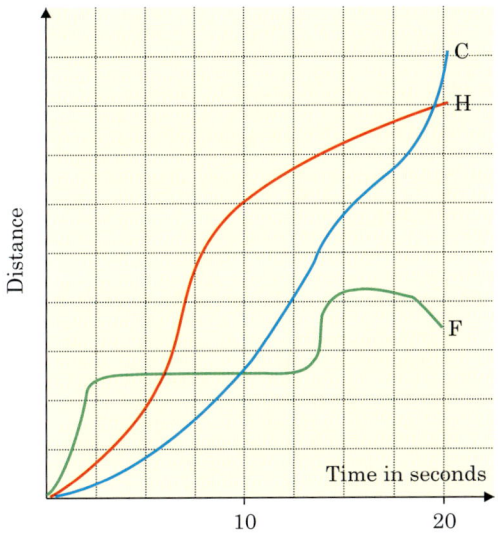

6. The graph shows the speed of the baton during a 4 × 100 m relay race.

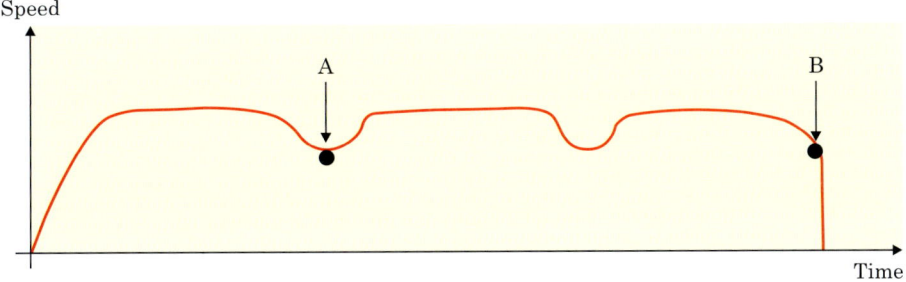

   a) Describe what is happening at point A.

   b) Describe what is happening at point B.

Revision exercise 8

1. In the diagram, the equations of the lines are $y = 3x$, $y = 6$, $y = 10 - x$ and $y = x - 3$.

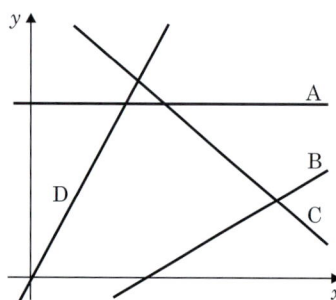

   Match each line with its equation.

2. A factory cafeteria contains a vending machine which sells drinks. On a typical day:

   - the machine starts half full

   - no drinks are sold before 9 a.m. and after 5 p.m.

   - drinks are sold at a slow rate throughout the day, except during the morning and lunch breaks (10.30–11 a.m. and 1–2 p.m.) when there is a greater demand

   - the machine is filled up just before the lunch break. (It takes about 10 minutes to fill.)

   Sketch a graph showing how the number of drinks in the machine may vary from 8 a.m. to 6 p.m.

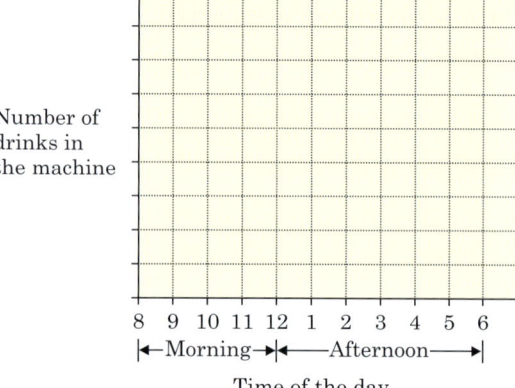

3. The distance–time graphs for several objects are shown. Decide which line represents each of the following:

   - car from Abu Dhabi

   - cyclist from Abu Dhabi

   - someone walking from Abu Dhabi

   - market stall on the road from Abu Dhabi

   - train from Abu Dhabi

   - cyclist travelling to Abu Dhabi

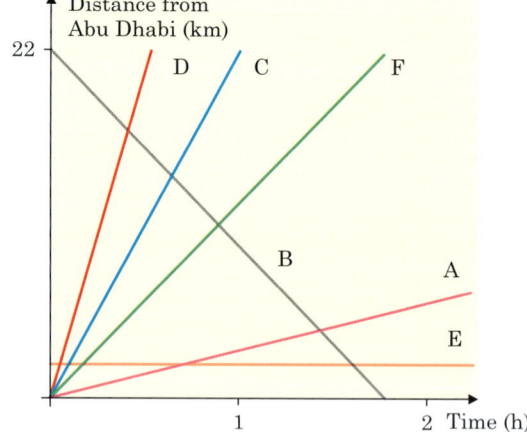

4. This graph shows a car journey from Gateshead to Middlesbrough and back again.

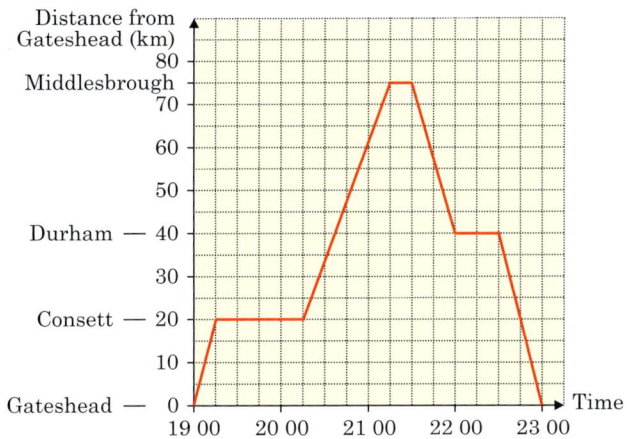

a) Where is the car

i) at 19 15

ii) at 22 15

iii) at 22 45?

b) How far is it

i) from Consett to Middlesbrough

ii) from Durham to Gateshead?

c) At what speed does the car travel

i) from Gateshead to Consett

ii) from Consett to Middlesbrough

iii) from Middlesbrough to Durham

iv) from Durham to Gateshead?

d) For how long is the car stationary during the journey?

5. a) Using a table of values, or otherwise, draw the graph of $y = x^2 + 2x - 3$ for $-4 \leqslant x \leqslant 2$.

b) Write down the roots of the equation $x^2 + 2x - 3 = 0$.

c) Write down the equation of the line of symmetry of the graph.

d) On the same set of axes, draw the graph of $y = 3x - 1$.

e) Write down the coordinates of the points of intersection of the line and the curve.

Examination-style exercise 8

NON-CALCULATOR

1. a) The equation of a straight line is $y = mx + c$.

Which letter in this equation represents the y-intercept? [1]

b) Write down the equation of the line shown on the grid below. [2]

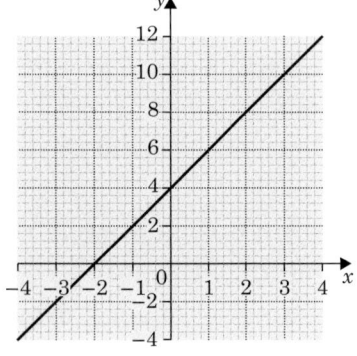

c) Copy and complete the table of values for $y = 10 - x^2$.

x	-4	-3	-2	-1	0	1	2	3	4
y	-6	1		9		9	6		-6

[3]

d) On a copy of the grid above, draw the graph of $y = 10 - x^2$. [3]

e) Write down the coordinates of the points of intersection of the straight line with your curve. [2]

2. a) i) Copy and complete the table of values for $y = x^2 - 2x - 2$.

x	-3	-2	-1	0	1	2	3	4	5
y	13		1		-3	-2	1	6	

[3]

ii) Draw the graph of $y = x^2 - 2x - 2$. [4]

iii) Use your graph to find the roots of the equation $x^2 - 2x - 2 = 0$.

Give your answers to 1 decimal place. [2]

iv) Use your graph to find the solutions to $x^2 - 2x - 2 = -1$. Give your answers to 1 decimal place. [2]

b) i) Copy and complete the table of values for the equation $y = \dfrac{2}{x}$.

x	0.25	0.5	1	2	3	4	5
y		4		1	0.7	0.5	0.4

[1]

ii) On the same grid draw the graph of $y = \dfrac{2}{x}$ for $0.25 \leqslant x \leqslant 5$. [3]

iii) Write down the x-coordinate of the point of intersection of your two graphs. [1]

3. a) Write down the equation of the straight line through $(0, -3)$ which is parallel to $y = 2x + 1$. [2]

b) Work out the equation of the straight line through $(-1, 5)$ which is parallel to $y = 6 - 3x$. [3]

4. A straight line, l, crosses the x-axis at $(4, 0)$ and the y-axis at $(0, 8)$.

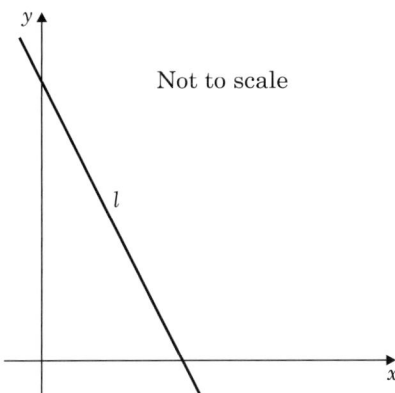

Not to scale

a) Work out the gradient of the line l. [1]

b) Write down the equation of the line l, in the form $y = mx + c$. [2]

5. a) The width of a carpet is x metres. The length of the carpet is 3 metres more than the width. Write down an expression, in terms of x, for

i) the length of the carpet [1]

ii) the area of the carpet. [1]

b) The area of the carpet is 7 square metres.

Show that $x^2 + 3x - 7 = 0$. [1]

c) i) Copy and complete the table of values for the equation $y = x^2 + 3x - 7$. [3]

x	-5	-4	-3	-2	-1	0	1	2
y	3		-7	-9		-7		3

ii) Draw axes using a scale of 1 cm to 1 unit for x and 2 cm to 1 unit for y.

Draw the graph of $y = x^2 + 3x - 7$ for $-5 \leqslant x \leqslant 2$. [4]

d) i) Use your graph to find the roots of the equation $x^2 + 3x - 7 = 0$. [2]

ii) Find the length of the carpet in part **(a)**. [1]

e) Mark the point $A(1, -1)$ on the grid.

i) Draw a straight line through A with a gradient of 2. [1]

ii) Write down the equation of this line in the form $y = mx + c$. [2]

6.

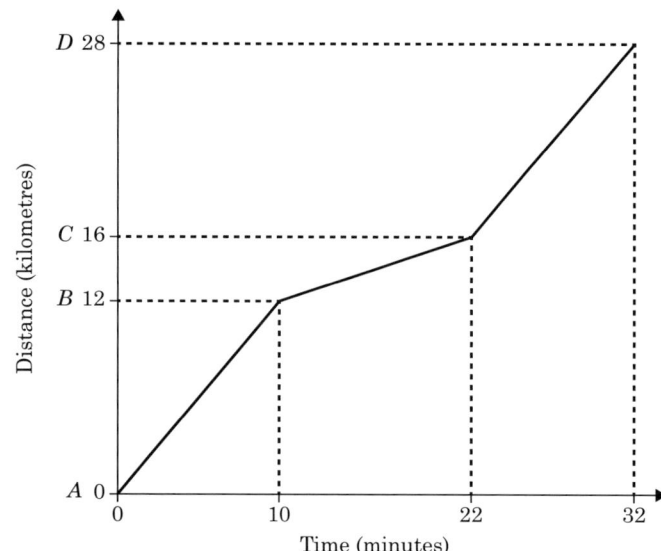

The diagram shows the graph of Mark's journey to visit his mother for her birthday.

Starting at A, he drove 12 kilometres to B at a constant speed.

Between B and C, he had to drive slowly, because of roadworks.

From C, he drove a further distance D at his original speed.

a) For how many minutes was he driving slowly due to the roadworks? [1]

b) At what speed did he drive during this slower part of his journey? [2]

c) What is the total distance from A to D? [1]

7.

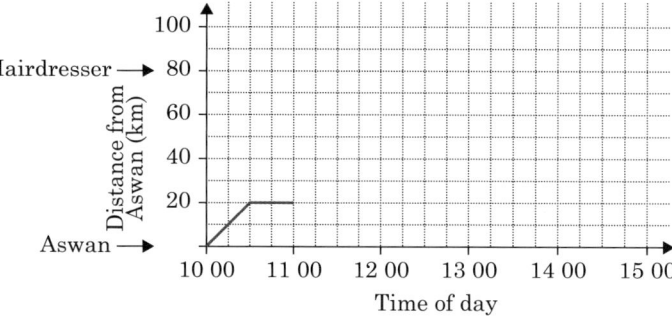

a) Chaturu drives from Aswan to a nail salon.

At 11 00 she continues her journey from the nail salon to the hairdresser, driving at 80 km/h. The first part of the journey is shown on the grid above.

i) How many minutes is Chaturu at the nail salon? [1]

ii) On a copy of the grid above, draw the rest of her journey to the hairdresser. [1]

b) Chaturu spends 1 hour in the hairdresser and then drives back to Aswan, at a constant speed, arriving at 14 30.

Show this information on the grid. [2]

Blaise Pascal (1623–1662) suffered the most appalling ill-health throughout his short life. He is best known for his work with Fermat on probability. This followed correspondence with a gentleman who was puzzled by the different results he got when throwing a dice. Pascal's work on probability became of enormous importance and showed for the first time that absolute certainty is not a necessity in mathematics and science. He also studied physics, but his last years were spent in religious meditation and illness.

- Understand and use Venn diagrams and set notation.
- Understand and use probability including the language, notation and the scale from 0 to 1. Calculate the probability of an event and understand the probability of an event not happening is (1 – the probability of the event happening).
- Understand and use relative frequency to estimate probability and find expected frequencies.
- Find the probability of combined events using these diagrams: sample space diagrams, Venn diagrams and tree diagrams.

9.1 Probability of an event

The probability of an event is a measure of the chance of it happening.

In probability you can ask questions such as . . .

How likely is it?

What are the chances of . . . ?

Here are some questions where you do not know the answer.

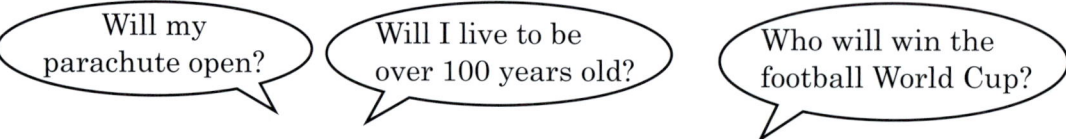

Will my parachute open?

Will I live to be over 100 years old?

Who will win the football World Cup?

Some events are certain. Some events are impossible.

Some events are in between certain and impossible.

The probability (or chance) of an event occurring is measured on a scale like this.

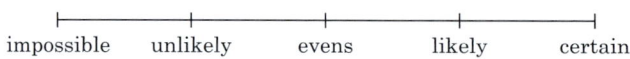

impossible   unlikely   evens   likely   certain

Exercise 9.1A

Draw a probability scale like this.

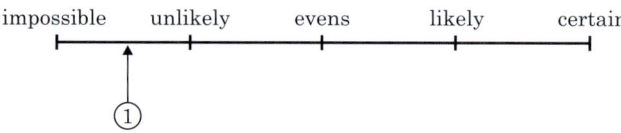

impossible   unlikely   evens   likely   certain

①

Draw an arrow to show the chance of the following events happening. [The arrow for Question **1** has been done for you.]

1. When a card is selected from a set numbered from 1 to 10, it is the number 1.
2. When a fair coin is flipped it will show a 'head'.
3. The letter 'a' appears somewhere on the next page of this book.
4. When a drawing pin is dropped it will land 'point up'.
5. There will be at least one baby born somewhere in China on the first day of next month.
6. The day after Monday will be Tuesday.
7. There will be a burst pipe in the school heating system next week and the school will have to close for three days.
8. You will blink your eyes in the next minute.
9. You will be asked to tidy your room this week.

Tip

Most coins have a 'head' side and a 'tail' side. You see coin tosses being used in cricket and football matches.

10. When a slice of toast is dropped, it will land on the floor buttered side down.

11. You will get maths homework this week.

12. England will win the next World Cup at football.

13. Your maths teacher has a black belt in Judo.

Probability as a number

As different countries have different words for saying how likely or unlikely any particular event is, and as this is not very precise, it is better to express probabilities using numbers on a scale instead of words.

The scale looks like this:

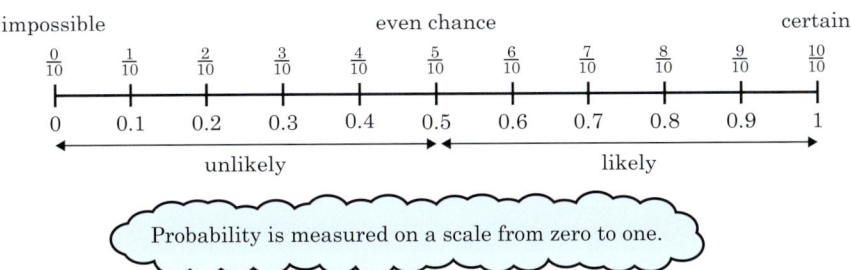

> **Tip**
>
> Do not use ratio to describe a probability.

Probabilities can be given as decimals, fractions or percentages.

Exercise 9.1B

Look at the events in Exercise 9.1A and, for each one, estimate the probability of it occurring using a probability from 0 to 1.

Copy each question and write your estimate of its probability next to it.

For example, in Question **1** you could write '0.1'.

Working out probabilities

The probability of an event occurring can be calculated using symmetry. For example, when you flip a coin you have an equal chance of getting a 'head' or a 'tail'. So the probability of flipping a 'head' is a half.

You can write this as 'P(flipping a head) = $\frac{1}{2}$'.

$$\text{Probability} = \frac{\text{number of favourable outcomes}}{\text{total number of outcomes}}$$

Example

A bag contains 6 blue counters and 2 red counters.

One counter is taken from the bag at random.

What is the probability that it is:

a) blue
b) red
c) green?

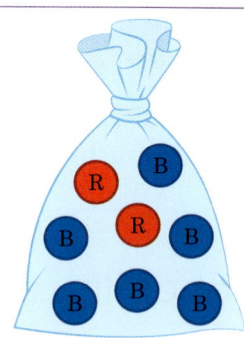

---

a) P(blue) = $\frac{6}{8}$ = $\frac{3}{4}$ There are 8 counters in total. 6 of these are blue.

b) P(red) = $\frac{2}{8}$ = $\frac{1}{4}$ There are 2 red counters out of 8.

c) P(green) = 0   There are no green counters so the probability is zero.

Tip

You can give your answers as unsimplified fractions in probability questions.

Exercise 9.1C

In this exercise, give your answers as fractions, unless otherwise stated.

1. A bag contains 3 white discs and 5 black discs. One disc is taken out at random. What is the probability that it is:

   a) white   **b)** black   **c)** purple?

2. Nine counters numbered 1, 2, 3, 4, 5, 6, 7, 8, 9 are placed in a bag. One is taken out at random. What is the probability that it is:

   a) a '5'   **b)** divisible by 3

   c) less than 5   **d)** divisible by 4

   e) greater than 10?

   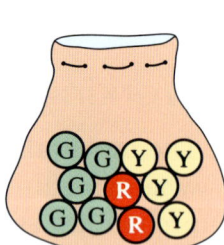

3. A bag contains 5 green balls, 2 red balls and 4 yellow balls. One ball is taken out at random. What is the probability that it is:

   a) green   **b)** red

   c) yellow   **d)** blue?

4. A bag contains 2 orange sweets, 4 lemon sweets, 5 strawberry sweets, 3 blueberry sweets and 3 melon sweets. Find the probability that one sweet selected at random is:

   a) a lemon sweet   **b)** a blueberry sweet   **c)** a grapefruit sweet.

5. A bag contains 8 orange balls, 5 green balls and 4 silver balls. Find the probability that a ball picked out at random is:

   a) silver   **b)** orange   **c)** green.

6. The numbers of drawing pins in ten boxes are as follows: 48, 46, 45, 49, 44, 46, 47, 48, 45, 46. One box is selected at random. Find the probability of the box containing:

   a) 49 drawing pins   **b)** 46 drawing pins   **c)** more than 47 drawing pins.

7. One ball is selected at random from those in this box. Find the probability of selecting:

   a) a white ball   **b)** a yellow or a black ball

   c) a ball which is not red.

8. **a)** A bag contains 5 red balls, 6 green balls and 2 black balls. Find the probability of selecting:

      i) a red ball   **ii)** a green ball.

   b) One black ball is removed from the bag. Find the new probability of selecting:

      i) a red ball   **ii)** a black ball.

9. A bag contains 12 white balls, 12 green balls and 12 purple balls. After 3 white balls, 4 green balls and 9 purple balls have been removed, what is the probability that the next ball to be selected will be white?

10. One person is chosen at random from a large group. What is the probability that this person's birthday is on a Monday in that year?

11. The numbering on a set of 28 dominoes is as follows:

6	6	6	6	6	6	6		5	5	5	5	5	5
6	5	4	3	2	1	0		5	4	3	2	1	0

4	4	4	4	4		3	3	3	3		2	2	2		1	1		0
4	3	2	1	0		3	2	1	0		2	1	0		1	0		0

   a) What is the probability of drawing a domino from a full set with:

      i) at least one six on it

      ii) at least one four on it

      iii) at least one two on it?

   b) What is the probability of drawing a 'double' from a full set?

   c) If I draw a double five which I do not return to the set, what is the probability of drawing another domino with a five on it?

> **Tip**
>
> A 'double' is a domino with the same number repeated.

12. The test results of 53 students are recorded in the two-way table below. One student is chosen at random.

	Grade			Total
	A	**B**	**C**	
Male	5	9	3	17
Female	10	12	14	36
Total	15	21	17	53

Find the probability that the student is:

a) male and achieved a grade B   **b)** female and achieved a grade A.

13. A bag contains blue, green and red counters. Six of the counters are blue.

The probability of randomly selecting a blue counter is $\frac{1}{4}$, of selecting a green counter is $\frac{1}{3}$ and of selecting a red counter is $\frac{5}{12}$.

What is the minimum number of counters in the bag?

Expected frequency

A probability experiment is **fair** if each of the possible outcomes is equally likely. For example, a six-sided dice is fair if the probability of rolling each of the numbers from 1 to 6 is $\frac{1}{6}$.

If the outcomes are *not* equally likely, then the experiment is said to be **biased**.

When you carry out an experiment, you can calculate the expected frequency of a given outcome occurring.

> Expected frequency = probability of success × number of trials

Example

A fair six-sided dice is rolled 240 times. How many times would you expect to roll a number greater than 4?

P(number greater than 4) = P(5 or 6) = $\frac{2}{6}$ = $\frac{1}{3}$ There are 6 equally likely outcomes.

Expected number of scores greater than 4   Expected number of successes

$$= \frac{1}{3} \times 240$$ = probability of a success

$$= 80$$ × number of trials

Exercise 9.1D

1. A fair six-sided dice is rolled 300 times. How many times would you expect to roll a 'six'?

2. The spinner shown has four equal sectors. How many 3s would you expect to get in 100 spins?

3. About one in eight of the population is left-handed. How many left-handed people would you expect to find in a company employing 400 people?

4. A bag contains a large number of marbles of which one in five is red. If I randomly select one marble on 200 occasions, how many of these marbles would I expect to be red?

5. The spinner shown is used for a simple game.

   a) What is the probability of spinning a zero?

   b) If the spinner is spun 200 times, how many times would you expect it to land on 50?

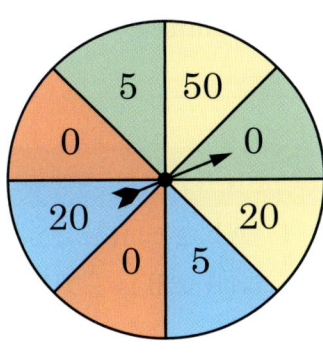

6. The numbered cards are shuffled and put into a pile.

One card is selected at random and not replaced. A second card is then selected.

If the first card is the '11', find the probability that the second card selected shows an even number.

9.2 Mutually exclusive events

Events are **mutually exclusive** if they cannot occur at the same time.

Examples
- Selecting an ace } from a pack of cards
  Selecting a ten

- Flipping a 'head'
  Flipping a 'tail'

- Getting a total of 5 on two dice
  Getting a total of 7 on two dice

The sum of the probabilities of mutually exclusive events is 1.

> The probability of an event not happening
> = 1 – the probability of the event happening.

Example

Every day Anna has the choice of going to work by bus, by train or by taxi. The probability of choosing to go by bus is 0.5 and the probability of choosing to go by train is 0.3.

Find the probability that Anna chooses:

a) to not go by train   **b)** to go by taxi.

a) P(not going by train) = 1 − P(going by train)   The three events 'going by bus', 'going by train' and 'going by taxi' are mutually exclusive.

$$= 1 - 0.3$$
$$= 0.7$$

b) P(going by taxi) = 1 − (0.5 + 0.3)   The sum of the probabilities is 1.
$$= 0.2$$

Exercise 9.2A

1. A bag contains a large number of coloured balls. The probability of selecting a red ball is $\frac{1}{5}$. What is the probability of selecting a ball which is not red?

2. A bag contains 7 white balls, 4 blue balls and 9 yellow balls. Find the probability of selecting:

   a) a white ball   **b)** a ball which is not white.

3. A motorist does a survey at some traffic lights on his way to work every day.

   He finds that the probability that the lights are 'red' when he arrives is 24%. What is the probability that the lights are not 'red'?

4. Government birth statistics show that the probability of a newborn baby being a boy is 0.506.

   What is the probability of a newborn baby being a girl?

5. The spinner has 8 equal sectors. Find the probability of:

   a) spinning a 5

   b) not spinning a 5

   c) spinning a 2

   d) not spinning a 2

   e) spinning a 7

   f) not spinning a 7.

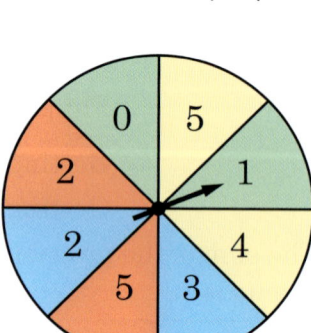

6. A bag contains a large number of balls coloured red, white, black or green. The probabilities of selecting each colour are as follows:

Colour	Red	White	Black	Green
Probability	0.3	0.1		0.3

Find the probability of selecting a ball:

a) which is black

b) which is not white.

7. In a survey the number of people in cars is recorded. When a car passes the school gates, the probability of having 1, 2, 3, ... occupants is as follows:

Number of people	1	2	3	4	more than 4
Probability	0.42	0.23		0.09	0.02

a) Find the probability that the next car past the school gates contains:

   i) three people   ii) fewer than 4 people.

b) One day, 2500 cars passed the gates. How many of the cars would you expect to have 2 people inside?

8. A box contains 30 balls which are yellow, black or white.

The probability of selecting a yellow ball is $\frac{1}{5}$, of selecting a black ball is $\frac{1}{2}$ and of selecting a white ball is $\frac{3}{10}$.

Some yellow balls are added to the bag.
The probability of randomly selecting a yellow ball is now $\frac{1}{4}$.

If the number of black balls and the number of white balls do not change, find how many yellow balls were added to the bag.

9. Bharat rolls a fair six-sided dice.

He says: 'The events rolling an even number and rolling a prime number are mutually exclusive'.

Is Bharat correct? Explain your answer.

9.3 Relative frequency

To work out the probability of a drawing pin landing point up, you can carry out an experiment in which a drawing pin is dropped many times. If the pin lands point up on x occasions out of a total number of N trials, the **relative frequency** of landing point up is $\dfrac{\text{number of successful trials}}{\text{total number of trials}}$ or $\dfrac{x}{N}$.

Example

Sayid thinks that he has a biased dice. To test this, he rolled the dice 300 times. For each set of 25 rolls, he recorded how many sixes he obtained. The following twelve numbers show how many sixes he rolled out of each set of 25.

5 4 6 6 6 5 3 7 6 5 6 5

Is Sayid's dice biased? Explain your answer.

> **Tip**
>
> When an experiment is repeated many times you can use the relative frequency as an estimate of the probability of the event occurring.

After 25 rolls, the relative frequency of rolling a six = $\dfrac{5}{25} = 0.2$

After 50 rolls the relative frequency of rolling a six = $\dfrac{5 + 4}{50} = 0.18$

After 75 rolls the relative frequency

of rolling a six = $\dfrac{5 + 4 + 6}{75} = 0.2$ and so on.

The results are plotted on this graph.

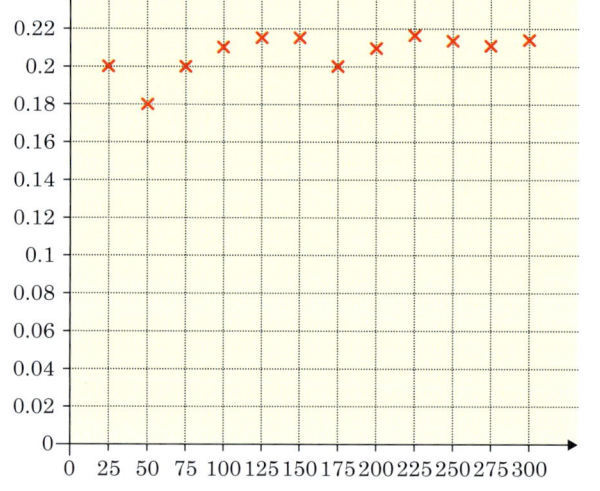

As more and more results are included, the average number of sixes per roll settles down at slightly over 0.21.

For this dice, you can say the relative frequency of sixes was just over 0.21.

If the dice was fair you would expect to

get a six on $\dfrac{1}{6}$ of the rolls. So the

relative frequency would be $0.1\dot{6}$.

Sayid's dice does appear to be biased because sixes occur more frequently than you would expect for a fair dice.

Exercise 9.3A

1. Conduct an experiment where you cannot predict the result. You could roll a dice with a piece of 'sticky tack' stuck to it. Or make a spinner where the axis is not quite in the centre. Or drop a drawing pin.

    Conduct the experiment many times and work out the relative frequency of a 'success' after every 10 or 20 trials.

    Plot a relative frequency graph like the one in the example to see if the results 'settle down' to a consistent value.

2. The spinner has an equal chance of giving any digit from 0 to 9. Four friends spun the pointer a different number of times and recorded the number of zeros they got.

Here are their results.

	Number of spins	Number of zeros	Relative frequency
Steve	10	2	0.2
Nick	150	14	0.093
Mike	200	41	0.205
Jason	1000	104	0.104

One of the four recorded his results incorrectly. Say who you think this was and explain why.

3. 50 buses arrived at the bus station this morning. Each bus was early, on time or late. The results are shown in the table below.

Outcome	Frequency
Early	14
On time	30
Late	6

What is the probability that the next bus will be:

a) late **b)** early **c)** on time?

4. A spinner has four sections, labelled A, B, C and D.

Daryl and Keren each spin the spinner a number of times.

The table shows some information.

	Number of spins	Number of Bs	Relative frequency of spinning a B
Daryl	20	7	
Keren	120		0.45

Copy and complete the table.

9.4 Venn diagrams

A **Venn diagram** shows all the possible ways in which something can be included in or excluded from several sets.

In the following diagram, $\mathscr{E}$ represents the **universal set**, which in this example is the set of all the whole numbers from 1 to 10.

P is the set of prime numbers, and E is the set of even numbers.

You can use set notation to describe sets. You use 'curly' brackets:

For example, P = {2, 3, 5, 7} or P = {prime numbers under 10}

You can also describe sets that are infinite.

For example, A = {x: x is a natural number} and
B = {x: $a \leqslant x \leqslant b$} describe the infinite set of natural numbers and the infinite set of real numbers between a and b (inclusive).

The **complement** (') of a set is the set of things *not* in the set.

P' = {1, 4, 6, 8, 9, 10}

You can also describe sets that are either the intersection ($\cap$) or the union ($\cup$) of two (or more) sets.

The intersection is the set of items in *both* sets: $P \cap E = \{2\}$

The union is the set of items in *either* (or both) of the sets: $P \cup E = \{2, 3, 4, 5, 6, 7, 8, 10\}$

n(P) is the number of elements in a set, so in this case n(P) = 4

1. In the Venn diagram,
   $\mathscr{E}$ = {people in a hotel}
   T = {people who like toast}
   E = {people who like eggs}

   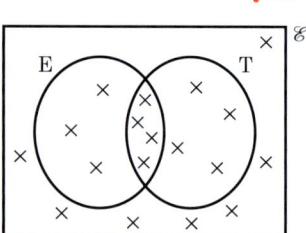

   a) Write down:

   i) n(T)

   ii) n(T' $\cap$ E)

   iii) n(T $\cap$ E).

   b) How many people are in the universal set?

   c) How many people like neither toast *nor* eggs?

2. In the Venn diagram,

 $\mathscr{E}$ = {students in a Year 10 class}

 R = {members of the rugby team}

 C = {members of the cricket team}

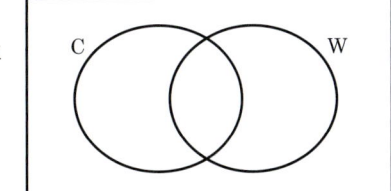

 a) Write down:

  i) n(R)

  ii) n(R $\cap$ C)

  iii) n(R $\cap$ C'}

  iv) n(R $\cup$ C}.

 b) How many students are there in the Year 10 class?

3. A class of 30 students travel to school one morning.

 12 students travel by car for part of the way, then walk the rest of the way.

 6 students travel all the way by car and 8 students walk all the way.

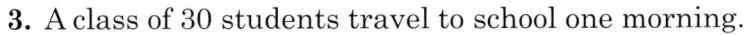

 $\mathscr{E}$ = {students in the class}

 C = {students who travel to school by car}

 W = {students who walk to school}

 a) Copy and complete the Venn diagram to illustrate this information.

 b) Write down:

  i) n(C)   **ii)** n(C $\cap$ W)

  iii) n(C $\cup$ W)   **iv)** n(C'$\cap$ W').

4. In a university mathematics class of 50 students, some students study geometry, some study logic, some study both and some study neither.

 27 study geometry, 19 study logic and 14 study neither.

 $\mathscr{E}$ = {students in the mathematics class}

 G = {students who study geometry}

 L = {students who study logic}

 a) Draw a Venn diagram to illustrate this information.

 b) Write down:

  i) n(G)   **ii)** n(G $\cap$ L)

  iii) n(G $\cap$ L')   **iv)** n(G'$\cup$ L').

Using Venn diagrams to find probabilities

Example

A group of 30 cats were given two different types of food. 7 cats liked only food A, 5 liked only food B, 16 liked both and 2 liked neither. This information is displayed in the following Venn diagram.

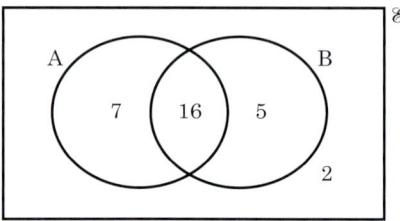

a) What is the probability that a randomly selected cat will like both foods A and B?

b) What is the probability that a randomly selected cat will like only food A?

c) What is the probability that a randomly selected cat does not like food B?

---

a) P(the cat likes both foods) $= \dfrac{16}{30} = \dfrac{8}{15}$ This is given by the intersection of the sets.

b) P(the cat likes only food A) $= \dfrac{7}{30}$ This is the number of cats in set A that are not in set B.

c) P(the cat does not like food B) $= \dfrac{7+2}{30} = \dfrac{9}{30}$ This is the total number of cats not in set B.

Exercise 9.4B

1. In a class of 25 students, some choose to study French, some choose to study German, some choose both and some choose neither. This information is illustrated in the following Venn diagram.

   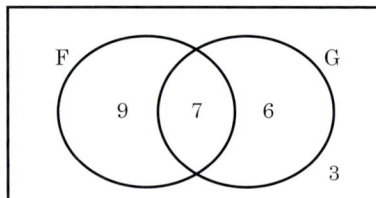

   $\mathscr{E} = \{$students in the class$\}$

   $F = \{$students who study French$\}$

   $G = \{$students who study German$\}$

   What is the probability that a student chosen randomly from the class:

   a) studies French

   b) studies French and German

   c) studies neither French nor German

   d) studies German but does not study French?

2. A group of 30 children were asked whether they liked eating carrots. Some said they liked eating them, some said they didn't like them but ate them anyway and some said they refused to eat them. This information is illustrated in the following Venn diagram.

   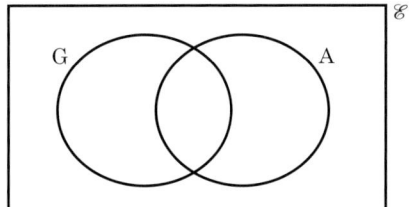

   $\mathscr{E}$ = {the group of children}

   C = {those who eat carrots}

   L = {those who like carrots}

   a) What is the probability that a randomly selected child:

   i) likes carrots

   ii) does not like carrots but does eat them

   iii) does not like carrots and refuses to eat them?

   b) What type of person does the number 4 in the diagram represent?

3. A sports club has 50 members. 20 of them do gymnastics and athletics. 18 of them do gymnastics but not athletics. 3 do neither.

   a) Copy and complete this Venn diagram to illustrate this information.

   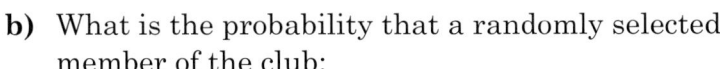

   $\mathscr{E}$ = {members of the club}

   G = {those who do gymnastics}

   A = {those who do athletics}

   b) What is the probability that a randomly selected member of the club:

   i) does athletics but not gymnastics

   ii) does gymnastics

   iii) does athletics?

4. A book club has 20 members. 13 of them like detective stories, 10 of them like horror stories and 4 of them like neither.

   a) Illustrate this information in a Venn diagram.

   b) What is the probability that a randomly selected member:

   i) likes both detective stories and horror stories

   ii) likes horror stories but not detective stories

   iii) does not like detective stories?

5. A group of 12 friends were discussing the various places they had visited. 2 said they had been to the USA but not Italy. 4 had been to Italy but not the USA. Twice the number of people who had been to neither had been to both.

   a) Illustrate this information in a Venn diagram.

   b) What is the probability that a randomly selected friend:

   i) had been to Italy and America

   ii) had been to Italy

   iii) had not been to Italy?

9.5 Probability diagrams

Other diagrams are sometimes useful for solving probability questions.

Sample space diagrams and **tree diagrams** are used to display all the possible outcomes of an event.

Sample space diagrams

Example

A black dice and a white dice are thrown at the same time and the total score recorded.

Draw a sample space diagram to display all the possible outcomes.

Find the probability of obtaining:

a) a total of 5

b) a total of 11

c) a 'two' on the black dice and a 'six' on the white dice.

		White dice					
		1	**2**	**3**	**4**	**5**	**6**
Black dice	**1**	2	3	4	⑤	6	7
	2	3	4	⑤	6	7	8
	3	4	⑤	6	7	8	9
	4	⑤	6	7	8	9	10
	5	6	7	8	9	10	|11|
	6	7	8	9	10	|11|	12

It is convenient to display all the possible outcomes in a grid.

There are 36 possible outcomes, shown by the numbers in each cell of the table.

a) Probability of obtaining a total of $5 = \dfrac{4}{36} = \dfrac{1}{9}$ There are four ways of obtaining a total of 5 on the two dice. They are circled on the diagram.

b) P(total of 11) $= \dfrac{2}{36} = \dfrac{1}{18}$ There are two ways of obtaining a total of 11. They are shown in rectangles on the diagram.

c) P(2 on black and 6 on white) $= \dfrac{1}{36}$ There is only one way of obtaining a 'two' on the black dice and a 'six' on the white dice.

Exercise 9.5A

1. The two sides of a coin are known as 'head' and 'tail'. A 10c and a 5c coin are flipped at the same time. List all the possible outcomes.

   Find the probability of obtaining:

   a) two heads

   b) a head and a tail.

2. A red six-sided dice and a blue six-sided dice are thrown at the same time. List all the possible outcomes in a systematic way. Find the probability of obtaining:

   a) a total of 10

   b) a total of 12

   c) a total less than 6

   d) the same number on both dice

   e) a total more than 9.

   What is the most likely total?

3. Two students are asked to choose a whole number between 3 and 7. Find the probability that:

   a) they choose the same number

   b) the sum of their numbers is 10

   c) the sum of their numbers starts with a 1

   d) they have chosen two consecutive numbers.

Tree diagrams

Example

A bag contains 5 red balls and 3 green balls. A ball is drawn at random and then replaced. Another ball is drawn.

a) Draw a tree diagram to illustrate this situation.

b) What is the probability that:

   i) both balls are green

   ii) one ball is red and one ball is green?

a)

Ball 1   Ball 2

$\frac{5}{8}$ R $\frac{5}{8}$ R

$\frac{3}{8}$ G

$\frac{3}{8}$ G $\frac{5}{8}$ R

$\frac{3}{8}$ G

> This branch represents selecting two red balls. The probability of each red is shown on its branch.

> This branch represents selecting a red ball followed by a green ball.

> This branch represents selecting a green ball followed by a red ball.

> This branch represents selecting two green balls.
> The probability of this event is obtained by multiplying the fractions on the two branches.

b) **i)** P(two green balls) $= \dfrac{3}{8} \times \dfrac{3}{8} = \dfrac{9}{64}$

    ii) P(one red, one green) = P(red, green OR green, red)

$$= \left(\frac{5}{8} \times \frac{3}{8} \right) + \left(\frac{3}{8} \times \frac{5}{8} \right) = \frac{30}{64} = \frac{15}{32}$$

Exercise 9.5B

1. A bag contains 10 discs; 7 are black and 3 white. A disc is selected, and then replaced. A second disc is selected.

   Copy and complete the tree diagram showing all the probabilities and outcomes.

   Find the probability of the following:

      a) both discs are black   **b)** both discs are white.

   $\frac{7}{10}$ B $\frac{7}{10}$ B

   W

   B

   $\frac{3}{10}$ W

   W

2. A bag contains 5 red balls and 3 green balls. A ball is drawn and then replaced before a ball is drawn again.

   Draw a tree diagram to show all the possible outcomes.

   Find the probability that:

      a) two green balls are drawn

      b) the first ball is red and the second is green.

3. A bag contains 7 green discs and 3 blue discs. A disc is drawn and then replaced. A second disc is drawn.

Copy and complete the tree diagram.

Find the probability that:

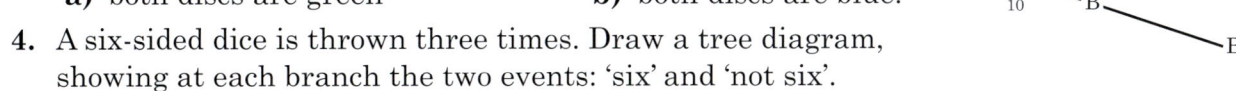

  a) both discs are green   **b)** both discs are blue.

4. A six-sided dice is thrown three times. Draw a tree diagram, showing at each branch the two events: 'six' and 'not six'.

What is the probability of throwing a total of:

a) three sixes

b) no sixes

c) one six

d) at least one six (use part **(b)**).

5. A bag contains 6 red marbles and 4 blue marbles. A marble is selected at random and then replaced. A second marble is selected and replaced, and then a third. Find the probability of selecting:

  a) three red marbles   **b)** three blue marbles

  c) no red marbles   **d)** at least one red marble.

6. When a cutting is taken from a geranium the probability that it grows is $\frac{3}{4}$. Three cuttings are taken. What is the probability that:

a) all three grow

b) none of them grow?

Revision exercise 9

1. A bag contains 3 red balls and 5 white balls. One ball is selected at random. Find the probability of selecting:

  a) a red ball   **b)** a white ball

  c) a blue ball.

2. A box contains 2 yellow discs, 4 blue discs and 5 green discs. A disc is selected at random. Find the probability of selecting:

  a) a yellow disc   **b)** a green disc

  c) a blue or a green disc

  d) not getting a green disc.

3. When two fair dice are thrown simultaneously, what is the probability of obtaining the same number on both dice?

4. A fair coin is flipped four times. What is the probability of obtaining at least three 'heads'?

5. Two fair six-sided dice are thrown. The product of the numbers on top is recorded.

  a) Draw a sample space diagram to show the possible outcomes.

  b) What is the probability that the product of the numbers on top is:

    i) 12   **ii)** 4   **iii)** 11?

6. A bag contains a large number of discs of which one in six is gold. If you randomly select one disc on 300 occasions, and put the disc back in the bag after each selection, how many times would you expect to select a gold disc?

7. a) What is the probability of getting a 6 with this spinner?

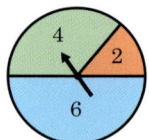

b) Draw a spinner like this with 8 equal sectors. Shade some sectors so that the chance of getting a shaded sector is three times the chance of getting a white sector.

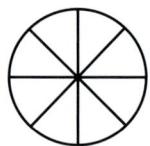

c) This spinner has some 1s, 2s and 3s written in the sectors. The chance of getting a 2 is twice the chance of getting a 3. The chance of getting a 1 is three times the chance of getting a 3. Draw the spinner and replace the question marks with the correct number of 1s, 2s and 3s.

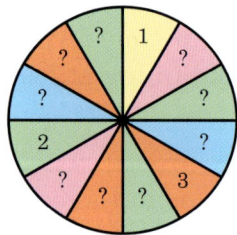

8. Cards with numbers 1, 2, 3, 4, 5, 6, 7, 8, 9, 10 are shuffled and then placed face down in a line. The cards are then turned over one at a time from the left.

In this example, the first card is a '4'.

a) Find the probability that the next card turned over will be:

   i) 7   **ii)** a number higher than 4.

b) Suppose the second card is a 1.

Find the probability that the next card will be:

   i) 6   **ii)** an even number

 iii) higher than 1.

9. A coin is biased so that the probability of a 'head' is $\frac{3}{4}$. Find the probability that, when flipped three times, it shows:

a) three tails

b) two heads and one tail

c) one head and two tails

d) no tails.

Write down the sum of the probabilities in parts **(a)**, **(b)**, **(c)** and **(d)**.

10. An 8-sided spinner has sectors numbered from 1 to 8.

a) Giovanna says: 'The events spinning an odd number and spinning an even number are mutually exclusive'.

Is Giovanna correct? Explain your answer.

b) Malik says: 'The events spinning an even number and spinning a prime number are mutually exclusive'.

Is Malik correct? Explain your answer.

Examination-style exercise 9

NON-CALCULATOR

1. A bag of 60 sweets contains 16 chocolates, 26 jellies and 18 toffees. A sweet is selected at random.

   What is the probability that it is a toffee? [1]

2. **a)** A bowl of fruit contains 3 mangos, 4 papayas, 2 pomegranates and 1 plum. Aminata chooses one piece of fruit at random. What is the probability she chooses:

   i) a papaya [1]

   ii) an apple? [1]

   b) The probability that it will rain in New Delhi on 1st September is $\dfrac{5}{12}$.

   State the probability that it will **not** rain in New Delhi on 1st September. [1]

3. **a)** 85% of the seeds in a packet will produce yellow flowers.

   One seed is chosen at random.

   What is the probability that it will **not** produce a yellow flower? [1]

   b) A box of 15 pencils contains 5 purple, 4 green and 6 pink pencils. One pencil is chosen at random from the box. Find the probability that it is:

   i) green [1]

   ii) green or pink [1]

   iii) yellow. [1]

4. **a)** There are 12 rabbits and 16 cats in an animal hospital.
   The vet chooses one animal at random.

   What is the probability that this is a rabbit?

   Write your answer as a fraction in its simplest form. [1]

   b) The probability that Fluffy the cat arrives at the hospital before 09 00 is $\dfrac{13}{18}$.

   What is the probability that Fluffy does not arrive before 09 00?

   Write your answer as a fraction. [1]

5. The diagram shows a six-sided spinner.

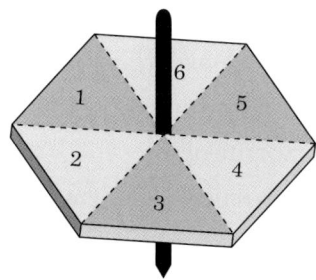

  a) Amara spins a biased spinner and the probability she gets a two is $\dfrac{5}{36}$. Find the probability she:

    i) does not get a two [1]

    ii) gets a seven [1]

    iii) gets a number on the spinner less than seven. [1]

  b) Jenny spins her spinner 198 times and gets a two 34 times. Write down the relative frequency of getting a two with Jenny's spinner. [1]

  c) The relative frequency of getting a two with Paulo's spinner is $\dfrac{21}{102}$.

    Which of the three spinners, Amara's, Jenny's or Paulo's, is most likely to give a two? [1]

6. Carlos is in a class of 12 students.

He compares the results of the students in a physics test with their results in a chemistry test. The table shows these results.

Student	A	B	C	D	E	F	G	H	I	J	K	L
Physics mark	17	8	11	15	14	19	9	12	19	18	13	15
Chemistry mark	10	13	10	8	11	7	14	11	10	11	11	10

A student is chosen at random.

What is the probability that the student scored *more than* 10 marks:

a) in physics [1]

b) in physics and in chemistry [1]

c) in at least one subject? [1]

7. A play is being performed twice; first on a Friday night and again on a Saturday night.

100 people who expressed an interest in seeing the play were asked which of the two nights they would be able to attend. 58 people said they could attend the Friday performance, 74 said they could attend the Saturday performance and 4 said they could attend neither.

  a) Draw a Venn diagram to illustrate this information. [3]

    $\mathscr{E}$ = {the 100 people asked}

    F = {those who could attend on Friday}

    S = {those who could attend on Saturday}

  b) What is the probability that a randomly selected person from the 100 asked:

    i) could attend either performance [1]

    ii) could only attend the Saturday performance [1]

    iii) could not attend the Saturday performance? [1]

Girolamo Cardan (1501–1576) was a colourful character who became Professor of Mathematics at Milan. As well as being a distinguished academic, he was an astrologer, a physician and a philosopher. His mathematical genius enabled him to develop the general theory of cubic and quartic equations, although a method for solving cubic equations which he claimed as his own was pirated from Niccolo Tartaglia.

Terence Tao (1975–) was born to ethnic Chinese immigrant parents and raised in Australia. He is generally regarded as one of the greatest living mathematicians and has been called the 'Mozart of mathematics' by his colleagues. He has authored over 300 research papers on topics such as differential equations, probability, number theory and an area of mathematics called combinatorics, for which he won the prestigious Fields Medal in 2006.

- Substitute into expressions and formulae.
- Construct and solve linear equations and simultanuous equations, and rearrange a formula to change the subject.

10.1 Forming and solving equations

Example

The length of a rectangle is twice the width.
If the perimeter is 36 cm, find the width.

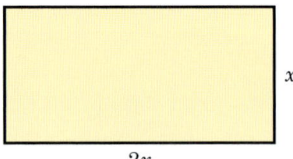

Let the width of the rectangle be x cm.

Then the length of the rectangle is $2x$ cm. — The length is twice the width.

$x + 2x + x + 2x = 36$ — Form an equation. The perimeter is the total distance around the edge.

$6x = 36$ — Simplify the left-hand side by adding like terms.

$x = 6$ — Solve the equation. Divide both sides of the equation by 6.

The width of the rectangle is 6 cm.

Exercise 10.1A

Answer each of these questions by forming an equation and then solving it.

1. Find the value of x if the perimeter is 7 cm.

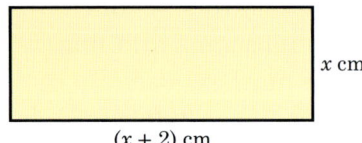

2. Find the value of x if the perimeter is 5 cm.

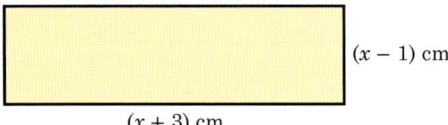

3. The length of a rectangle is 3 times its width. If the perimeter of the rectangle is 11 cm, find its width.

4. The length of a rectangle is 4 cm more than its width. If its perimeter is 13 cm, find its width.

5. The width of a rectangle is 5 cm less than its length. If the perimeter of the rectangle is 18 cm, find its length.

Tip

Let the width be x cm.

6. Find the value of x in the following rectangles.

a)
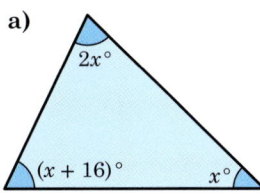

Area = 18 cm^2

x cm

5 cm

b)
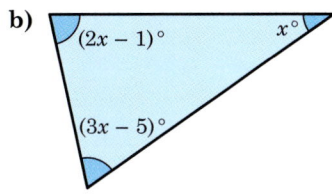

Area = 15 cm^2

$(x + 3)$ cm

4 cm

7. Find the value of x in the following triangles.

a)

$2x°$

$(x + 16)°$

$x°$

b)
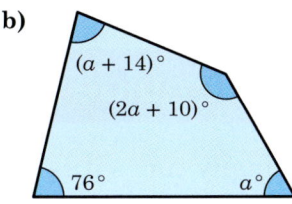

$(2x − 1)°$

$x°$

$(3x − 5)°$

8. The angles of a triangle are $32°$, $x°$ and $(4x + 3)°$.

Find the value of x.

9. Find the value of a in the diagrams below.

a)

$a°$

$a°$

$(2a − 32)°$

b)

$(a + 14)°$

$(2a + 10)°$

$76°$

$a°$

10. The sum of three consecutive whole numbers is 168. Let the first number be x. Form an equation and hence find the three numbers.

11. The sum of four consecutive whole numbers is 170. Find the numbers.

12. In this triangle AB = x cm.

BC is 3 cm shorter than AB.

AC is twice as long as BC.

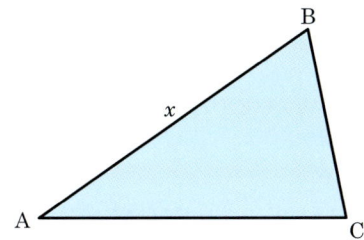

B

x

A

C

> **Tip**
>
> Consecutive numbers are numbers that are next to each other, for example 1, 2 and 3.

a) Write down, in terms of x, the lengths of:

  i) BC

  ii) AC

The perimeter of the triangle is 41 cm.

b) Write down an equation in x and solve it to find x.

13. The area of rectangle A is twice the area of rectangle B. Find the value of x.

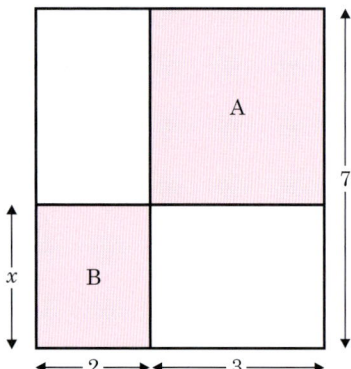

10.2 Substituting into formulae

When a calculation is repeated many times it is often helpful to use a **formula**, which is a rule that connects two or more quantities that may vary. For example, when a building society calculates how much money they can offer two people as a

mortgage it may use a formula such as '$2\frac{1}{2}$ times the main salary plus the second salary'.

Publishers use a formula to work out the selling price of a book based on the production costs and the expected sales of the book. There are many formulae used in maths and science.

Example 1

A formula connecting velocity with acceleration and time is $v = u + at$.

Find the value of v when $u = 3$, $a = 4$ and $t = 6$.

$v = u + at$	Write the formula.
$v = 3 + (4 \times 6)$	Replace each letter by its value.
$v = 27$	Calculate the answer. Remember to use BIDMAS.

Example 2

A formula for the tension in a spring is $T = \dfrac{kx}{a}$.

Find the value of T when $k = 13$, $x = 5$ and $a = 2$.

$T = \dfrac{kx}{a}$	Write the formula.
$T = \dfrac{13 \times 5}{2}$	Replace each letter by its value.
$T = 32\dfrac{1}{2}$ or 32.5	Calculate the answer.

Exercise 10.2A

1. A formula involving force, mass and acceleration is $F = ma$. Find the value of F when $m = 12$ and $a = 3$.

2. The height of a growing tree is given by the formula $h = 2t + 15$. Find the value of h when $t = 7$.

3. The time required to cook a joint of meat is given by the formula
   $T = (\text{mass of joint}) \times 3 + \dfrac{1}{2}$. Find the value of T when $(\text{mass of joint}) = 2\dfrac{1}{2}$.

4. An important formula in physics states that $I = mu - mv$. Find the value of I when $m = 6$, $u = 8$ and $v = 5$.

5. The distance travelled by an accelerating car is given by the formula $s = \left(\dfrac{u + v}{2}\right)t$. Find the value of s when $u = 17$, $v = 25$ and $t = 4$.

6. Einstein's famous formula states that $E = mc^2$. Find the value of E when $m = 0.0001$ and $c = 3 \times 10^8$.

7. The height of a stone thrown upwards is given by $h = ut - 5t^2$. Find the value of h when $u = 70$ and $t = 3$.

8. The speed of an accelerating particle is given by the formula $v^2 = u^2 + 2as$. Find the exact value of v when $u = 11$, $a = 5$ and $s = 6$.

9. The time period T of a simple pendulum is given by the formula $T = 2\pi\sqrt{\left(\dfrac{l}{g}\right)}$,

   where l is the length of the pendulum and g is the gravitational acceleration. Find T when $l = 0.65$, $g = 9.81$ and $\pi = 3.142$.

10. The sum S of the squares of the integers from 1 to n is given by
    $S = \dfrac{1}{6}n(n + 1)(2n + 1)$. Find S when $n = 12$.

10.3 Changing the subject of a formula

Consider the formula $F = ma$. F is called the **subject** of the formula because it is on its own on one side of the equals sign.

Sometimes you need to change the subject of the formula. For example, if you want to calculate the value of a, given F and m, you can rearrange the formula so that a is the subject.

In this case, divide both sides by m so that $\dfrac{F}{m} = a$. Now a is the subject of the formula.

(You usually rewrite the formula so the new subject is on the left: in this case, $a = \dfrac{F}{m}$.)

Example

Make x the subject of the formulae below.

a) $ax - p = t$

  $ax = t + p$ Add p to both sides.

  $x = \dfrac{t + p}{a}$ Divide both sides by a.

b) $y(x + y) = v^2$

  $yx + y^2 = v^2$ Expand the brackets.

  $yx = v^2 - y^2$ Subtract y^2 from both sides.

  $x = \dfrac{v^2 - y^2}{y}$ Divide both sides by y.

Exercise 10.3A

Make x the subject.

1. $x + b = e$

2. $x - t = m$

3. $x - f = a + b$

4. $x + h = A + B$

5. $x + t = y + t$

6. $a + x = b$

7. $k + x = m$

8. $v + x = w + y$

9. $ax = b$

10. $hx = m$

11. $mx = a + b$

12. $kx = c - d$

13. $vx = e + n$

14. $3x = y + z$

15. $xp = r$

16. $xm = h - m$

17. $ax + t = a$

18. $mx - e = k$

19. $ux - h = m$

20. $ex + q = t$

21. $kx - u^2 = v^2$

22. $gx + t^2 = s^2$

23. $xa + k = m^2$

24. $xm - v = m$

25. $a + bx = c$

26. $t + sx = y$

27. $y + cx = z$

28. $a + hx = 2a$

29. $mx - b = b$

30. $kx + ab = cd$

31. $a(x - b) = c$

32. $c(x - d) = e$

33. $m(x + m) = n^2$

34. $k(x - a) = t$

35. $h(x - h) = k$

36. $m(x + b) = n$

37. $a(x - a) = a^2$

38. $c(a + x) = d$

39. $m(b + x) = e$

Formulae involving fractions

Example

Make x the subject in the formulae below.

a) $\dfrac{x}{a} = p$

$x = ap$ Multiply both sides by a.

b) $\dfrac{m}{x} = t$

$m = xt$ Multiply both sides by x.

$\dfrac{m}{t} = x$ Divide both sides by t.

c) $v = \dfrac{a^2}{x}$

$vx = a^2$ Multiply both sides by x.

$x = \dfrac{a^2}{v}$ Divide both sides by v.

Exercise 10.3B

Make x the subject.

1. $\dfrac{x}{t} = m$

2. $\dfrac{x}{e} = n$

3. $\dfrac{x}{p} = a$

4. $am = \dfrac{x}{t}$

5. $bc = \dfrac{x}{a}$

6. $e = \dfrac{x}{y^2}$

7. $\dfrac{x}{a} = (b + c)$

8. $\dfrac{x}{t} = (c - d)$

9. $\dfrac{x}{m} = s + t$

10. $\dfrac{x}{k} = h + i$

11. $\dfrac{x}{b} = \dfrac{a}{c}$

12. $\dfrac{x}{m} = \dfrac{z}{y}$

13. $\dfrac{x}{h} = \dfrac{c}{d}$

14. $\dfrac{m}{n} = \dfrac{x}{e}$

15. $\dfrac{b}{e} = \dfrac{x}{h}$

16. $\dfrac{x}{(a + b)} = c$

17. $\dfrac{x}{(h + k)} = m$

18. $\dfrac{x}{u} = \dfrac{m}{y}$

19. $\dfrac{x}{(h - k)} = t$

20. $\dfrac{x}{(a + b)} = (z + t)$

21. $t = \dfrac{e}{x}$

22. $a = \dfrac{e}{x}$

23. $m = \dfrac{h}{x}$

24. $\dfrac{a}{b} = \dfrac{c}{x}$

25. $\dfrac{u}{x} = \dfrac{c}{d}$

26. $\dfrac{m}{x} = t^2$

27. $\dfrac{a^2}{b^2} = \dfrac{c^2}{x}$

10.4 Simultaneous equations

Simultaneous equations are used when you have two unknown quantities that you want to find and two pieces of information connecting these quantities.

You can solve simultaneous equations in different ways.

Graphical solution

Example

Layla and Rashid are children. Layla is 5 years older than Rashid.

The sum of their ages is 12 years.

How old is each child?

Let Layla be x years old and Rashid be y years old.		Define the two variables.
[sum = 12]	$x + y = 12$	Write each piece of information as an equation.
[difference = 5]	$x - y = 5$	

Now draw on the same set of axes the graphs of $\quad x + y = 12$

$$\text{and} \quad x - y = 5$$

$x + y = 12$ goes through $(0, 12)$, $(2, 10)$, $(6, 6)$, $(12, 0)$.

$x - y = 5$ goes through $(5, 0)$, $(7, 2)$, $(10, 5)$.

The point $(8.5, 3.5)$ lies on both lines. The lines cross at this point.

Hence $x = 8.5$, $y = 3.5$ are the solutions of the simultaneous equations
$x + y = 12$, $x - y = 5$.

Layla is $8\frac{1}{2}$ years old and

Rashid is $3\frac{1}{2}$ years old.

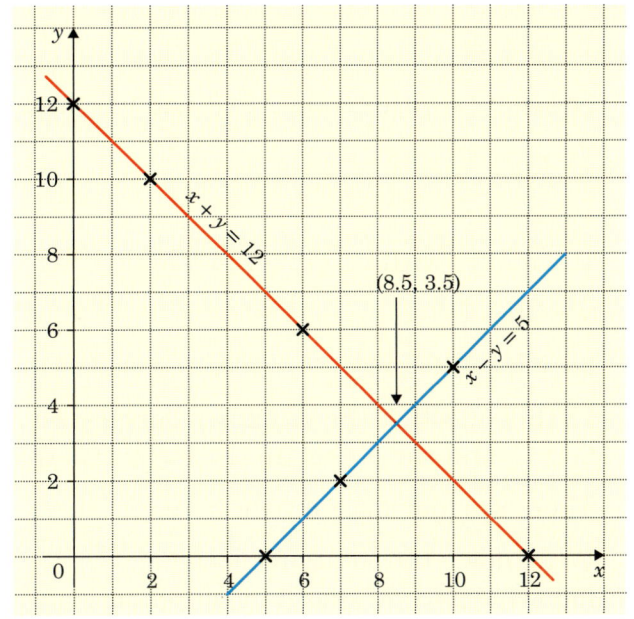

Exercise 10.4A

1. Use the graphs to solve these pairs of simultaneous equations.

   a) $x + y = 10$

   $\quad y - 2x = 1$

   b) $2x + 5y = 17$

   $\quad y - 2x = 1$

   c) $x + y = 10$

   $\quad 2x + 5y = 17$

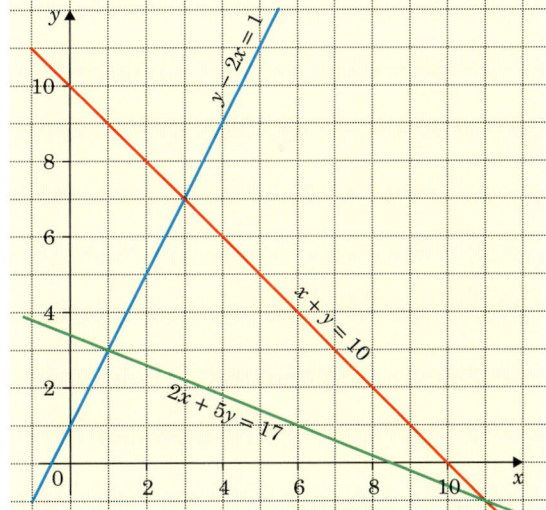

In Questions **2** to **6**, solve the simultaneous equations by drawing graphs. Use a scale of 1 cm to 1 unit on both axes.

2. $x + y = 6$

$\quad 2x + y = 8$

Draw axes with x and y from 0 to 8.

3. $x + 2y = 8$

$\quad 3x + y = 9$

Draw axes with x and y from 0 to 9.

4. $x + 3y = 6$

$x - y = 2$

Draw axes with x from 0 to 8 and y from 0 to 10.

5. $5x + y = 10$

$x - y = -4$

Draw axes with x from -4 to 4 and y from -2 to 4.

6. $a + 2b = 11$

$2a + b = 13$

In this question the unknowns are a and b. Draw the a-axis across the page from 0 to 13 and the b-axis up the page also from 0 to 13.

7. There are four lines drawn here.

Write down the solutions to the following pairs of simultaneous equations.

a) $x - 2y = 4$ **b)** $x + y = 7$

     $x + 4y = 4$ $y - 3x = 3$

c) $y - 3x = 3$ **d)** $x + 4y = 4$

     $x - 2y = 4$ $x + y = 7$

e) $x + 4y = 4$

     $y - 3x = 3$

(For part **(e)** give x and y correct to 1 d.p.)

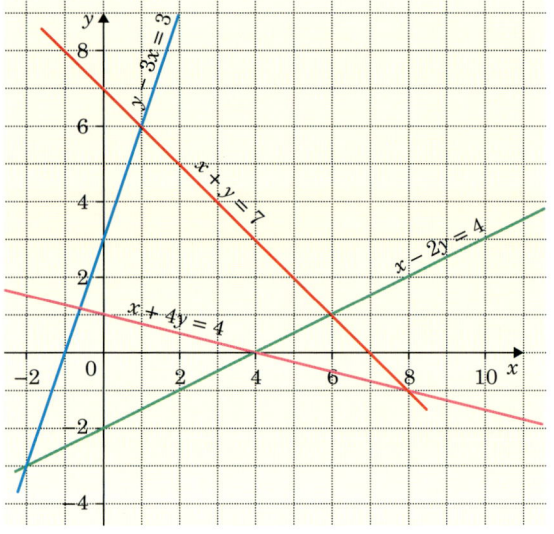

Algebraic solution

You can also solve simultaneous equations without drawing graphs. There are two methods: substitution and elimination.

You may need to choose for yourself which method to use in a question.

Substitution method

This method is best used when one equation is in the form $y = \dots$ or $x = \dots$, as in equation [2] of the following example.

Example

Solve the simultaneous equations

$$3x - 2y = 0 \quad [1]$$
$$y = 7 - 2x \quad [2]$$

$3x - 2(7 - 2x) = 0$	Substitute for y in equation [1].
$3x - 14 + 4x = 0$	Expand the brackets and simplify.
$7x = 14$	

Hence $\quad\quad\quad x = 2$

$$y = 7 - 2 \times 2 \quad\quad \text{Substitute for } x \text{ in equation [2].}$$

Hence $\quad\quad\quad y = 3$

The solutions are $x = 2$ and $y = 3$.

Check your solution using equation [1]:   Check using the equation that you didn't use to

$$3 \times 2 - 2 \times 3 = 0 \quad\quad ✓ \quad\quad \text{find } x \text{ or } y.$$

These values of x and y are the only pair which simultaneously satisfy both equations.

Exercise 10.4B

Use the substitution method to solve the following pairs of simultaneous equations.

1. $2x + y = 5$
$\quad x = 5 - 3y$

2. $x = 8 - 2y$
$\quad 2x + 3y = 14$

3. $y = 10 - 3x$
$\quad x - y = 2$

4. $y = -3 - 2x$
$\quad x - y = -3$

5. $y = 14 - 4x$
$\quad x + 5y = 13$

6. $x = 1 - 2y$
$\quad 2x + 3y = 4$

7. $y = 5 - 2x$
$\quad 3x - 2y = 4$

8. $y = 13 - 2x$
$\quad 5x - 4y = 13$

9. $7x + 2y = 19$
$\quad x = 4 + y$

10. $b = a - 5$
$\quad a + b = -1$

11. $a = 6 - 4b$
$\quad 8b - a = -3$

12. $b = 4 - a$
$\quad 2a + b = 5$

For Questions **13** to **18**, you will need to rearrange one of the equations.

13. $3m = 2n - 6\frac{1}{2}$
$\quad 4m + n = 6$

14. $2w + 3x - 13 = 0$
$\quad x + 5w - 13 = 0$

15. $x + 2(y - 6) = 0$
$\quad 3x + 4y = 30$

16. $2x = 4 + z$
$\quad 6x - 5z = 18$

17. $3m - n = 5$
$\quad 2m + 5n = 7$

18. $5c - d - 11 = 0$
$\quad 4d + 3c = -5$

It is useful, at this point, to revise the operations of addition and subtraction with negative numbers.

Example

Simplify:

a) $-7 + -4$

b) $-3x + (-4x)$

c) $4y - (-3y)$

d) $3a + (-3a)$

a) $-7 + -4 = -7 - 4 = -11$ — Adding a negative number is the same as subtracting the positive number.

b) $-3x + (-4x) = -3x - 4x = -7x$

c) $4y - (-3y) = 4y + 3y = 7y$ — Subtracting a negative value is the same as adding the positive value.

d) $3a + (-3a) = 3a - 3a = 0$

Elimination method

The elimination method involves making the coefficient of both x terms or both y terms the same (ignoring signs). Do this by multiplying every term in each equation by an appropriate value.

You usually use this method when the two equations are in the form $ax + by = c$.

Example 1

Solve the simultaneous equations

$$3x + y = 7 \qquad [1]$$
$$5x - y = 9 \qquad [2]$$

[1] + [2]: $8x = 16$ — The number of ys are the same.

The signs are different so ADD the two equations

since $y + (-y) = y - y = 0$.

$x = 2$ — Find x.

$3 \times 2 + y = 7$ — Substitute x into equation [1].

$y = 1$ — Find y.

Check using equation [2]: — Check using the equation that you didn't use to find x or y.

$5 \times 2 - 1 = 9 \qquad \checkmark$

Hence the solutions are $x = 2$ and $y = 1$.

Sometimes you need to multiply one of the equations to make the number of ys (or the number of xs) the same.

Example 2

$$2x + 3y = 14 \quad [1]$$

$$4x + y = 18 \quad [2]$$

$[2] \times 3$:	$12x + 3y = 54 \quad [3]$	Make the number of ys the same in both equations.
	$\underline{2x + 3y = 14} \quad [1]$	Rewrite equation [1].
$[3] - [1]$:	$10x = 40$	The signs are the same, so SUBTRACT the two equations
		since $3y - 3y = 0$.
	$x = 4$	Find x.
	$4 \times 4 + y = 18$	Substitute x into equation [2].
	$y = 2$	Find y.
Check using equation [1]:		Check using the equation that you didn't use to find x or y.

$$2 \times 4 + 3 \times 2 = 14 \quad ✓$$

The solutions are $x = 4$ and $y = 2$.

Sometimes you need to multiply *both* equations to make the number of xs or ys the same.

In this example, it is the number of xs that have been made the same.

Example 3

Solve the simultaneous equations

$$2x + 3y = 5 \quad [1]$$

$$5x - 2y = -16 \quad [2]$$

$[1] \times 5$:	$10x + 15y = 25 \quad [3]$	Multiply equation [1] by 5 to get $10x$.
$[2] \times 2$:	$\underline{10x - 4y = -32} \quad [4]$	Multiply equation [2] by 2 to get $10x$.
$[3] - [4]$:	$15y - (-4y) = 25 - (-32)$	The signs of the x terms are the same so SUBTRACT to eliminate x.
	$19y = 57$	
	$y = 3$	Find y.

$$2x + 3 \times 3 = 5 \qquad \text{Substitute for } y \text{ in equation [1].}$$

$$2x = 5 - 9 = -4$$

$$x = -2 \qquad \text{Find } x.$$

Check using equation [2]:

Check using the equation that you didn't use to find x or y.

$$5 \times (-2) - 2 \times 3 = -16 \quad \checkmark$$

The solutions are $x = -2$ and $y = 3$.

Exercise 10.4C

Use the elimination method to solve the following pairs of simultaneous equations.

1. $2x + 5y = 24$
$4x + 3y = 20$

2. $5x + 2y = 13$
$2x + 6y = 26$

3. $3x + y = 11$
$9x + 2y = 28$

4. $x + 2y = 17$
$8x + 3y = 45$

5. $3x + 2y = 19$
$x + 8y = 21$

6. $2a + 3b = 9$
$4a + b = 13$

7. $2x + 3y = 11$
$3x + 4y = 15$

8. $3x + 8y = 27$
$4x + 3y = 13$

9. $2x + 7y = 17$
$5x + 3y = -1$

10. $5x + 3y = 23$
$2x + 4y = 12$

11. $7x + 5y = 32$
$3x + 4y = 23$

12. $3x + 2y = 4$
$4x + 5y = 10$

13. $3x + 2y = 11$
$2x - y = -3$

14. $3x + 2y = 7$
$2x - 3y = -4$

15. $x - 2y = -4$
$3x + y = 9$

16. $5x - 7y = 27$
$3x - 4y = 16$

17. $3x - 2y = 7$
$4x + y = 13$

18. $x - y = -1$
$2x - y = 0$

19. $y - x = -1$
$3x - y = 5$

20. $x - 3y = -5$
$2y + 3x + 4 = 0$

Problems solved by simultaneous equations

Example

The price of 3 cups of coffee and 4 cups of tea is $6.95.

The price of 2 cups of coffee and 5 cups of tea is $7.20.

Find the cost of one cup of coffee and one cup of tea.

Let x be the cost of coffee and y be the cost of tea.   Define your variables.

$$3x + 4y = 6.95 \quad [1]$$

$$2x + 5y = 7.2 \quad [2] \qquad \text{Write down two equations.}$$

$[1] \times 5:$ $\quad 15x + 20y = 34.75 \quad [3]$

$[2] \times 4:$ $\quad \underline{8x + 20y = 28.8} \quad [4] \qquad$ Make the number of ys the same.

$[3] - [4]:$ $\qquad 7x = 5.95 \qquad$ The signs are the same so subtract.

$$x = 0.85 \qquad \text{Find } x.$$

$$3 \times 0.85 + 4y = 6.95 \qquad \text{Substitute for } x \text{ in equation [1].}$$

$$4y = 4.4$$

$$y = 1.1 \qquad \text{Find } y.$$

Check using equation [2]:

$$2 \times 0.85 + 5 \times 1.1 = 7.2 \quad ✓$$

The price of a cup of coffee is \$0.85 and the price of a cup of tea is \$1.10.

Exercise 10.4D

Solve each problem by forming a pair of simultaneous equations.

1. Find two numbers with a sum of 15 and a difference of 4.

   [Let the numbers be x and y.]

2. Twice one number added to three times another gives 21.

   Find the possible pairs of numbers, if the difference between them is 3.

3. The average of two numbers is 7, and the difference between them is 6. Find the numbers.

4. Here is a number puzzle. The ? and * stand for numbers which you need to find. The totals of each row and column are given.

   Write down two equations involving ? and * and solve them to find the values of ? and *.

?	*	?	*	36
?	*	*	?	36
*	?	*	*	33
?	*	?	*	36
39	33	36	33	

5. The line with equation $y + ax = c$ passes through the points (1, 5) and (3, 1). Find a and c.

> **Tip**
>
> For the point (1, 5) substitute $x = 1$ and $y = 5$ into $y + ax = c$.

6. The line $y = mx + c$ passes through (2, 5) and (4, 13).

Find m and c.

7. A stone is thrown into the air and its height, h metres above the ground, is given by the equation $h = at - bt^2$.

From an experiment we know that $h = 40$ when $t = 2$ and that $h = 45$ when $t = 3$.

Show that $a - 2b = 20$ and $a - 3b = 15$.

Solve these equations to find a and b.

8. A shop owner can buy either two televisions and three DVD players for $1750 or four televisions and one DVD player for $1250. Find the cost of one of each. Let the cost of a television be x and the cost of a DVD player be y.

9. A spider can lay either white or brown eggs. Three white eggs and two brown eggs have a mass of 13 grams, while five white eggs and four brown eggs have a mass of 24 grams. Find the mass of one brown egg and the mass of one white egg.

10. A bag contains forty counters. All of the counters have either a 2 or a 5 written on them. If the total value of the counters is 155, find the number of each kind.

11. A meter in a car park takes only 10c and 50c coins and contains twenty-one coins altogether. If the value of the coins is $4.90, find the number of coins of each value.

Revision exercise 10

1. This is a rectangle. Work out the value of x and hence find the perimeter of the rectangle.

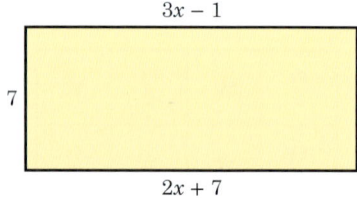

2. Find the length of the sides of this equilateral triangle.

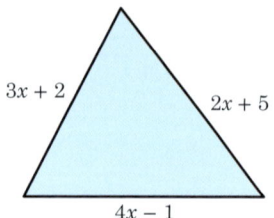

3. Petra has $12 and Suki has nothing. They both receive the same money for doing a delivery job. Now Petra has three times as much as Suki. How much did they each receive for the job?

4. If $H = \dfrac{1}{2}\left(\dfrac{1}{x} + \dfrac{1}{y}\right)$ find H when $x = 4$ and $y = 6$.

5. Given that $s - 3t = rt$, express:

  a) s in terms of r and t

  b) r in terms of s and t.

6. Solve these simultaneous equations.

  a) $7c + 3d = 29$ **b)** $2x - 3y = 7$
       $5c - 4d = 33$ $2y - 3x = -8$

Examination-style exercise 10

NON-CALCULATOR SECTION

1. **a)** Simplify the expression $6p - 3q - (2p + 2q)$. [2]

    b) Solve the equation $3(2x - 7) = 21$. [3]

    c) A kite has sides of length j cm and k cm.

    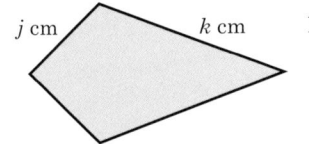

    j cm k cm   Not to scale

    i) Write down an expression in terms of j and k cm for the perimeter of the kite. [1]

    ii) The perimeter of the kite is 144 centimetres. Write down an equation in j and k. [1]

    iii) If $k = 2j$, find the value of k. [2]

    d) **i)** Use the formula $w = \dfrac{p - t}{q}$ to find the value of w

    when $p = \dfrac{5}{6}$, $t = \dfrac{2}{3}$ and $q = \dfrac{1}{2}$.

    Show all your working clearly. [3]

    ii) Rearrange the formula in part **(d)(i)** to find p in terms of w, q and t. [2]

2. Solve the simultaneous equations

    $$3x - 2y = 3$$
    $$x + 4y = 8$$ [3]

3. **a)**

    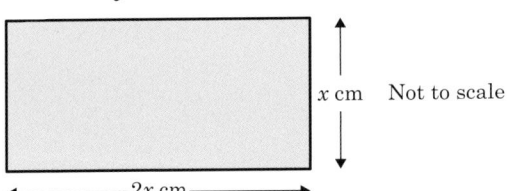

    x cm   Not to scale

    —$2x$ cm—

    The perimeter of the rectangle in the diagram above is 72 centimetres.

    i) Find the value of x. [2]

    ii) Using this value of x, calculate the area of the rectangle. Give your answer in cm^2. [2]

b)

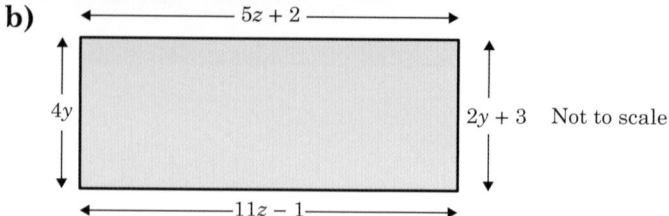

The diagram above shows another rectangle.

i) In this rectangle $4y = 2y + 3$. Solve the equation to find y. [2]

ii) Write down an equation in z. [1]

iii) Solve the equation in part **(b)(ii)** to find z. [3]

c)

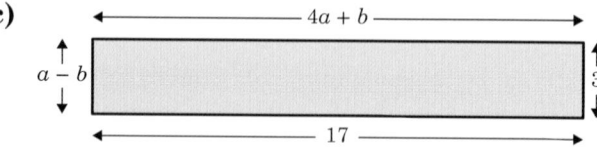

The diagram above shows another rectangle.

i) Write down two equations in a and b. [2]

ii) Solve these two equations simultaneously to find a and b. [3]

CALCULATOR SECTION

4. **a)** Kinetic energy, E, is related to mass, m, and velocity, v, by the formula

$$E = \frac{1}{2} mv^2$$

i) Calculate E when $m = 6$ and $v = 11$. [2]

ii) Calculate v when $m = 16$ and $E = 450$. [2]

iii) Make m the subject of the formula. [2]

b) Solve the simultaneous equations

$$4x + y = 13$$
$$2x + 3y = 9$$

[3]

Leonard Euler (1707–1783) was born near Basel in Switzerland but moved to St Petersburg in Russia and later to Berlin. He was a gifted mathematician and philosopher and he made significant contributions to the development of geometry, trigonometry, calculus and number theory. For example, in trigonometry he introduced the use of small letters a, b, c for the sides of a triangle and capital letters A, B, C for the angles, a notation that is widely used today. He also used r, R and s to represent the radius of the inscribed circle, the radius of the circumscribed circle and the semi-perimeter of a triangle, giving the beautiful formula $4rRs = abc$.

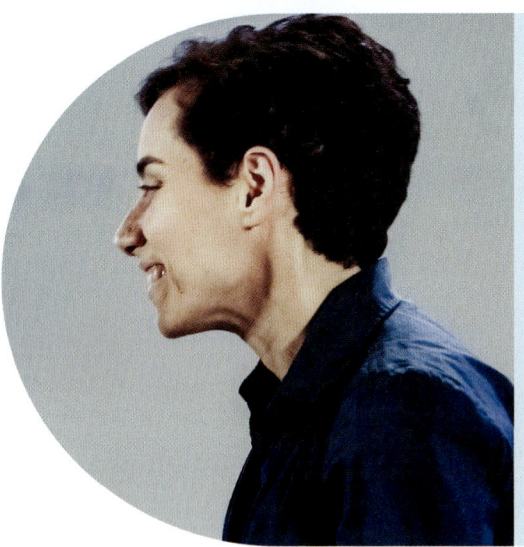

Maryam Mirzakhani (1977–2017) was the first woman ever to be awarded the Fields Medal, the most prestigious award in mathematics, for outstanding work on the geometry of curved surfaces. On a curved surface, the shortest distance between two points is not a straight line but is part of a curve called a geodesic. Mirzakhani became fascinated with surfaces which are saddle shaped, and in 2004 she solved a problem related to how many geodesics these kinds of surfaces have.

- Understand similar shapes including finding missing lengths.
- Calculate unknown sides and angles of a right-angled triangle using trigonometry or Pythagoras' theorem including solving problems in two dimensions.
- Draw and describe reflections, rotations, enlargements and translations.

11.1 Reflections and rotations

Reflection

A′B′C′D′ is the **image** of ABCD after reflection in the broken line (the mirror line).

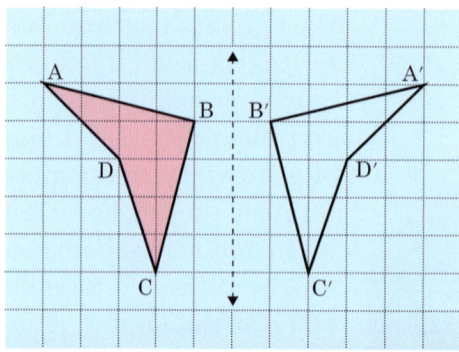

Exercise 11.1A

On squared paper draw the object and its image after reflection in the dotted line. You could use tracing paper to help you.

1.

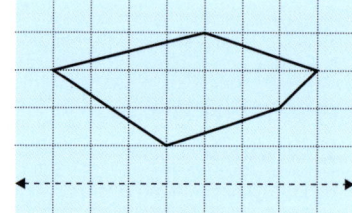

2.

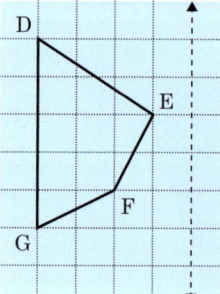

3.

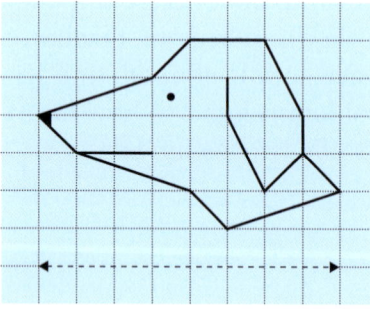

4.

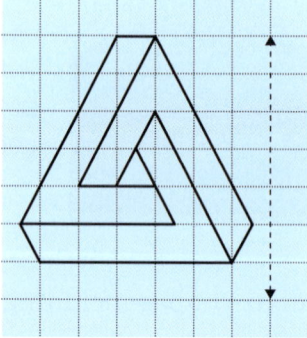

At Core level, you only need to reflect in mirror lines that are horizontal or vertical.

You could try these questions by turning your page round so that the mirror lines are all vertical.

5.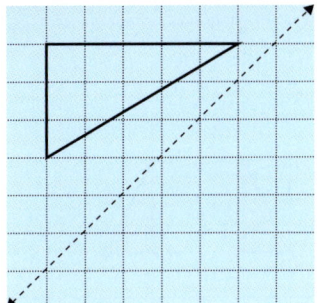

6.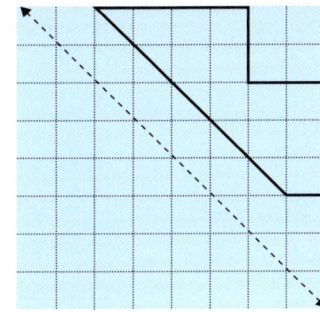

7.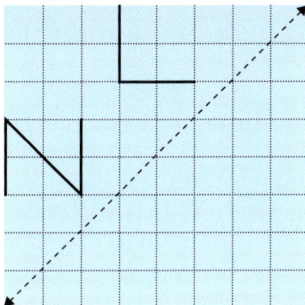

8.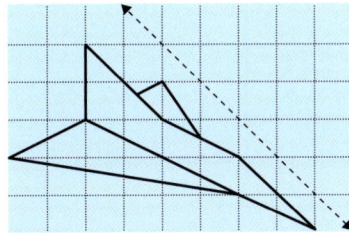

9.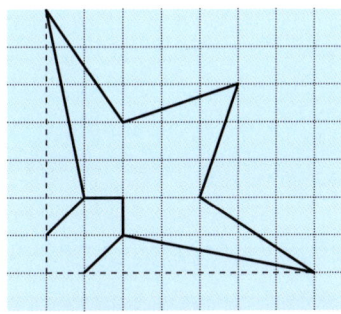

Reflect this shape in *both* of the broken lines.

Example

Reflect triangle A in

a) the *x*-axis

b) the line $x = 1$.

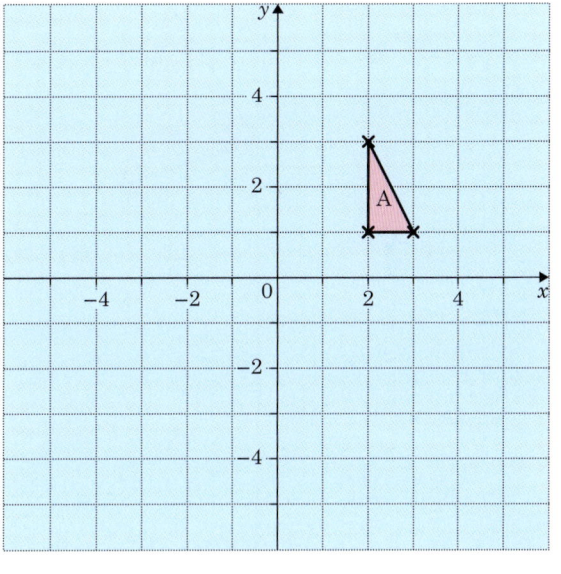

Triangle B is a reflection in the x-axis.

The line $x = 1$ is a vertical line through $x = 1$.

Triangle C is a reflection in the line $x = 1$.

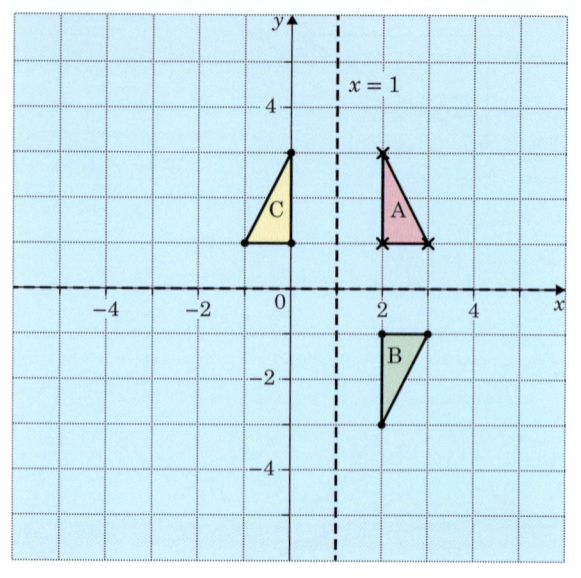

Exercise 11.1B

1. Copy the diagram below.

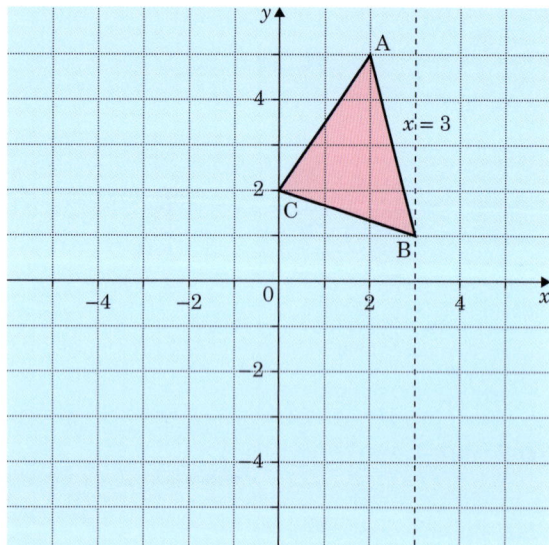

Draw the image of $\triangle ABC$ after reflection in:

a) the x-axis. Label it $\triangle 1$.

b) the y-axis. Label it $\triangle 2$.

c) the line $x = 3$. Label it $\triangle 3$.

For Questions **2** to **5** draw a pair of axes so that both x and y can take values from -7 to $+7$.

2. a) Plot and label P(7, 5), Q(7, 2), R(5, 2).

  b) Draw the lines $y = -1$, $x = 1$ and $y = x$. Use dotted lines.

  c) Draw the image of $\triangle$PQR after reflection in:
  - **i)** the line $y = -1$. Label it $\triangle$1.
  - **ii)** the line $x = 1$. Label it $\triangle$2.
  - **iii)** the line $y = x$. Label it $\triangle$3.

  d) Write down the coordinates of the image of point P in each case.

3. a) Plot and label L(7, -5), M(7, -1), N(5, -1).

  b) Draw the lines $y = -4$ and $x = 3$. Use dotted lines.

  c) Draw the image of $\triangle$LMN after reflection in:
  - **i)** the x-axis. Label it $\triangle$1.
  - **ii)** the line $x = 3$. Label it $\triangle$2.
  - **iii)** the line $y = -4$. Label it $\triangle$3.

  d) Write down the coordinates of the image of point L in each case.

4. a) Draw the line $x + y = 7$. [It passes through (0, 7) and (7, 0).]

  b) Draw $\triangle$1 at (-3, -1), (-1, -1), (-1, -4).

  c) Reflect $\triangle$1 in the y-axis onto $\triangle$2.

  d) Reflect $\triangle$2 in the x-axis onto $\triangle$3.

  e) Reflect $\triangle$3 in the line $x + y = 7$ onto $\triangle$4.

  f) Reflect $\triangle$4 in the y-axis onto $\triangle$5.

  g) Write down the coordinates of $\triangle$5.

5. a) Draw the lines $y = 2$, $x = -1$ and $y = x$.

  b) Draw $\triangle$1 at (1, -3), (-3, -3), (-3, -5).

  c) Reflect $\triangle$1 in the line $y = x$ onto $\triangle$2.

  d) Reflect $\triangle$2 in the line $y = 2$ onto $\triangle$3.

  e) Reflect $\triangle$3 in the line $x = -1$ onto $\triangle$4.

  f) Reflect $\triangle$4 in the line $y = x$ onto $\triangle$5.

  g) Write down the coordinates of $\triangle$5.

Tip

For part **(c)(iii)**, turn your page round so the mirror line is either horizontal or vertical. Note that you wouldn't be expected to do this type of reflection at Core level.

Tip

For part **(e)**, turn your page round so the mirror line is either horizontal or vertical. Note that you wouldn't be expected to do this type of reflection in your exam.

6. Find the equation of the mirror line for the reflection:

a) $\triangle 1$ onto $\triangle 2$

b) $\triangle 1$ onto $\triangle 3$

c) $\triangle 1$ onto $\triangle 4$.

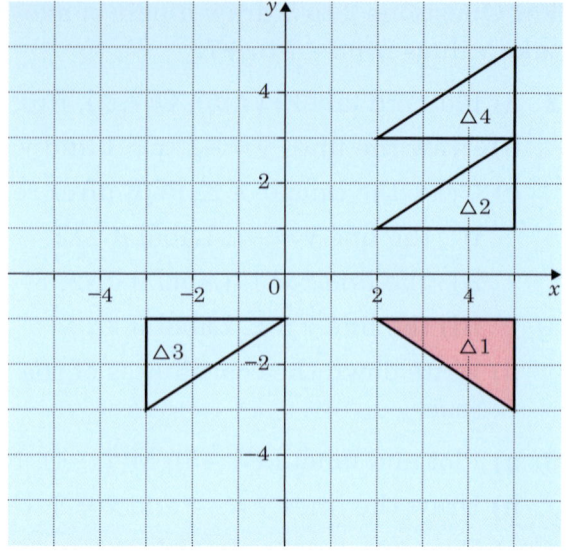

Rotation

Example

Rotate $\triangle ABC$ 90° clockwise about centre C.

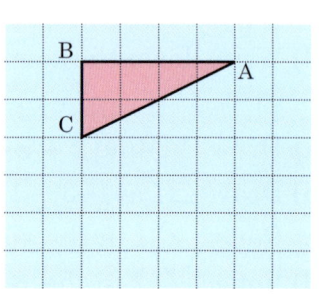

Draw $\triangle ABC$ on tracing paper and then put the tip of your pencil on C.

Turn the tracing paper 90° clockwise about C. The tracing paper now shows the position of $\triangle A'B'C'$.

$\triangle A'B'C'$ is the image of $\triangle ABC$ after a 90° clockwise rotation about centre C.

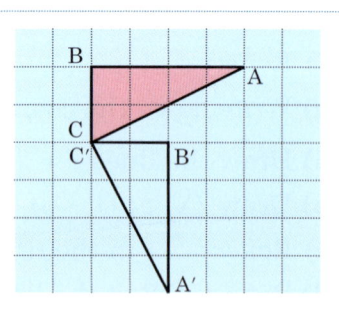

To describe a rotation, you need to give:
- the centre
- the angle
- the direction (e.g. clockwise).

Exercise 11.1C

Draw the object and its image under the rotation given.

Take O as the centre of rotation in each case.

1.

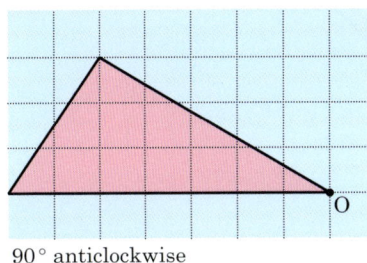

90° anticlockwise

2.

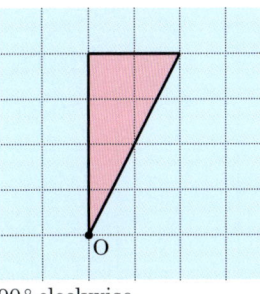

90° clockwise

3.

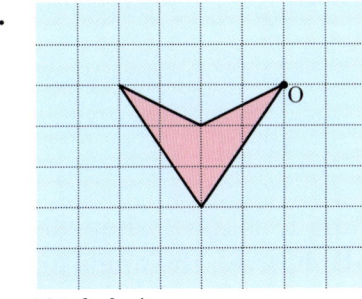

90° clockwise

4.

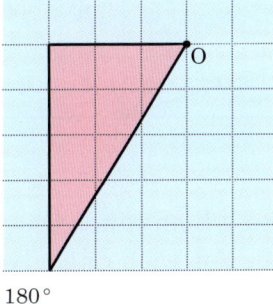

180°

5.

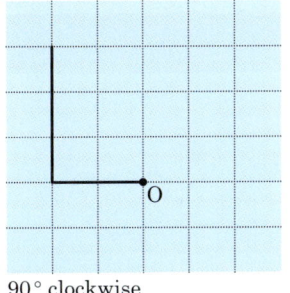

90° clockwise

6.

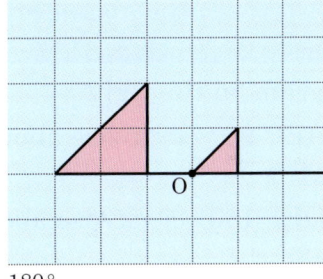

180°

You need to be able to rotate shapes about the origin, a vertex of the shape, or the midpoint of one of the sides.

If you want to extend yourself and think about other centres of rotation, try Question **7**.

7. The shape on the right has been rotated about several different centres to form the pattern below.

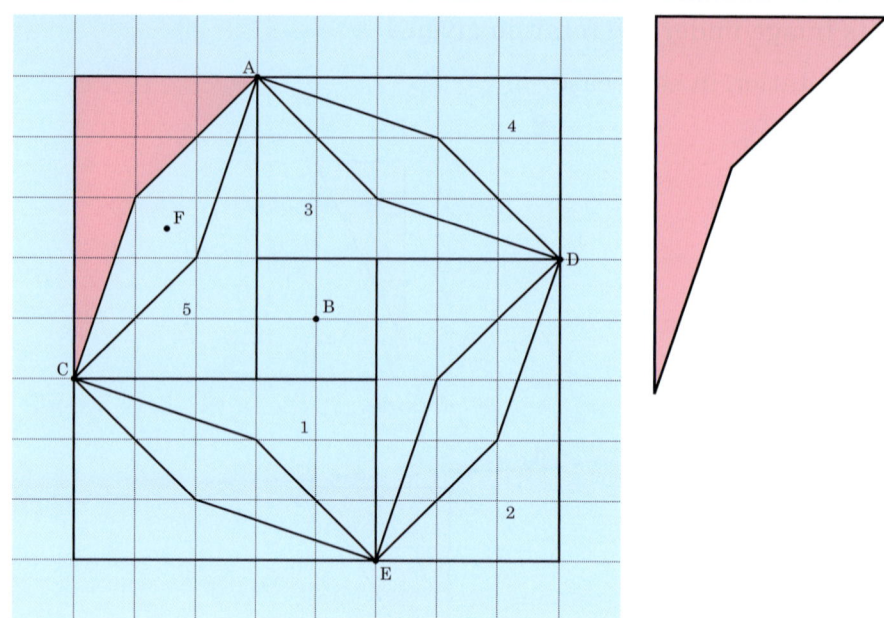

Describe the rotation which takes the shaded shape onto shape 1, shape 2, shape 3, shape 4 and shape 5. For each one, give the centre (A, B, C, D, E or F), the angle and the direction of the rotation. [e.g. 'centre C, 90°, clockwise'.]

Exercise 11.1D

1. Copy the diagram on the right.

a) Rotate △ABC 90° clockwise about (0, 0). Label it △1.

b) Rotate △DEF 180° clockwise about (0, 0). Label it △2.

c) Rotate △GHI 90° clockwise about (0, 0). Label it △3.

For Questions **2** and **3** draw a pair of axes with values of x and y from -7 to $+7$.

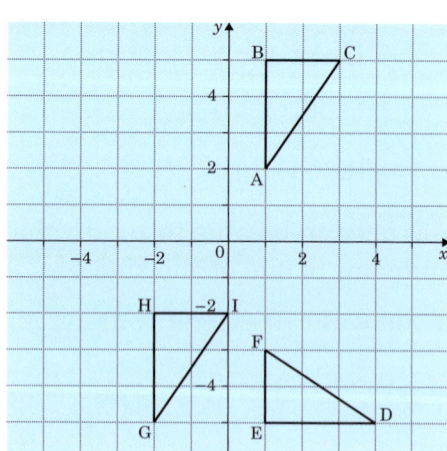

2. a) Plot △1 at (2, 3), (6, 3), (3, 6).

b) Rotate △1 90° clockwise about (2, 3) onto △2.

c) Rotate △1 180° about (0, 0) onto △3.

d) Rotate △1 90° anticlockwise about (0, 0) onto △4.

3. a) Plot $\triangle$1 at (4, 4), (6, 6), (2, 6).

   b) Rotate $\triangle$1 90° anticlockwise about (4, 4) onto $\triangle$2.

   c) Rotate $\triangle$1 180° about (0, 0) onto $\triangle$3.

   d) Rotate $\triangle$1 90° anticlockwise about (5, 5) onto $\triangle$4.

Finding the centre of a rotation

Example

Find the centre of rotation which takes the shaded shape onto the unshaded shape.

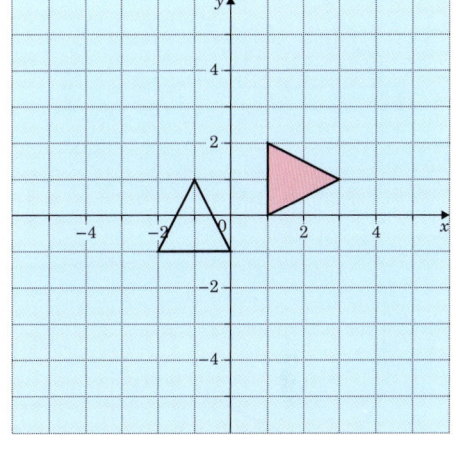

Use a piece of tracing paper to draw the shaded shape.

Place your pencil on a coordinate point and rotate the tracing paper so your drawn shape fits over the unshaded shape.

Once you have found the correct coordinate, mark it with a cross.

The centre of rotation is (1, −1).

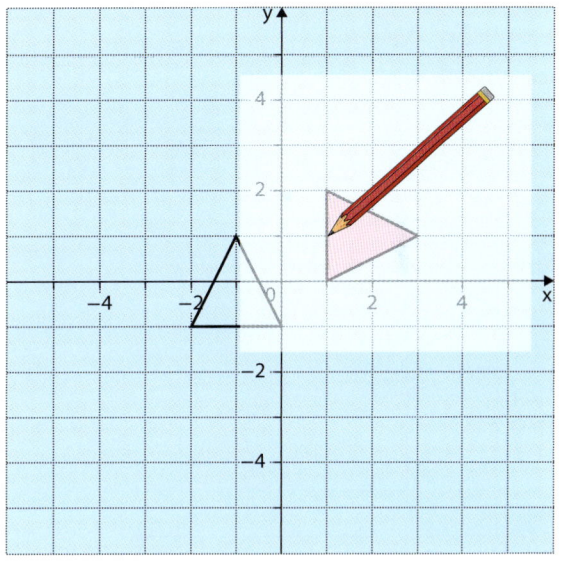

Exercise 11.1E

In each of these questions, find the coordinates of the centre of the rotation which takes the shaded shape onto the unshaded shape.

1.

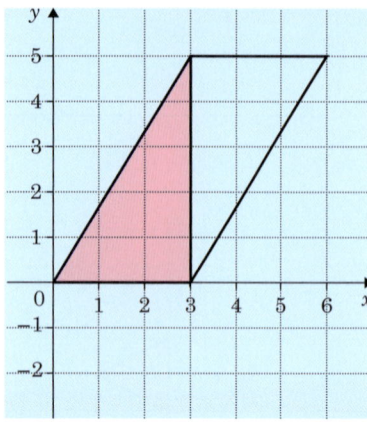

2.

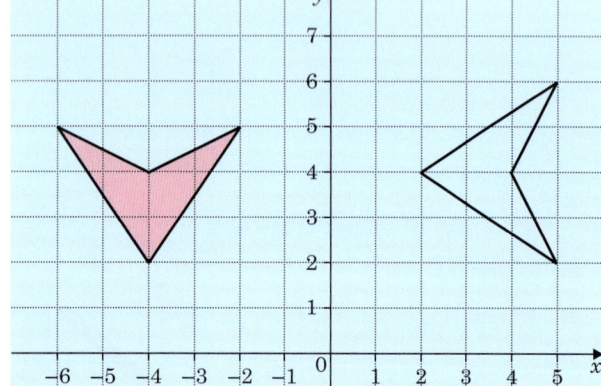

3.

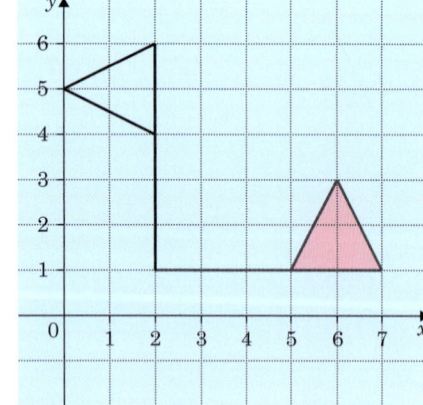

11.2 Enlargements

A

B

C

Photo A has been enlarged to give photos B and C.

The shape of the face is exactly the same in all the pictures.

The length and the width must both be multiplied by the same number so that the image isn't distorted.

Pictures, and other shapes, where this is true are called **similar**.

Photo A measures 22 mm by 27 mm.

Photo B measures 44 mm by 54 mm.

Photo C measures 66 mm by 81 mm.

From A to B both the width and the height have been multiplied by 2:

$$22 \times 2 = 44$$
$$27 \times 2 = 54$$

You say B is an **enlargement** of A with a **scale factor** of 2.

Similarly C is an enlargement of A with a scale factor of 3 since

$$\frac{66}{22} = \frac{81}{27} = 3$$

Also C is an enlargement of B with a scale factor of $1\frac{1}{2}$.

The scale factor of an enlargement can be found by dividing corresponding lengths on two pictures.

In this enlargement the scale factor is $\dfrac{21}{14} = 1.5$

You can work out missing lengths in an enlarged shape if you know the scale factor of enlargement.

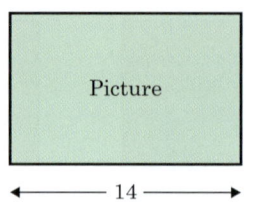

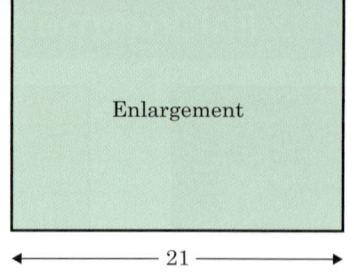

Example

The diagram shows two triangles, A and B.

B is an enlargement of A with scale factor $\dfrac{7.5}{3} = 2.5$.

Find the value of x and the value of y.

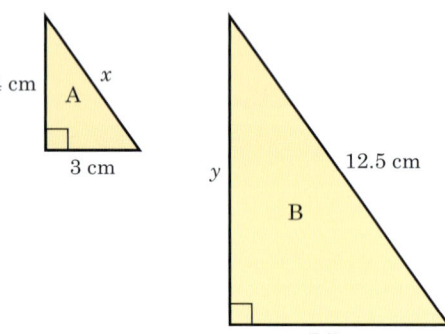

> **Tip**
>
> To find an unknown length in the smaller shape you divide by the scale factor.

$\dfrac{12.5}{2.5} = 5$ so $x = 5$ cm

$4 \times 2.5 = 10$ so $y = 10$ cm

The scale factor is 2.5 so to find x, divide 12.5 by 2.5.

To find y, multiply 4 by 2.5.

> **Tip**
>
> To find an unknown length in the larger shape you multiply by the scale factor.

Exercise 11.2A

1. This picture is to be enlarged and we want the enlargement to fit exactly in a frame. Which of the following frames will the picture fit?

   Write 'yes' or 'no'.

   a) 100 mm by 76 mm

   b) 110 mm by 76 mm

   c) 150 mm by 114 mm

   d) 75 mm by 57 mm

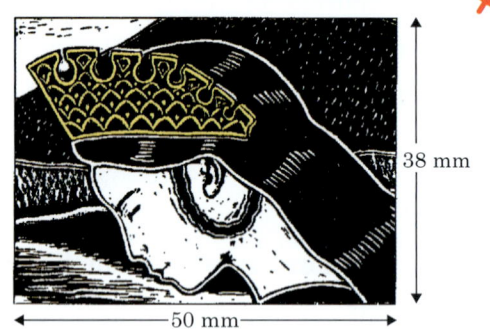

2. This picture is to be enlarged so that it fits exactly into the frame. Find the length x.

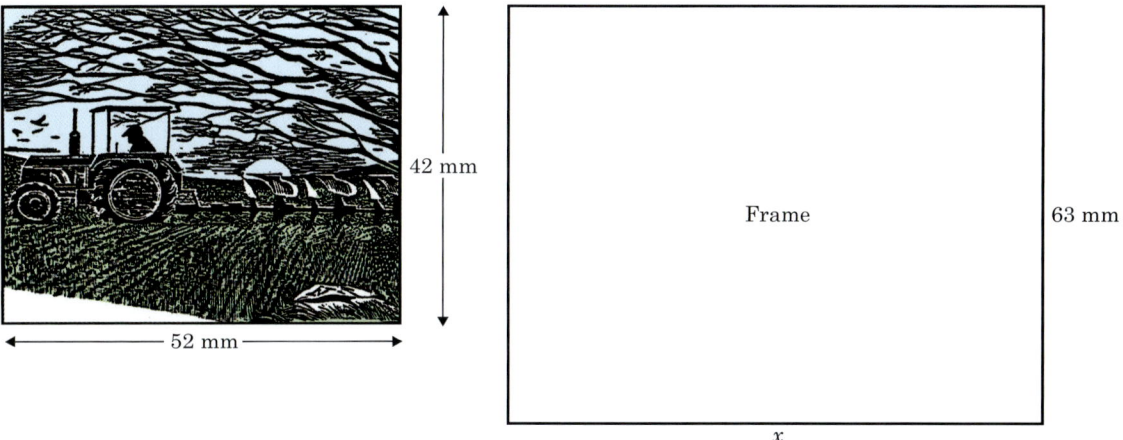

3. This picture is enlarged or reduced to fit into each of the frames shown. Calculate y and z.

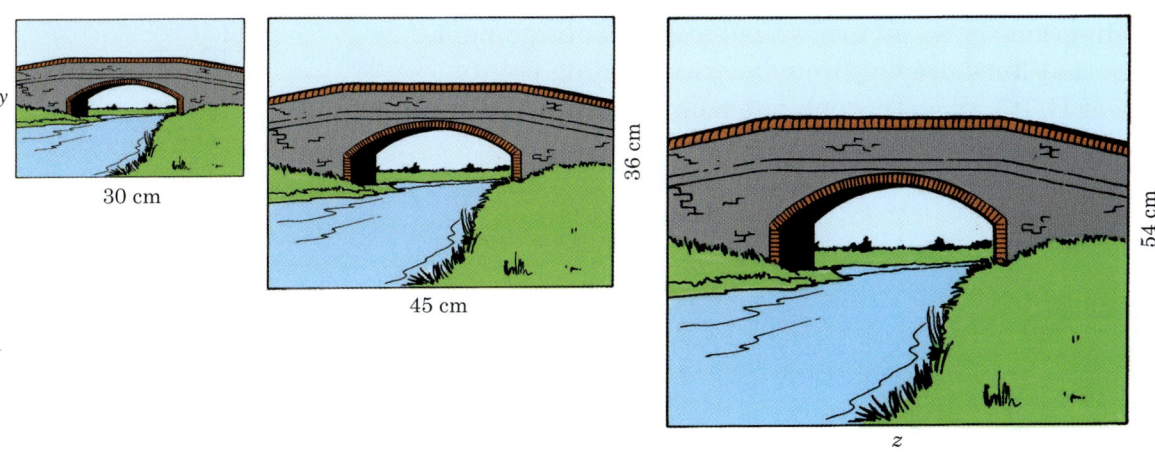

4. Here is the start of an enlargement with scale factor 2 of a house.

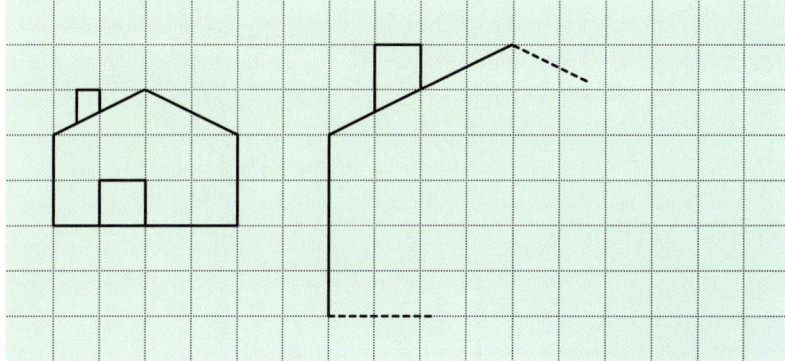

Copy and complete the enlargement in your book (use squared paper).

5. Draw an enlargement of this figure with scale factor 3.

6. Draw an enlargement of this shape with:

a) scale factor 2

b) scale factor 3.

Measure the angles a and b on each enlargement.

Write the correct version of this sentence: 'In an enlargement, the angles in a shape are changed/unchanged'.

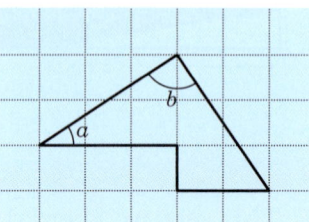

7. This diagram shows an arrowhead and its enlargement. Notice that lines drawn through corresponding points (A, A′ or B, B′) all go through one point O.

This point is called the **centre of enlargement**.

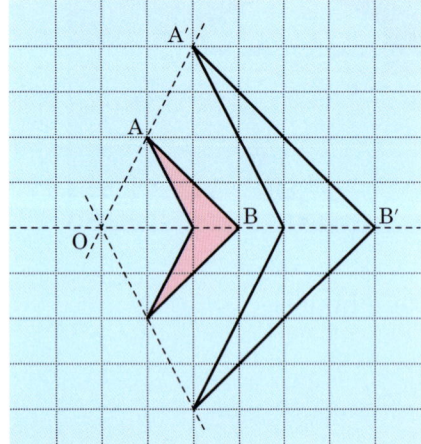

> **Tip**
>
> The dotted lines in the diagram in Question **7** are called **construction lines**.

Copy and complete:

OA′ = _____ × OA

OB′ = _____ × OB.

8. Copy this shape and its enlargement. Draw construction lines to find the centre of enlargement.

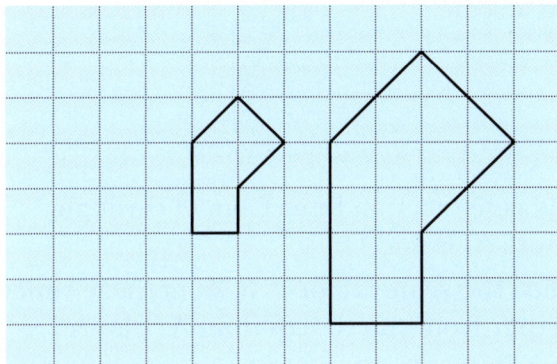

9. In this diagram, △1 is an enlargement with scale factor 2 of the shaded triangle, with O_1 as centre of enlargement.

Also △2 is an enlargement with scale factor 3 of the shaded triangle, with O_2 as centre of enlargement.

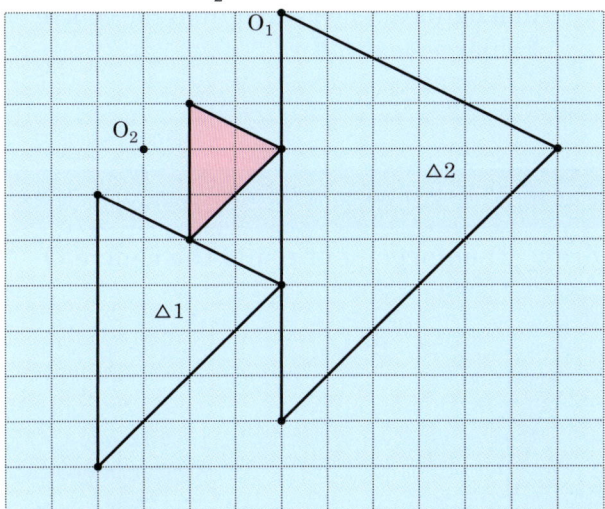

a) Copy the diagram and draw construction lines to find the centre of enlargement from △1 onto △2.

b) What is the scale factor for the enlargement △1 onto △2?

Tip

Leave space on the left side of the diagram.

To give a mathematical description of an enlargement you need to state:

• the scale factor
• the centre of enlargement.

Example

Enlarge triangle ABC onto triangle A'B'C' with a scale factor of 3 and centre O.

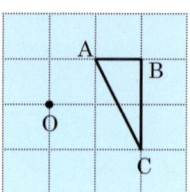

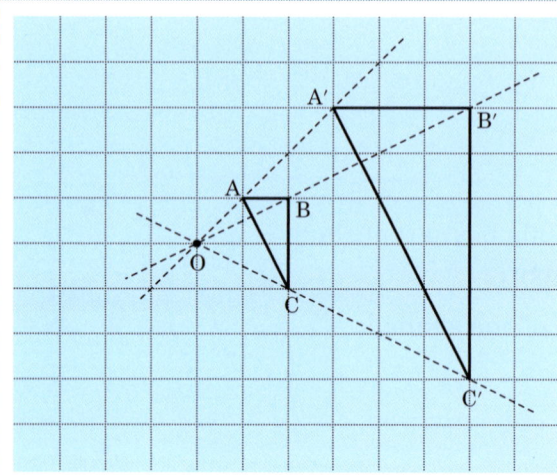

Add construction lines from O through points A, B and C.

Since the scale factor is 3, A' is three times farther from the centre than A. Plot A'.

Repeat this for B' and C'.

Then draw in the completed triangle.

> **Tip**
>
> OA' = 3 × OA; OB' = 3 × OB; OC' = 3 × OC.
>
> All lengths are measured from the centre of enlargement.

Exercise 11.2B

In Questions **1** to **6** copy the diagram and draw an enlargement using the centre O and the scale factor given.

1.

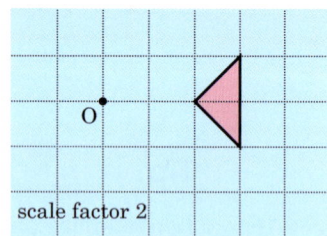

scale factor 2

2.

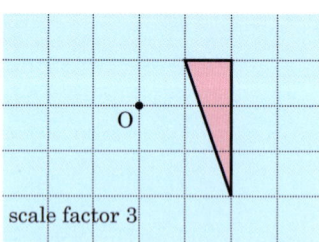

scale factor 3

3.

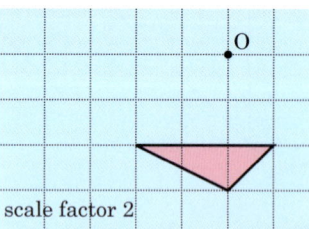

scale factor 2

4.

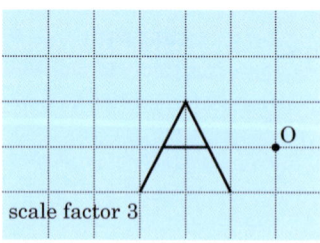

scale factor 3

5.

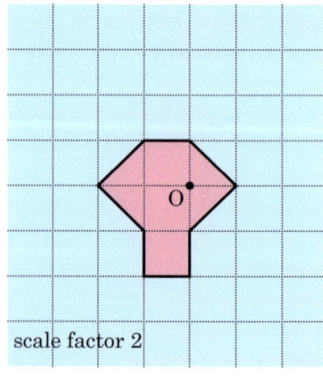

scale factor 2

6.

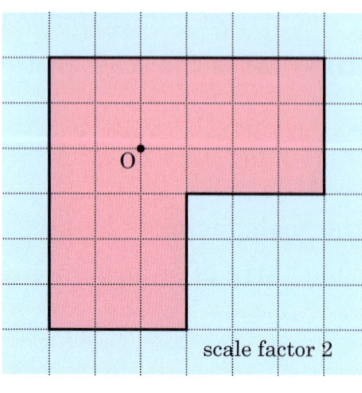

scale factor 2

7. a) Copy the diagram on the right.

  b) Draw the image of △1 after
  enlargement with scale factor 3,
  centre (0, 0). Label the image △4.

  c) Draw the image of △2 after
  enlargement with scale factor 2,
  centre (−1, 3). Label the image △5.

  d) Draw the image of △3 after
  enlargement with scale factor 2,
  centre (−1, −5). Label the
  image △6.

  e) Write down the coordinates of the
  'pointed ends' of △4, △5 and △6.

  [The 'pointed end' is the vertex of
  the triangle with the smallest
  angle.]

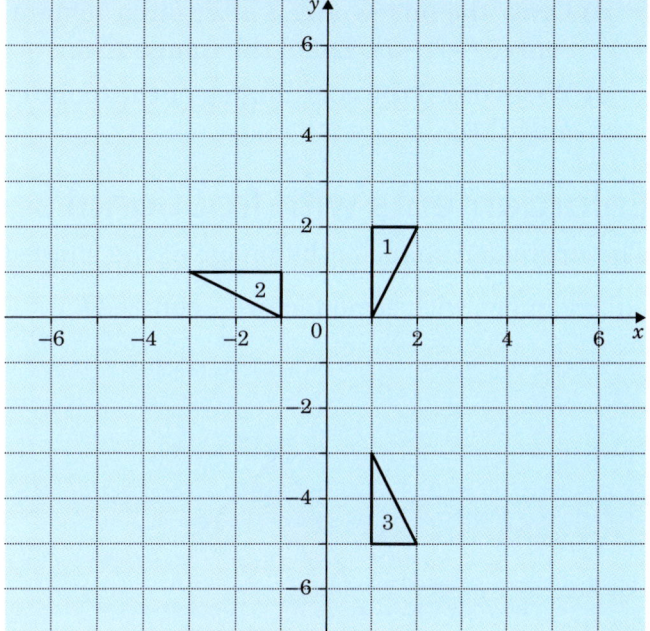

For Questions **8** and **9** draw a pair of axes with values from −7 to +7.

8. a) Plot and label the triangles with the following vertices:

  △1: (5, 5), (5, 7), (4, 7)
  △2: (−6, −5), (−3, −5), (−3, −4)
  △3: (1, −4), (1, −6), (2, −6).

  b) Draw the image of △1 after enlargement with scale factor 2,
  centre (7, 7). Label the image △4.

  c) Draw the image of △2 after enlargement with scale factor 3,
  centre (−6, −7). Label the image △5.

  d) Draw the image of △3 after enlargement with scale factor 2,
  centre (−1, −5). Label the image △6.

  e) Write down the coordinates of the 'pointed ends' of △4, △5
  and △6.

9. a) Plot and label the triangles with the following vertices:

  △1: (5, 3), (5, 6), (4, 6)
  △2: (4, −3), (1, −3), (1, −2)
  △3: (−4, −7), (−7, −7), (−7, −6).

  b) Draw the image of △1 after enlargement with scale factor 2,
  centre (7, 7). Label the image △4.

  c) Draw the image of △2 after enlargement with scale factor 3,
  centre (5, −4). Label the image △5.

d) Draw the image of △3 after enlargement with scale factor 4, centre (−7, −7). Label the image △6.

e) Write down the coordinates of the 'pointed ends' of △4, △5 and △6.

Enlargements with fractional scale factors

The unshaded shape is the image of the shaded shape after an enlargement with scale factor $\frac{1}{2}$, centre O.

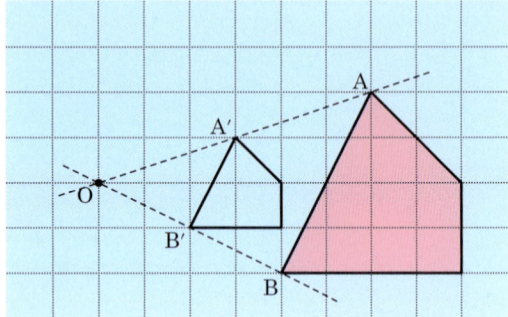

Note: $OA' = \frac{1}{2} \times OA$

$OB' = \frac{1}{2} \times OB$

Even though the shape has undergone a reduction, mathematicians prefer to call it an enlargement with a scale factor that is a proper fraction.

Example

Enlarge this shape by scale factor $\frac{1}{3}$ using the centre of enlargement marked O.

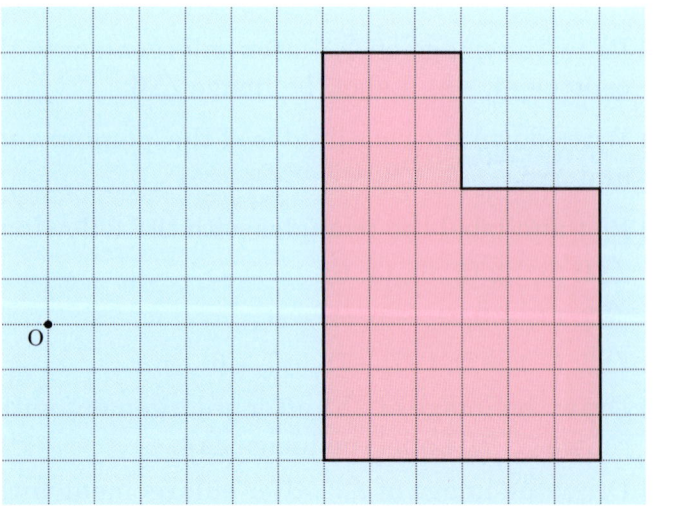

Draw construction lines from the centre of enlargement to the vertices of the given shape.

Each vertex on the image is $\frac{1}{3}$ of the distance from the centre of enlargement to the original vertex.

Once you have plotted the vertices of the image, join them up.

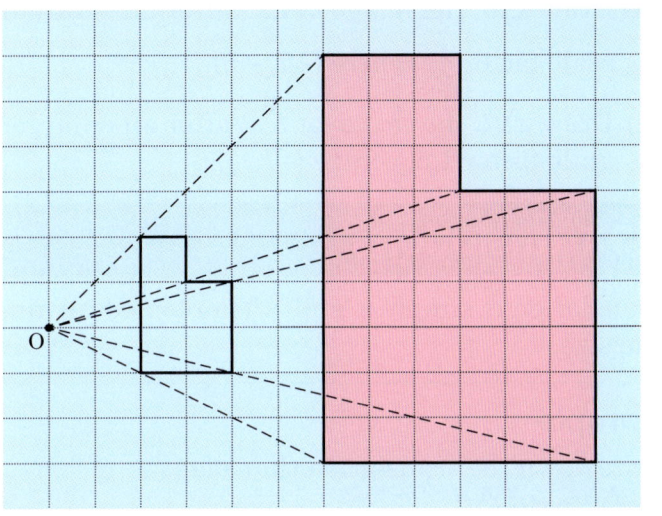

Exercise 11.2C

In Questions **1** to **3** copy the diagram and draw an enlargement using the centre O and the scale factor given.

1.

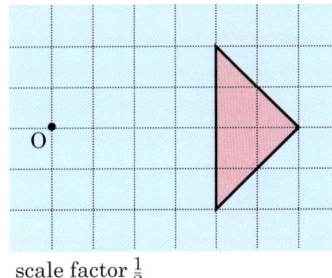

scale factor $\frac{1}{2}$

2.

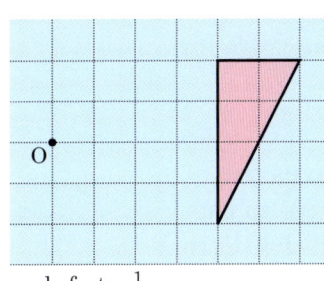

scale factor $\frac{1}{2}$

3.

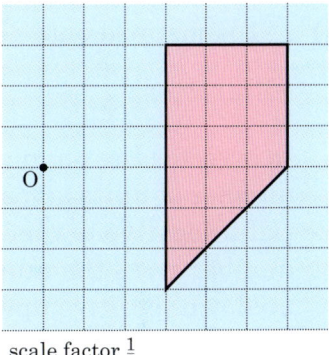

scale factor $\frac{1}{3}$

For Question **4** draw a pair of axes with values from -7 to $+7$.

4. a) Plot and label the triangles with the following vertices:

  $\triangle$1: (7, 6), (1, 6), (1, 3)

  $\triangle$2: (7, −1), (7, −7), (3, −7)

  $\triangle$3: (−5, 7), (−5, 1), (−7, 1).

 b) Draw $\triangle$4, the image of $\triangle$1 after an enlargement with scale factor $\frac{1}{3}$, centre (−2, 0).

 c) Draw $\triangle$5, the image of $\triangle$2 after an enlargement with scale factor $\frac{1}{2}$, centre (−5, −7).

d) Draw △6, the image of △3 after an enlargement with scale factor $\frac{1}{2}$, centre $(-7, -5)$.

e) Draw △7, the image of △3 after an enlargement with scale factor 1.5, centre $(-7, -7)$.

11.3 Translations

A translation is simply a 'shift'. There is no turning or reflection and the object stays the same size.

Example

Write down the translation that maps
a) △1 onto △2
b) △2 onto △3
c) △3 onto △2.

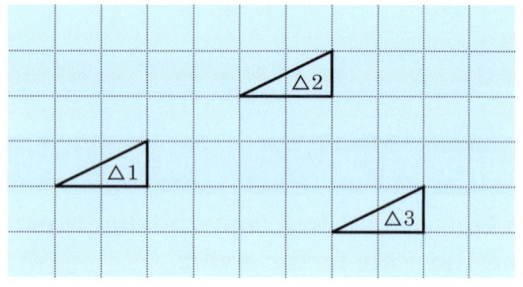

a) △1 is mapped onto △2 by the translation with vector $\begin{pmatrix} 4 \\ 2 \end{pmatrix}$.

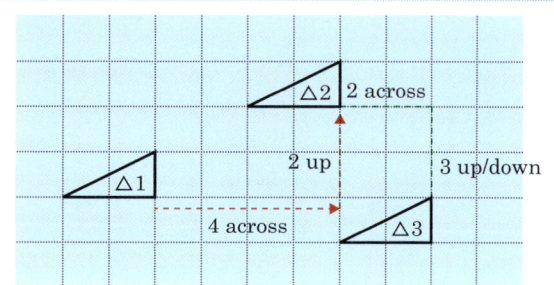

In a vector the top number gives the number of units across (positive to the right/negative to the left) and the bottom number gives the number of units up/down (positive upwards/negative downwards).

So $\begin{pmatrix} 4 \\ 2 \end{pmatrix}$ is $\begin{array}{l} 4 \text{ across } \rightarrow \\ 2 \text{ up } \uparrow \end{array}$

b) △2 is mapped onto △3 by the translation with vector $\begin{pmatrix} 2 \\ -3 \end{pmatrix}$. $\begin{pmatrix} 2 \\ -3 \end{pmatrix}$ is $\begin{array}{l} 2 \text{ across } \rightarrow \\ 3 \text{ down } \downarrow \end{array}$

c) △3 is mapped onto △2 by the translation with vector $\begin{pmatrix} -2 \\ 3 \end{pmatrix}$. $\begin{pmatrix} -2 \\ 3 \end{pmatrix}$ is $\begin{array}{l} 2 \text{ across } \leftarrow \\ 3 \text{ up } \uparrow \end{array}$

Exercise 11.3A

1. Look at the diagram shown.

   Write down the vector for each of the following translations:

   a) $\triangle 1 \rightarrow \triangle 2$ **b)** $\triangle 1 \rightarrow \triangle 3$ **c)** $\triangle 1 \rightarrow \triangle 4$ **d)** $\triangle 1 \rightarrow \triangle 5$

   e) $\triangle 1 \rightarrow \triangle 6$ **f)** $\triangle 6 \rightarrow \triangle 5$ **g)** $\triangle 1 \rightarrow \triangle 7$ **h)** $\triangle 2 \rightarrow \triangle 3$

   i) $\triangle 2 \rightarrow \triangle 4$ **j)** $\triangle 2 \rightarrow \triangle 5$ **k)** $\triangle 2 \rightarrow \triangle 6$ **l)** $\triangle 2 \rightarrow \triangle 7$

   m) $\triangle 3 \rightarrow \triangle 5$ **n)** $\triangle 7 \rightarrow \triangle 2$

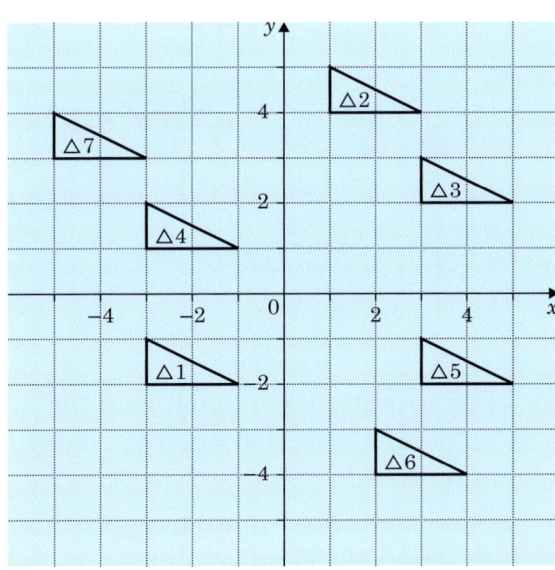

For Question **2** draw a pair of axes with values from −7 to +7.

2. **a)** Draw $\triangle 1$ with coordinates (1, 1), (1, 4) and (3, 1).

   b) Translate $\triangle 1$ using vector $\begin{pmatrix} 3 \\ 2 \end{pmatrix}$. Label the image $\triangle 2$.

   c) Translate $\triangle 1$ using vector $\begin{pmatrix} -4 \\ -1 \end{pmatrix}$. Label the image $\triangle 3$.

   d) Translate $\triangle 1$ using vector $\begin{pmatrix} -5 \\ 3 \end{pmatrix}$. Label the image $\triangle 4$.

   e) Translate $\triangle 1$ using vector $\begin{pmatrix} 2 \\ -3 \end{pmatrix}$. Label the image $\triangle 5$.

Describing transformations

Example

Describe fully the transformation that maps △1 onto:

a) △2

b) △3

c) △4

d) △5.

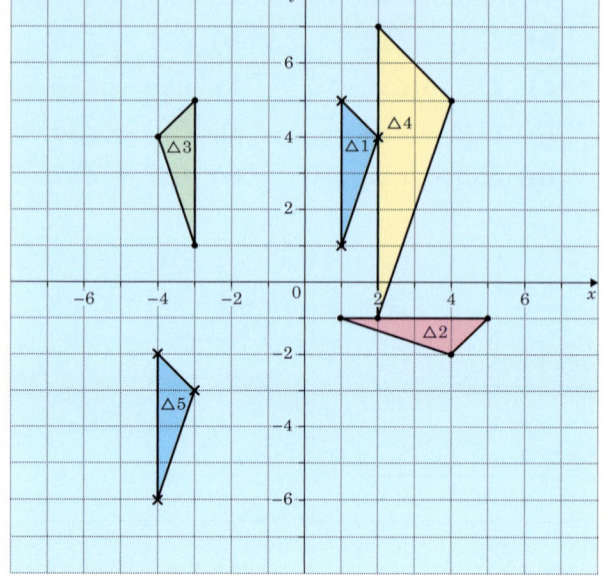

a) △2 is a rotation of 90° clockwise about (0, 0).   For a rotation, you must give the angle, the direction and the centre of rotation.

b) △3 is a reflection in the line $x = -1$.   For a reflection, you must give the equation of the mirror line.

c) △4 is an enlargement, scale factor 2, centre (0, 3).   For an enlargement, you must give the scale factor and centre of enlargement.

d) △5 is a translation using vector $\begin{pmatrix} -5 \\ -7 \end{pmatrix}$.   For a translation, you must give the vector.

Exercise 11.3B

1. **a)** Reflect shape A in line 1 onto shape B.

   b) What transformation will map shape B onto shape C?

   c) What transformation will map shape A onto shape C?

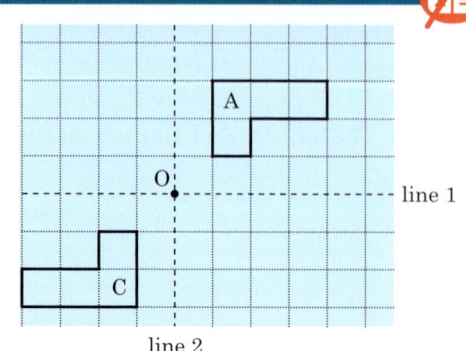

2. **a)** Rotate △D 90° clockwise about (0, 0). Label the
    image △E.
   b) What transformation will map △E onto △F?
   c) What transformation will map △D onto △F?

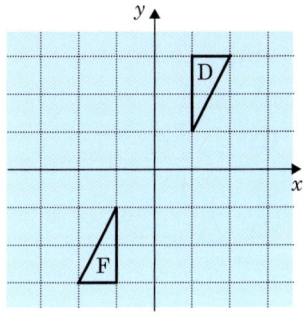

3. **a)** Draw △X.
   b) Translate △X using vector $\begin{pmatrix} 4 \\ 0 \end{pmatrix}$
    onto △Y.
   c) What transformation maps △Y
    onto △Z?
   d) What transformation maps △X
    onto △Z?

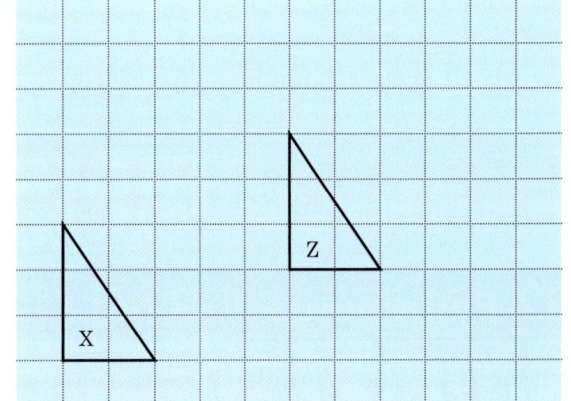

4. Mariette says that the transformation that maps △A
   onto △D is the same as the transformation that maps △B
   onto △C.

   Is Mariette correct? Explain your answer.

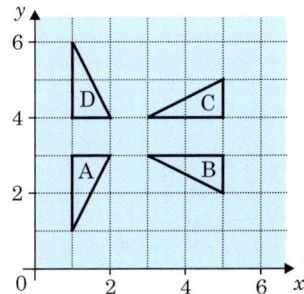

5. Describe fully the following transformations:
   a) △A onto △B
   b) △B onto △C
   c) △C onto △D
   d) △C onto △E.

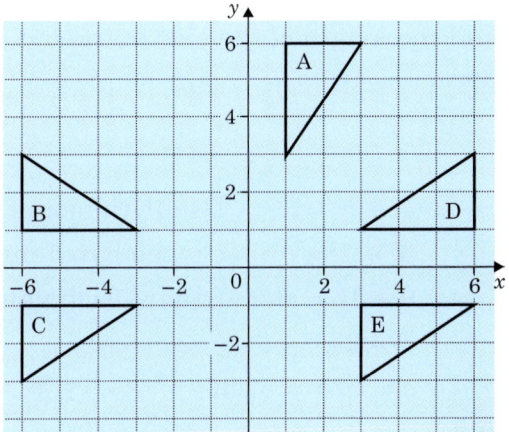

6. Describe fully the transformation that maps $\triangle 1$ onto:

   a) $\triangle 2$

   b) $\triangle 3$

   c) $\triangle 4$

   d) $\triangle 5$.

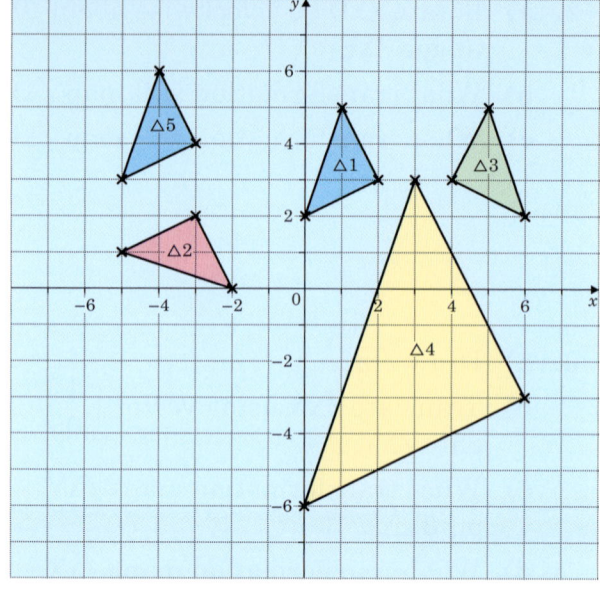

11.4 Pythagoras' theorem

In a right-angled triangle the square on the **hypotenuse** is equal to the sum of the squares on the other two sides.

$a^2 + b^2 = c^2$

The *converse* is also true: 'If the square on one side of a triangle is equal to the sum of the squares on the other two sides, then the triangle is right-angled'.

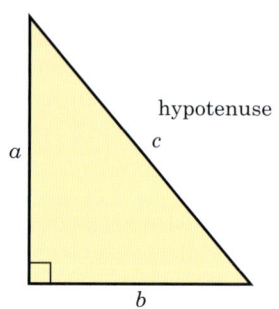

Example 1

Find the length of the side marked c.

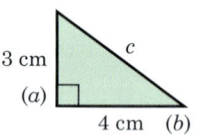

$a^2 + b^2 = c^2$ Use Pythagoras' theorem.

$3^2 + 4^2 = c^2$ c is the hypotenuse.

$9 + 16 = c^2$ Square each value you know.

$c^2 = 25$

$c = \sqrt{25} = 5$ cm

Example 2

Find the length of the side marked d.

Give your answer correct to 3 significant figures.

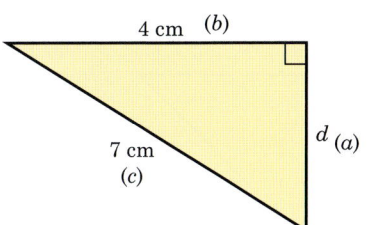

$a^2 + b^2 = c^2$ — Use Pythagoras' theorem.

$d^2 + 4^2 = 7^2$ — d is one of the shorter sides.

$d^2 = 49 - 16$ — Square each value you know and rearrange.

$d = 33$

$d = \sqrt{33}$ — Find the square root.

$= 5.74$ cm (3 s.f.) Round as required.

You might be asked to give your answer as an exact square root, or round it to a given number of significant figures.

Example 3

A triangle has sides of length 12 cm, 5 cm and 13 cm.
Show that the triangle has a right angle.

Work out $a^2 + b^2$:

$5^2 + 12^2 = 25 + 144 = 169$ — a and b are the two shorter sides.

Work out c^2:

$13^2 = 169$ — c is the longest side, the hypotenuse.

Since $a^2 + b^2 = c^2$, the triangle is right-angled.

Exercise 11.4A

In Questions **1** to **4**, find x. Give your answers as exact values.

1.

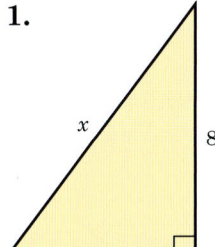

2.

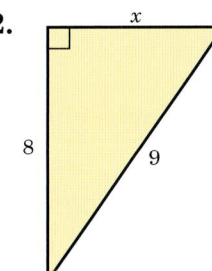

3.

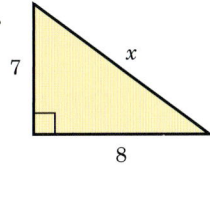

4.

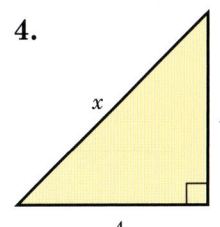

In Questions **5** to **8**, find x. Give your answers correct to 3 significant figures.

5. 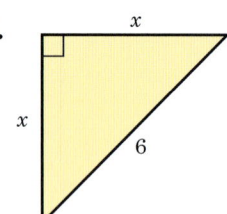   **6.** 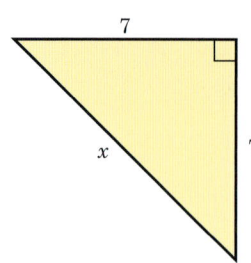   **7.** 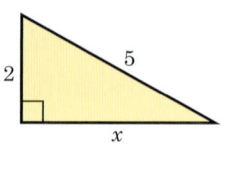   **8.**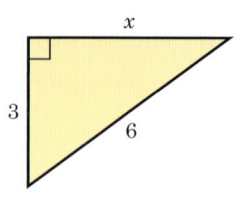

9. Find the length of a diagonal of a rectangle of length 9 cm and width 4 cm.

10. A square has diagonals of length 10 cm. Find the length of the sides of the square.

11. A 4 m ladder rests against a vertical wall with its foot 2 m from the wall. How far up the wall does the ladder reach?

12. A ship sails 20 km due north and then 35 km due east. How far is it from its starting point?

13. A thin wire of length 18 cm is bent in the shape shown.

Calculate the direct distance from A to B.

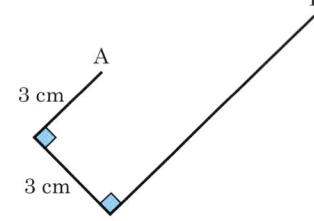

14. In the diagram A is (1, 2) and B is (6, 4).

Work out the length AB. [First find the length of AN and BN.]

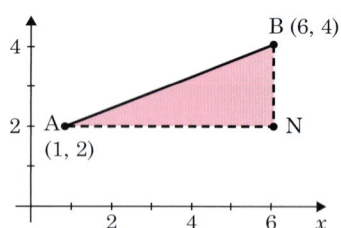

15. On squared paper plot P(1, 3), Q(6, 0), R(6, 6). Find the lengths of the sides of triangle PQR. Is the triangle isosceles?

Questions **16** to **21** are more difficult. In each case, find the value of x.

16. 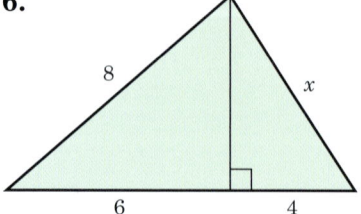   **17.** 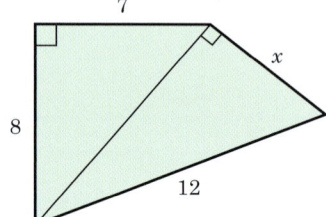   **18.**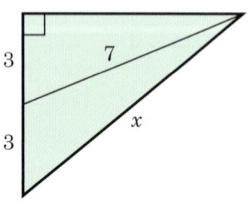

19.

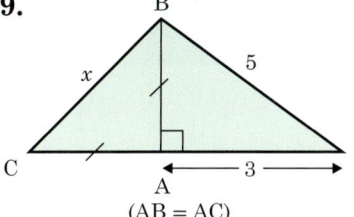

20.

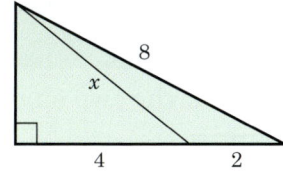

21.

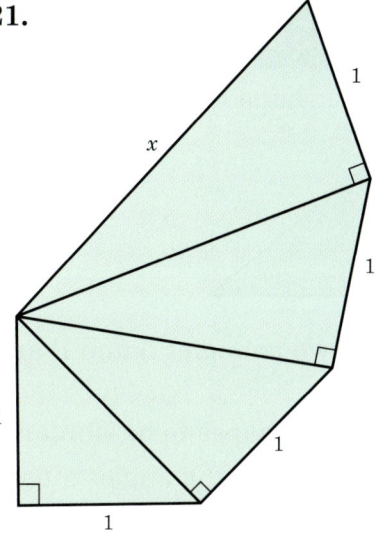

22. A well-known right-angled triangle is the 3, 4, 5 triangle [$3^2 + 4^2 = 5^2$]. It is interesting to look at other right-angled triangles where all the sides are whole numbers.

a) i) Find c if $a = 5$, $b = 12$.

   ii) Find c if $a = 7$, $b = 24$.

   iii) Find a if $c = 41$, $b = 40$.

b) Write the results in a table.

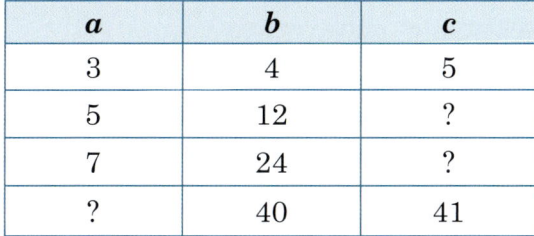

a	b	c
3	4	5
5	12	?
7	24	?
?	40	41

c) Look at the sequences in the 'a' column and in the 'b' column. Also write down the connection between b and c for each triangle.

d) Predict the next three sets of values of a, b, c. Check to see if they really do form right-angled triangles.

23. Alexis and Philip were arguing about a triangle that had sides 10 cm, 11 cm and 15 cm. Alexis said the triangle had a right angle, and Philip said that it did not. Who was correct? Show working to explain.

11.5 Similar shapes

If one shape is an enlargement of another, the two shapes are mathematically **similar**.

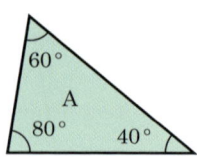
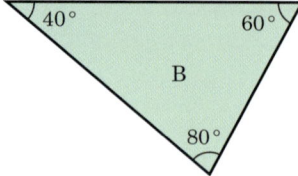

The two triangles A and B are similar because they have the same angles.

For two shapes to be similar:

- corresponding angles must be equal
- corresponding edges must be in the same proportion, i.e. enlarged with the same scale factor.

Example 1

The two quadrilaterals C and D are similar.
Calculate the length of the side marked x.

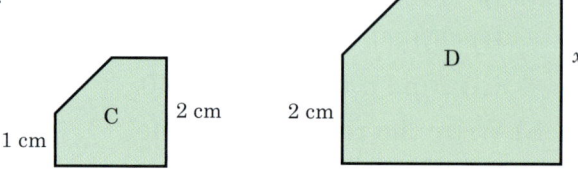

Scale factor of enlargement = 2 ÷ 1 = 2

So, $x = 2 \times 2 = 4$ cm

Calculate the scale factor of enlargement from the information given.

All the edges of shape D are twice as long as the edges of shape C.

Example 2

Explain why rectangles E and F are not similar.

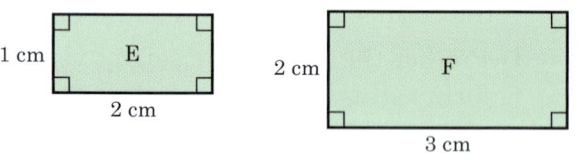

Scale factor for widths: 2 ÷ 1 = 2

Work out the scale factors for the given pairs of sides.

Scale factor for lengths: 3 ÷ 2 = 1.5

Since the given pairs of sides have been enlarged with different scale factors, the rectangles are not similar.

Example 3

Triangles A and B are similar.

All measurements are in centimetres.

Find the value of x.

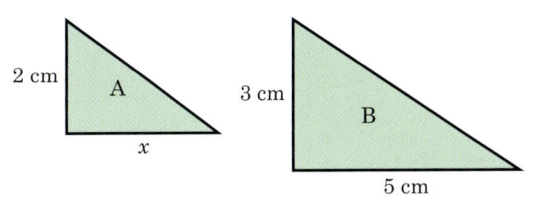

The scale factor of the enlargement is

$$\frac{3}{2} = 1.5$$

So, $x = \dfrac{5}{1.5} = 3\dfrac{1}{3}$ cm

Triangle B is an enlargement of triangle A.

Exercise 11.5A

1. Which of the shapes B, C, D is/are similar to shape A?

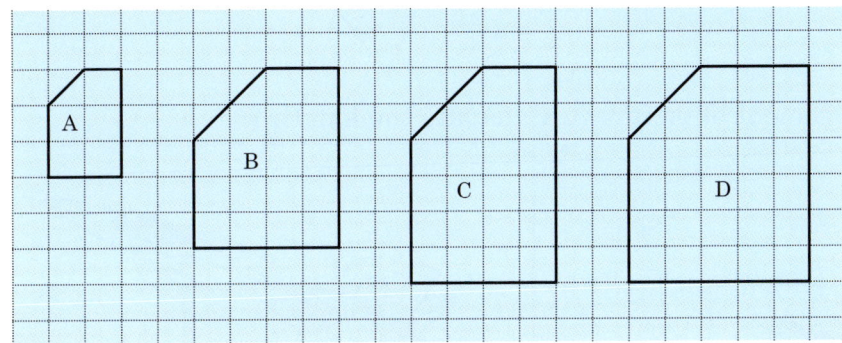

In Questions **2** to **7**, find the sides marked with letters; all lengths are given in cm. The pairs of shapes are similar.

2.

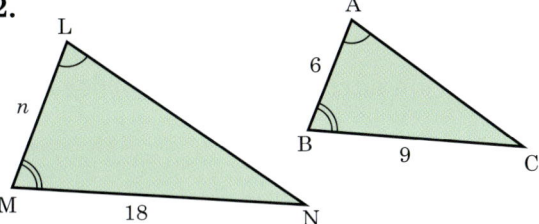

3.

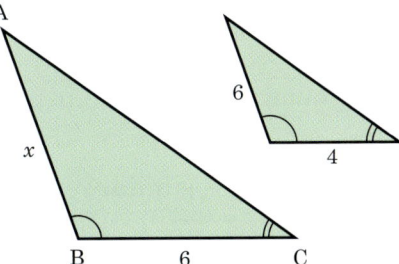

4.

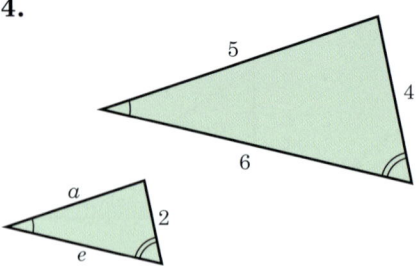

5.

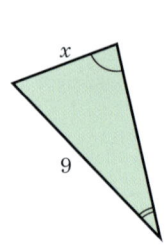

6.

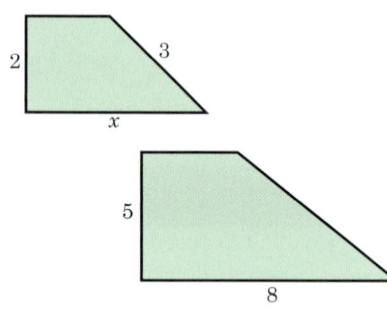

7.

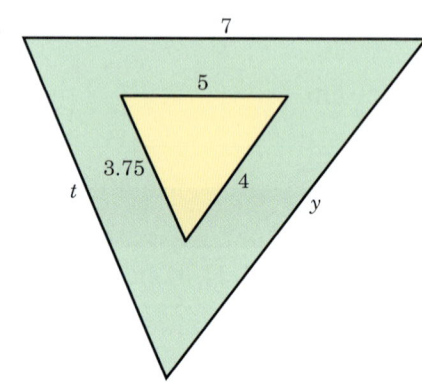

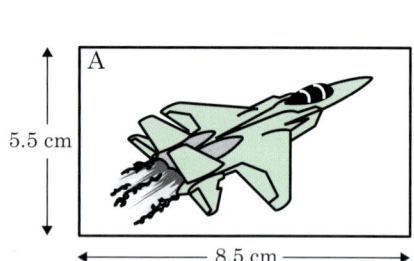

8. Picture B is an enlargement of picture A. Calculate the length x.

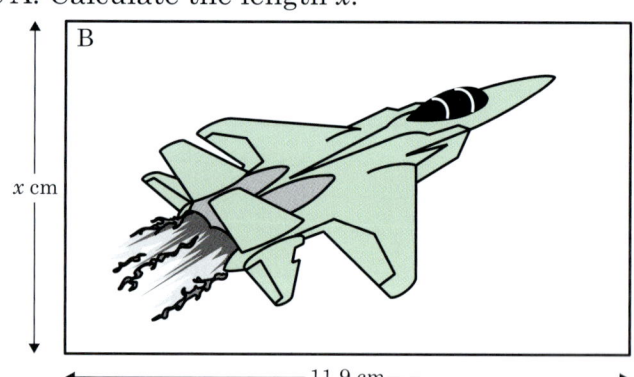

9. The drawing shows a rectangular picture 16 cm × 8 cm surrounded by a border of width 4 cm.

Are the two rectangles similar?

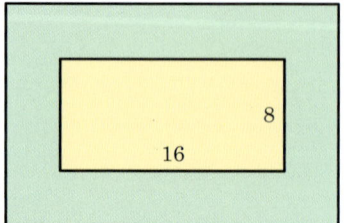

10. Which of the following *must* be similar to each other?

a) Two equilateral triangles   b) Two rectangles   c) Two isosceles triangles

d) Two squares   e) Two regular pentagons f) Two kites

g) Two rhombuses   h) Two circles

11. a) Explain why triangles ABC and EBD are similar.

b) Given that EB = 7 cm, calculate the length AB.

c) Write down the length AE.

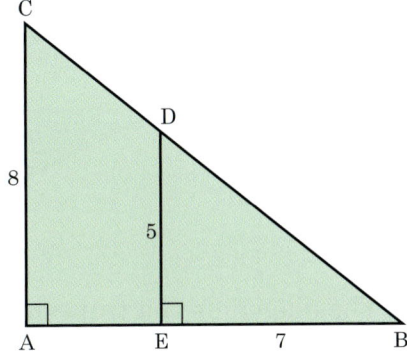

> **Tip**
>
> Draw the two triangles separately and mark on any measurements that you know.

In Questions **12**, **13** and **14** use similar triangles to find the sides marked with letters. All lengths are in cm.

12.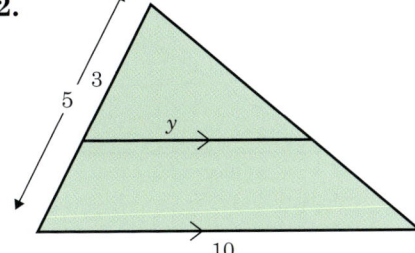

13.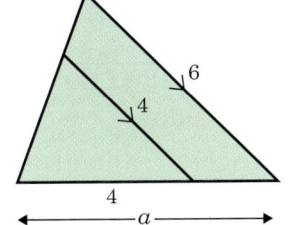

14.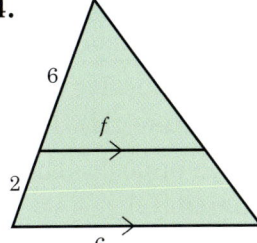

15. A tree of height 4 m casts a shadow of length 6.5 m. Find the height of a house casting a shadow 26 m long.

16. a) Explain why triangles ABC and CDE are similar.

b) Find the value of:

i) *a*

ii) *b*.

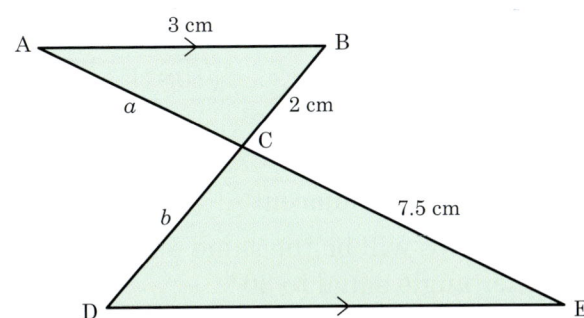

17. The diagram shows the side view of a swimming pool being filled with water. Calculate the length x.

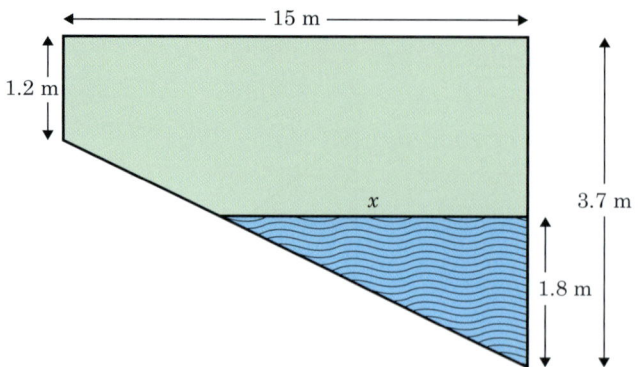

18. The diagonals of a trapezium ABCD intersect at O. AB is parallel to DC, AB = 3 cm and DC = 6 cm. Show that triangles ABO and CDO are similar. If CO = 4 cm and OB = 3 cm, find AO and DO.

11.6 Trigonometry in right-angled triangles

Trigonometry is used to calculate sides and angles in right-angled triangles.

The side opposite the right angle is called the **hypotenuse** (H). It is the longest side.

The side opposite the marked angle is called the **opposite** (O).

The other side is called the **adjacent** (A).

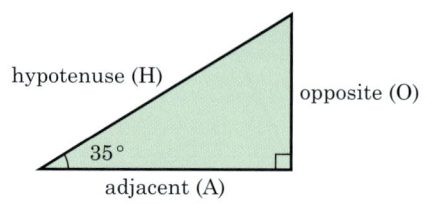

Consider two triangles, one of which is an enlargement of the other.

For the angle $30°$,

the ratio $= \dfrac{\text{opposite}}{\text{hypotenuse}} = \dfrac{6}{12} = \dfrac{2}{4} = \dfrac{1}{2}$

the ratio $= \dfrac{\text{adjacent}}{\text{hypotenuse}} = \dfrac{\sqrt{108}}{12} = \dfrac{\sqrt{12}}{4} = 0.866...$

and the ratio $= \dfrac{\text{opposite}}{\text{adjacent}} = \dfrac{6}{\sqrt{108}} = \dfrac{2}{\sqrt{12}} = 0.577...$

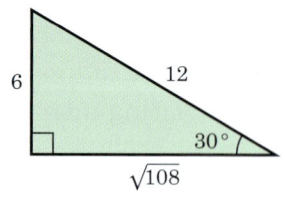

These ratios will be the same for all right-angled triangles with another angle equal to $30°$.

The ratios are called **sine (sin)**, **cosine (cos)** and **tangent (tan)**.

In a right-angled triangle, the ratios are defined for an angle x:

$$\sin x = \frac{O}{H} \qquad \cos x = \frac{A}{H} \qquad \tan x = \frac{O}{A}$$

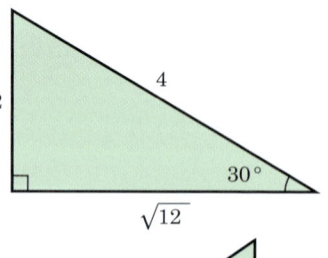

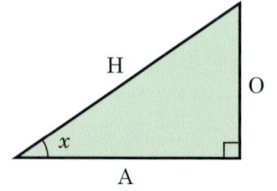

It is important to get the letters in the correct positions.

Some people find a good way of remembering the order of the letters is to use the mnemonic **SOHCAHTOA**.

e.g.   SOH: $\sin x = \dfrac{O}{H}$ CAH: $\cos x = \dfrac{A}{H}$ TOA: $\tan x = \dfrac{O}{A}$

You need to be able to find the sine, cosine or tangent for a given angle using your calculator. Try this short introductory exercise to practise doing this with your model of calculator.

Exercise 11.6A

Use your calculator to find each of these, giving your answers correct to 3 decimal places.

1. $\sin 40°$ **2.** $\cos 35°$ **3.** $\tan 25°$ **4.** $\sin 80°$ **5.** $\cos 75°$

6. $\tan 64°$ **7.** $\sin 51°$ **8.** $\cos 17°$ **9.** $\tan 12°$

Finding the length of a side

Example

Find the value of l. Give your answer to 1 decimal place.

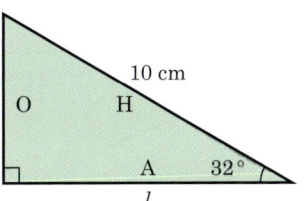

$\cos 32° = \dfrac{A}{H} = \dfrac{l}{10}$

$l = 10 \times \cos 32°$

$l = 8.5$ cm (to 1 d.p)

Label the sides O, A and H and choose the correct ratio.
Side H is given and A is required, so use cosine.

Substitute into the formula for cosine and rearrange.
Calculate l to the required degree of accuracy.

Exercise 11.6B

Find the lengths marked with letters. All lengths are in cm.

Give answers correct to 1 d.p.

1. 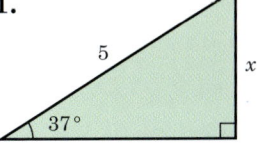   **2.** 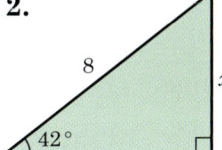   **3.** 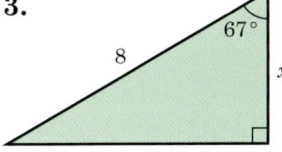   **4.**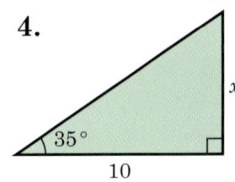

5.

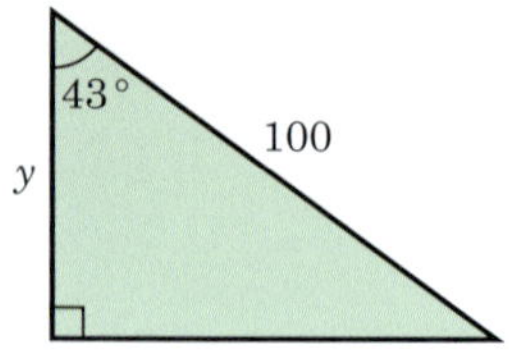

$43°$
100
y

6.

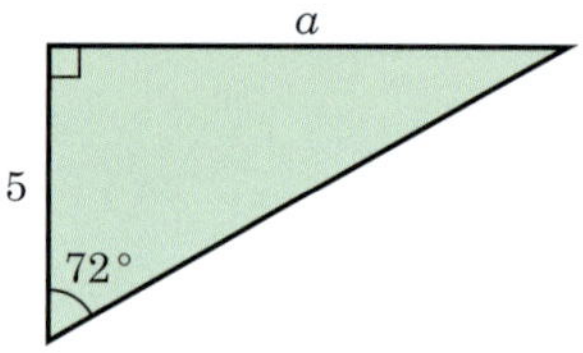

a
5
$72°$

7.

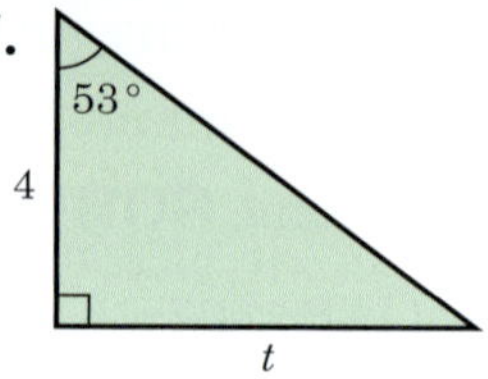

$53°$
4
t

8.

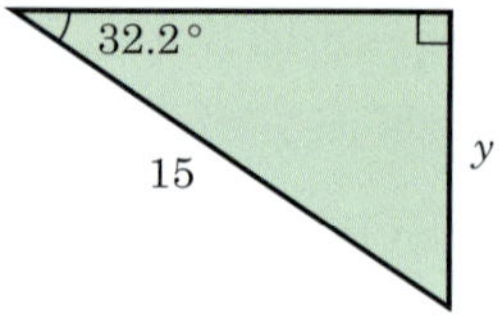

$32.2°$
15
y

9.

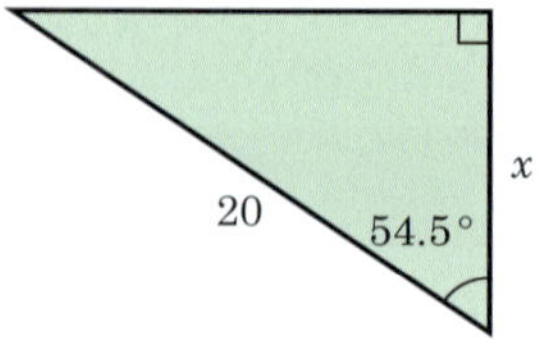

20
$54.5°$
x

10.

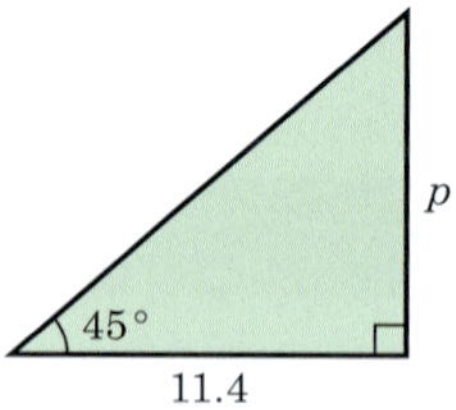

$45°$
11.4
p

11.

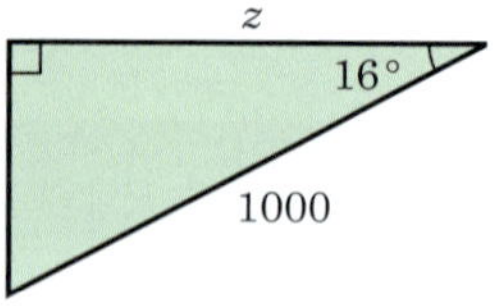

z
$16°$
1000

12.

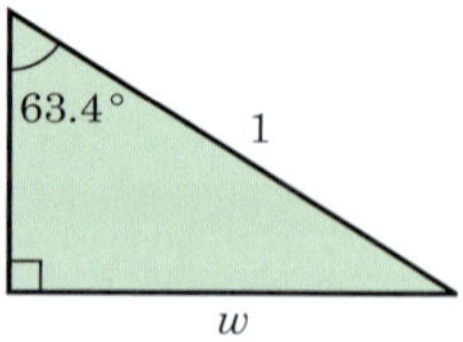

$63.4°$
1
w

13.

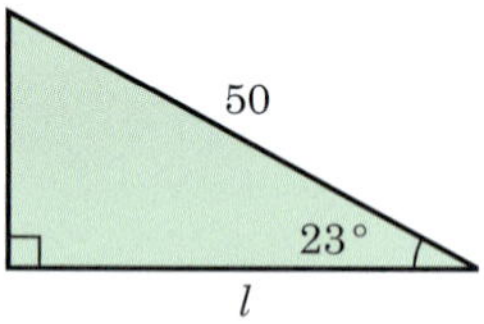

50
$23°$
l

14.

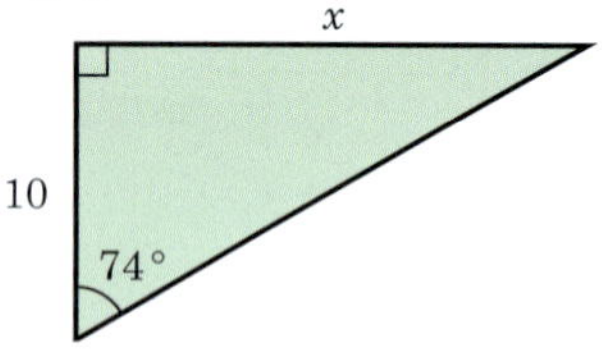

x
10
$74°$

15.

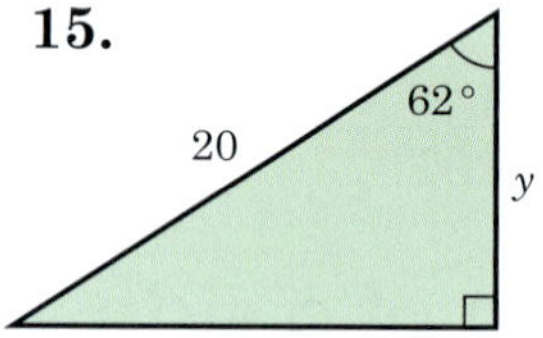

$62°$
20
y

16.

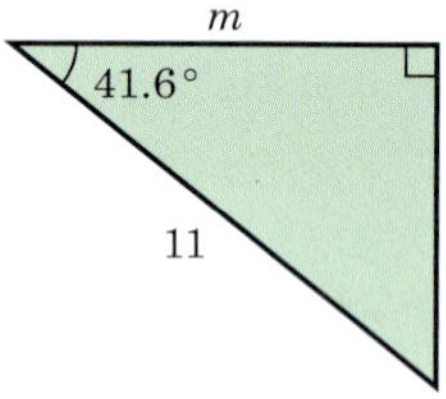

m
$41.6°$
11

17.

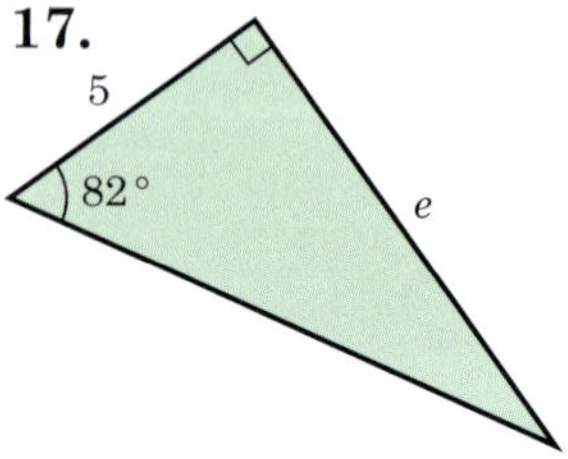

5
$82°$
e

18.

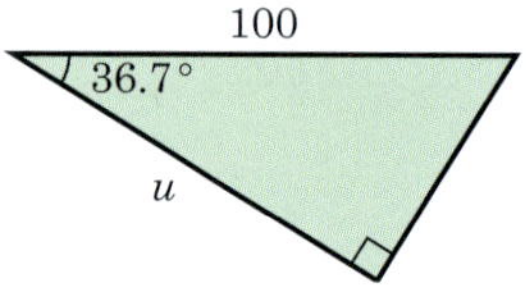

100
$36.7°$
u

19.

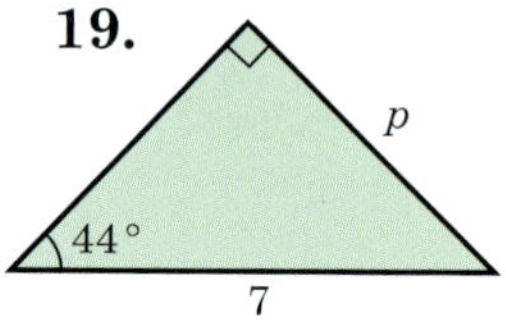

p
$44°$
7

20.

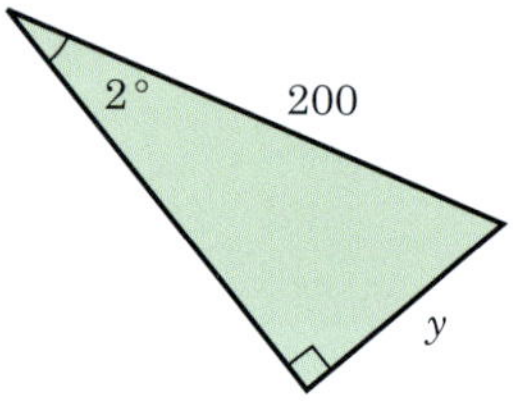

$2°$
200
y

Example

Find the value of x, giving your answer correct to
1 decimal place.

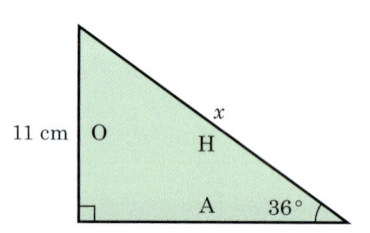

11 cm O H x A $36°$

Label the sides O, A and H and choose the
correct ratio.

$$\sin 36° = \frac{O}{H} = \frac{11}{x}$$

$$x \sin 36° = 11$$

Side O is given and H is required so use sine.

Substitute into the formula for sine and
rearrange.

$$x = \frac{11}{\sin 36°} = 18.7 \text{ cm (to 1 d.p.)}$$

Calculate x to the required degree of
accuracy.

Exercise 11.6C

This exercise is more difficult. Find the lengths marked with letters.

Give answers correct to 1 d.p.

1.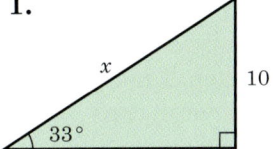

2.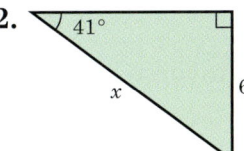

3.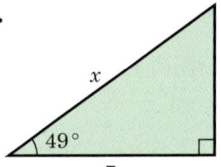

4.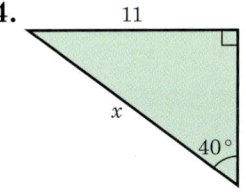

5.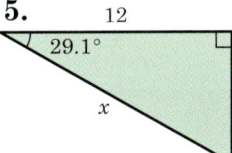

6.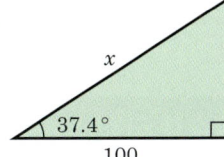

7.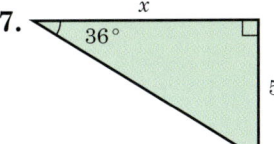

8.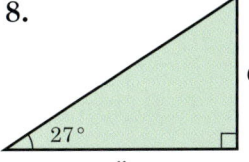

9.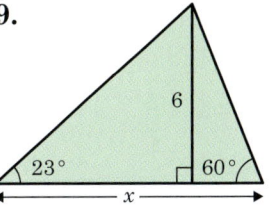

10.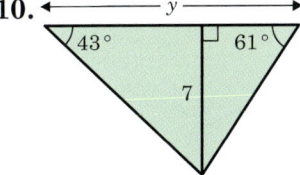

11.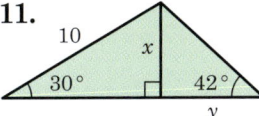

12.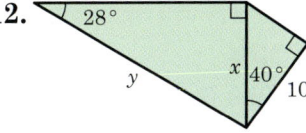

13.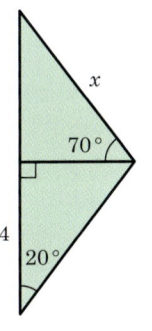

14.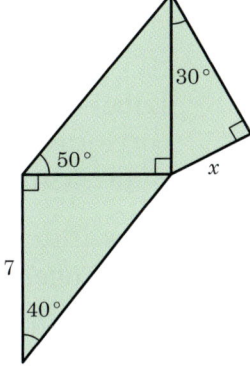

15.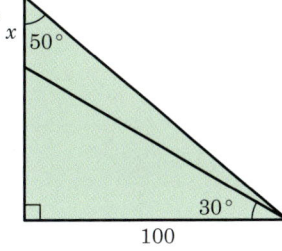

16.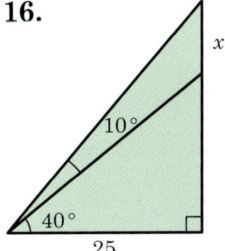

Finding angles

You can also use sine, cosine or tangent to find the size of an unknown angle in a right-angled triangle.

To do this, you use the **inverse** sine, cosine and tangent functions on your calculator.

For example, if $\sin x = 0.5$, you can use inverse sine to find the value of x.

$$x = \sin^{-1}(0.5) = 30°$$

Try this short exercise to practise finding the angles for the given ratios.

Exercise 11.6D

Use your calculator to find each of these angles, giving your answers correct to 1 decimal place.

1. $\sin x = 0.8$ 2. $\cos x = 0.7$ 3. $\tan x = 0.5$

4. $\sin x = 0.4$ 5. $\cos x = 0.2$ 6. $\tan x = 1.4$

7. $\sin x = 0.57$ 8. $\cos x = 0.84$ 9. $\tan x = 0.23$

Example

Find the size of angle x. Give your answer to 1 decimal place.

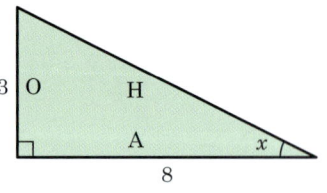

$$\tan x = \frac{O}{A} = \frac{3}{8}$$

$$\tan x = 0.375 \Rightarrow x = \tan^{-1}(0.375)$$

So, $x = 20.6°$ (1 d.p.)

Label the sides O, A and H and choose the correct ratio.
Sides O and A are given so use tangent.

Substitute into the formula for tangent and rearrange.

Use the inverse tangent button and round to 1 decimal place.

Exercise 11.6E

Find the angles marked with letters. Give the answers correct to 1 d.p.

1.

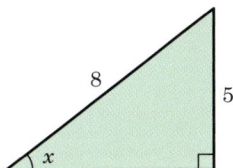

2.

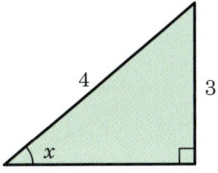

3.

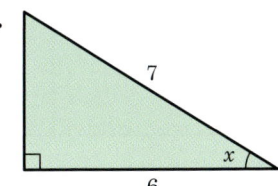

4.

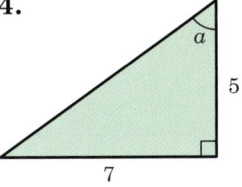

5.

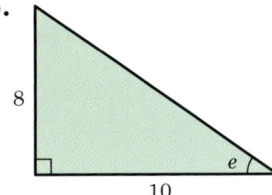

6.

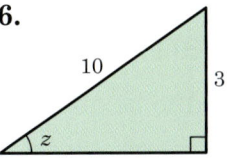

7.

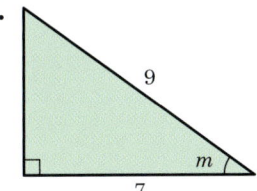

8.

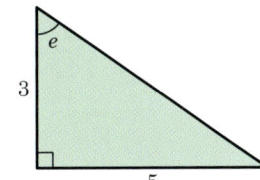

9.

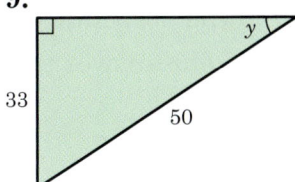

10.

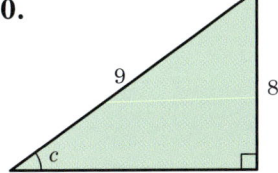

11.

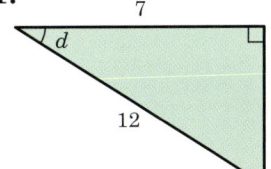

12.

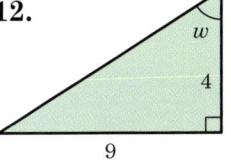

13.

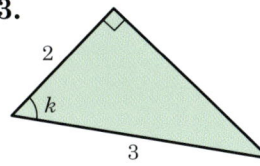

14.

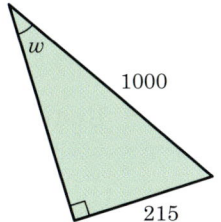

15.

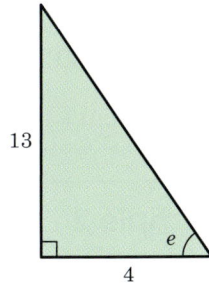

16.

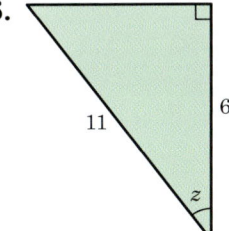

17.

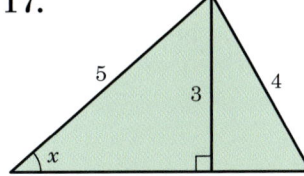

18.

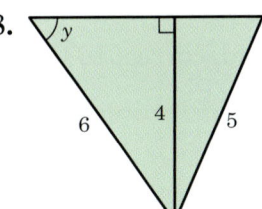

19.

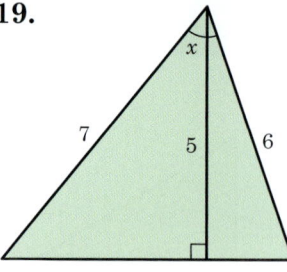

20.

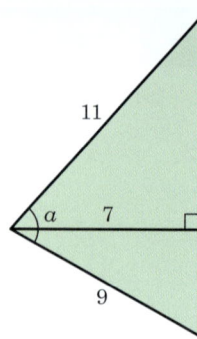

Problem-solving using trigonometry

You need to be able solve problems using trigonometry.

Example 1

An equilateral triangle has sides of 5 cm.

Find the perpendicular height of the triangle.

Give your answer to 3 significant figures.

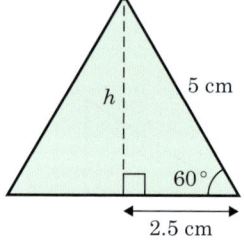

Draw a sketch diagram of the problem.

$$\sin 60° = \frac{h}{5}$$

h is the side opposite the angle and 5 is the hypotenuse.

$$h = 5 \times \sin 60°$$

Use sine.

$$h = 4.33 \text{ cm (3 s.f.)}$$

Rearrange and solve for h.

Example 2

A plane flies directly from airport A to airport B.

Airport B is 80 km north and 50 km east of airport A.

Find, correct to the nearest degree, the bearing on which the plane flies.

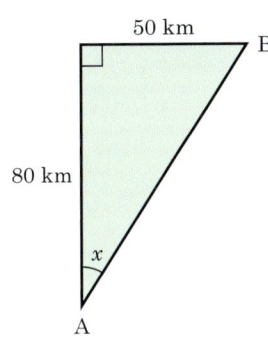

Draw a diagram to show this situation.

The angle marked x shows the required bearing.

$\tan x = \dfrac{50}{80}$

$\tan x = 0.625$

You are given O and A so use tangent.

Work out the ratio and then use inverse tangent to find x.

So, $x = \tan^{-1}(0.625) = 32°$ (nearest degree)

The plane flies on a bearing of 032°

Remember to give your bearing using three figures.

Exercise 11.6F

Begin each question by drawing a large clearly labelled diagram.

1. A ladder of length 4 m rests against a vertical wall so that the base of the ladder is 1.5 m from the wall.

   Calculate the angle between the ladder and the ground.

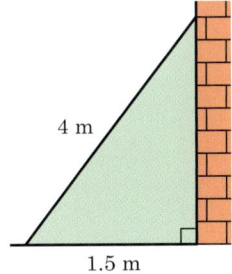

2. A ladder of length 4 m rests against a vertical wall so that the angle between the ladder and the ground is $66°$. How far up the wall does the ladder reach?

3. From a distance of 20 m the angle of elevation to the top of a tower is $35°$.

   How high is the tower?

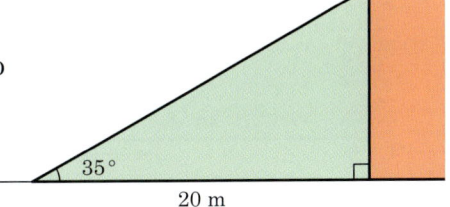

4. A point G is 40 m away from a building, which is 15 m high.

What is the angle of elevation to the top of the building from G (the angle marked x)?

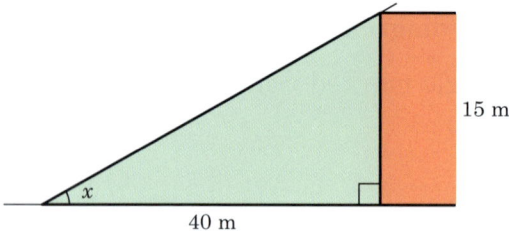

5. A boy is flying a kite from a string of length 60 m.

If the string makes an angle of $71°$ with the horizontal, what is the height of the kite?

Ignore the height of the boy.

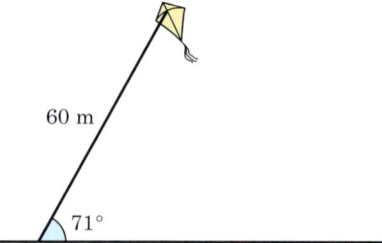

6. A straight tunnel is 80 m long and slopes downwards at an angle of $11°$ to the horizontal. Find the vertical drop in travelling from the top to the bottom of the tunnel.

7. The diagram shows the frame of a bicycle.

Find the length of the crossbar, marked x.

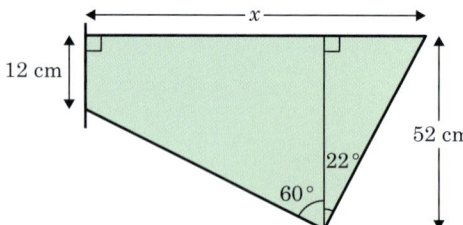

8. Calculate the value of x.

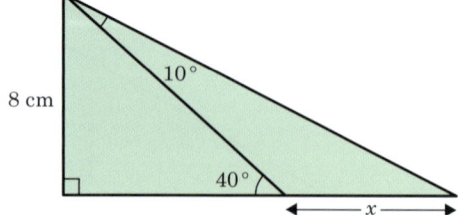

9. AB is a chord of a circle of radius 5 cm and centre O.

   The perpendicular bisector of AB passes through O, and also bisects the angle AÔB. If AÔB = 100°, calculate the length of the chord AB.

   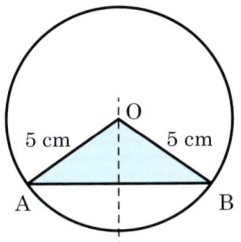

   > **Tip**
   >
   > The perpendicular bisector of a line intersects the line at right angles and cuts the line into two equal-length parts.

10. A ship is due south of a lighthouse L. It sails on a bearing of 055° for a distance of 80 km until it is due east of the lighthouse.

    How far is it now from the lighthouse?

    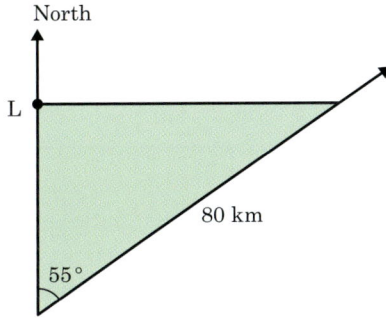

11. A ship is due south of a lighthouse. It sails on a bearing of 071° for a distance of 200 km until it is due east of the lighthouse. How far is it now from the lighthouse?

12. A ship is due north of a lighthouse. It sails on a bearing of 200° at a speed of 15 km/h for five hours until it is due west of the lighthouse. How far is it now from the lighthouse?

13. An isosceles triangle has sides of length 8 cm, 8 cm and 5 cm.

    Find the angle between the two equal sides.

    > **Tip**
    >
    > Draw the triangle and divide it into two right-angled triangles.

14. The angles of an isosceles triangle are 66°, 66° and 48°.

    If the shortest side of the triangle is 8.4 cm, find the length of one of the two equal sides.

15. A regular pentagon is inscribed in a circle of radius 7 cm.

Find the angle a and then the length of a side of the pentagon.

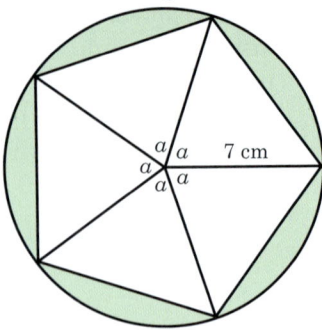

16. Find the acute angle between the diagonals of a rectangle whose sides are 5 cm and 7 cm.

Revision exercise 11

1.

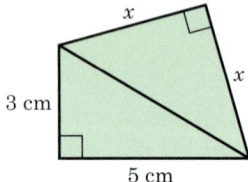

Find the value of x.

2. A regular octagon of side length 20 cm is cut out of a square piece of card.

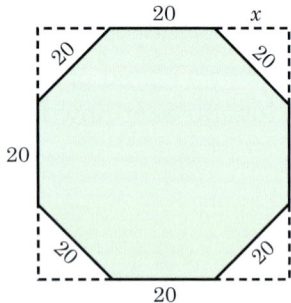

a) Find the value of x and hence find the size of the smallest square of card from which this octagon can be cut.

b) Calculate the area of the octagon. Give your answer correct to 3 s.f.

3. A photo 21 cm by 12 cm is enlarged as shown.

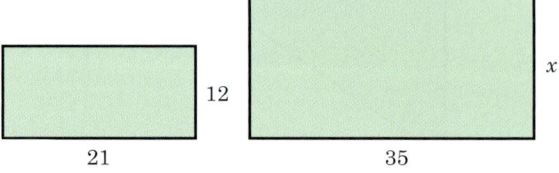

a) What is the scale factor of the enlargement?

b) Work out the length marked x.

4. Look at the diagram below.

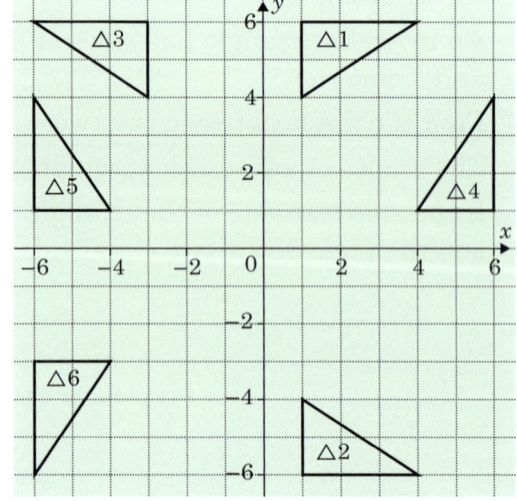

Describe fully the following transformations.

a) $\triangle 1 \to \triangle 2$ b) $\triangle 1 \to \triangle 3$

c) $\triangle 5 \to \triangle 4$ d) $\triangle 5 \to \triangle 1$

e) $\triangle 5 \to \triangle 6$

5. Plot and label the triangles with the following vertices.

$\triangle 1$: (−3, −6), (−3, −2), (−5, −2)

$\triangle 2$: (−5, −1), (−5, −7), (−8, −1)

$\triangle 3$: (8, 4), (8, 8), (6, 8)

$\triangle 4$: (−3, 1), (−3, 3), (−4, 3)

Describe fully the following transformations.

a) $\triangle 1 \to \triangle 2$ b) $\triangle 1 \to \triangle 3$

c) $\triangle 1 \to \triangle 4$ d) $\triangle 4 \to \triangle 2$

6. a) Plot and label:

$\triangle 1$: (−3, 4), (−3, 8), (−1, 8)

$\triangle 5$: (−8, −2), (−8, −6), (−6, −2)

b) Draw the triangles $\triangle 2$, $\triangle 3$, $\triangle 4$, $\triangle 6$ and $\triangle 7$ as follows:

i) $\triangle 1$ onto $\triangle 2$: translation with vector $\begin{pmatrix} 9 \\ -4 \end{pmatrix}$.

ii) $\triangle 1$ onto $\triangle 3$: reflection in the x-axis.

iii) $\triangle 1$ onto $\triangle 4$: rotation $180°$ about (0, 0).

iv) $\triangle 5$ onto $\triangle 6$: reflection in the line $x = -1$.

v) $\triangle 6$ onto $\triangle 7$: enlargement, scale factor 2, centre (−8, −8).

7. Calculate the value of x in each of these diagrams.

Give your answers correct to 2 significant figures.

a)
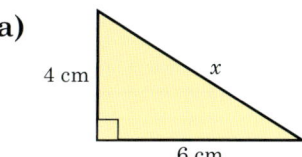

b)
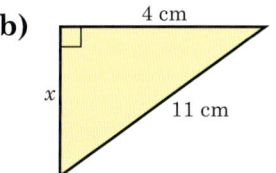

c)
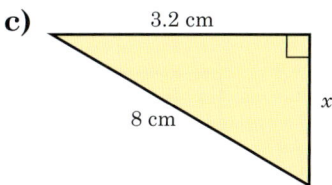

8. Given BD = 1 m, calculate the length AC.

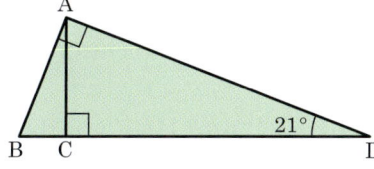

9. Calculate the value of the side or angle marked with a letter.

a)
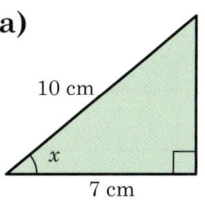

b)
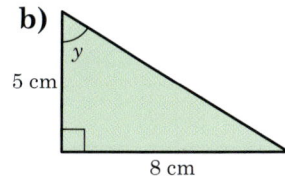

c)
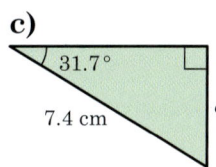

d)

10. Rectangle B is an enlargement of rectangle A.

Calculate the length marked x.

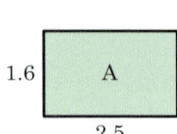
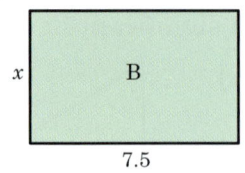

11. The pairs of shapes are similar. Find the lengths of the sides marked with letters.

a)

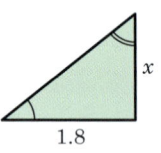
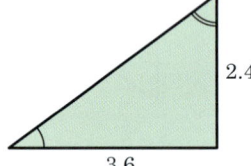

b)

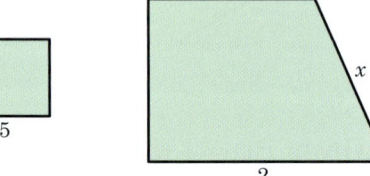

12. The sides of triangle ABC are each increased by 1 cm to form triangle DEF.

Are triangles ABC and DEF similar? Explain your answer.

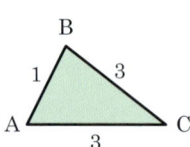
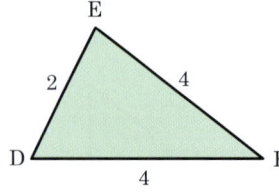

13. Calculate the length marked x.

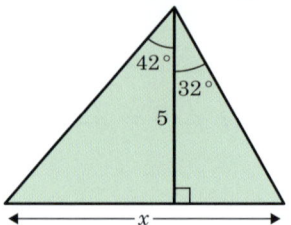

14. a) Explain why triangles PQR and PST are similar.

b) Given that PQ = 8 cm, calculate the length QS.

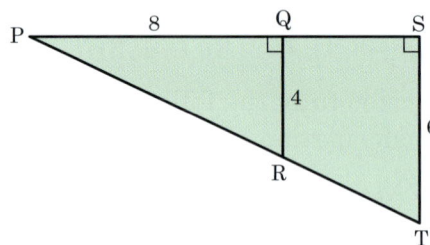

Examination-style exercise 11

NON-CALCULATOR SECTION

1.

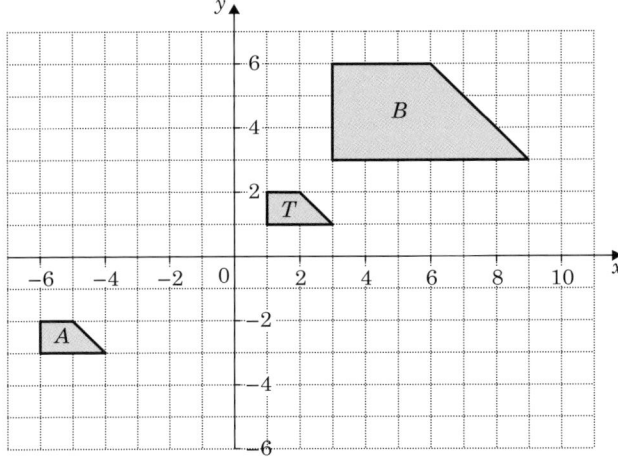

a) In each case, describe fully the single transformation which maps:

   i) *T* onto *A* [3]

   ii) *T* onto *B*. [3]

b) Draw on a copy of the grid the rotation of *T* by 90° anticlockwise about (0, 0).
Label your answer *R*. [2]

c) Draw on your grid the reflection of *T* in the line $y = -2$.
Label your answer *M*. [2]

2.

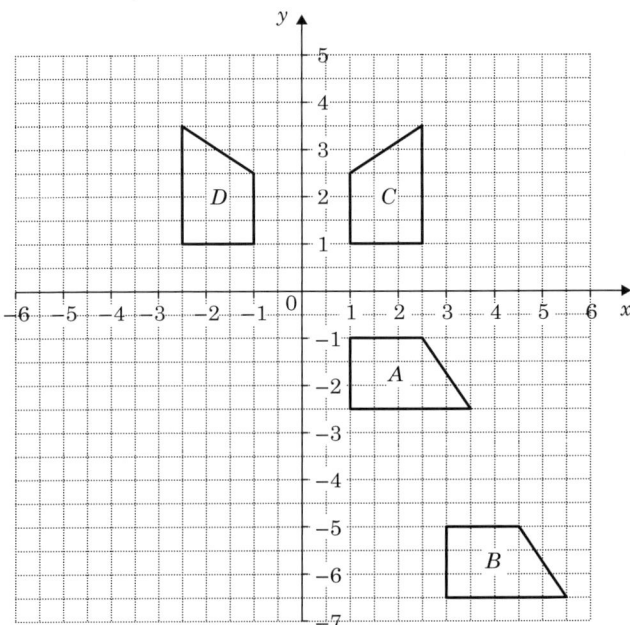

Describe fully the single transformation which maps:

a) *A* onto *B* [3]

b) *C* onto *D* [2]

c) *A* onto *C*. [3]

CALCULATOR SECTION

3. *ABC* is a right-angled triangle.

AB = 0.5 m and *AC* = 1.3 m.

Calculate the length of *BC*. [2]

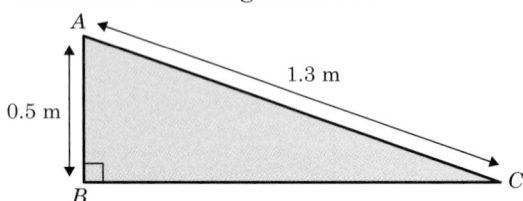

4.

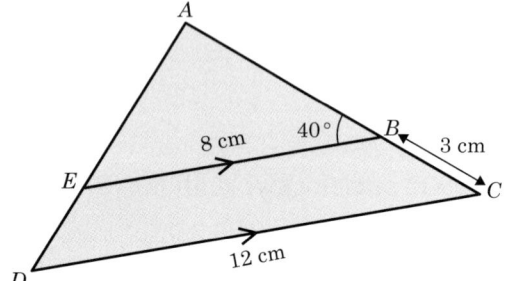

In the diagram *BE* is parallel to *CD*.

a) Copy and complete the following statement.

Triangle *ACD* is .... to triangle *ABE*. [1]

b) *BE* = 8 cm, *CD* = 12 cm and *BC* = 3 cm.

Calculate the length of *AB*. [2]

c) Angle *ABE* = 40°.

Calculate the size of the reflex angle at *C*. [2]

5.

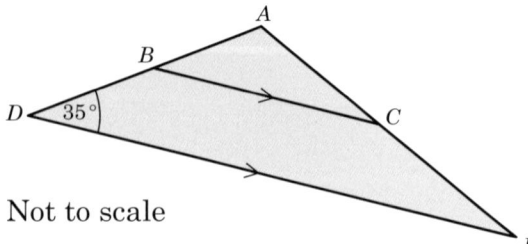

Not to scale

In the diagram BC is parallel to DE. ABD and ACE are straight lines.

a) Choose one of the following words to complete the statement.

congruent   equilateral   isosceles   similar

Triangle ABC and triangle ADE are ....   [1]

b) Angle $BDE = 35°$.
Calculate the size of angle DBC.   [1]

6.

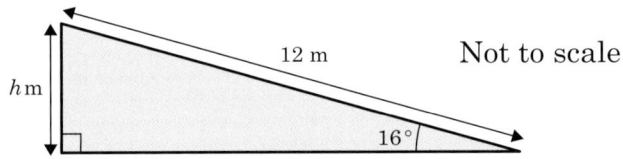

Not to scale

A ramp from the garden to the door of a house slopes upwards at an angle of 16° to the horizontal.

The length of the ramp is 12 metres.

Calculate the difference in height, h metres, between the door of the house and the garden.   [2]

7.

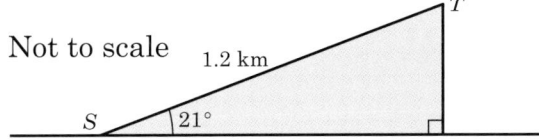

The diagram shows the flight path, ST, of an aircraft on takeoff.

The flight path is 1.2 kilometres long and slopes at an angle of 21° to the horizontal.

Calculate the height of the plane at the end of the flight path, showing all your working.

Give your answer in metres.   [3]

8. In the diagram below ABD is a straight line. $AB = 8$ m and $AC = 12$ m.
Angle $BAC = 90°$.

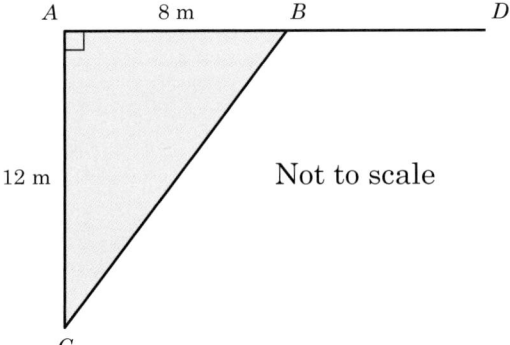

Not to scale

    a) i) Use trigonometry to calculate angle *ABC*. [2]

      ii) Find angle *CBD*. [1]

    b) Calculate the length of *BC*. [2]

    c) Work out the perimeter and area of triangle *ABC*. Give the correct units for each. [3]

9.

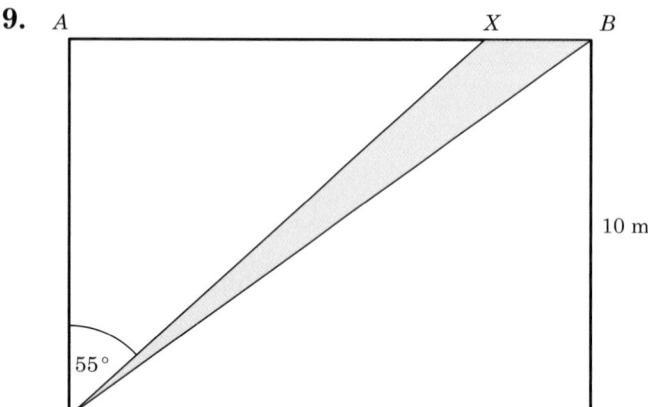

Not to scale

The diagram shows a rectangular garden *ABCD* which has a flower bed *DXB*.
DC = 18 metres, *BC* = 10 metres and angle *ADX* = 55°.

    a) Calculate the area of triangle *BDC*. [2]

    b) Calculate the length of *AX*. [2]

    c) Calculate the area of the flower bed. [3]

    d) Calculate the length of *BD*. [2]

10.

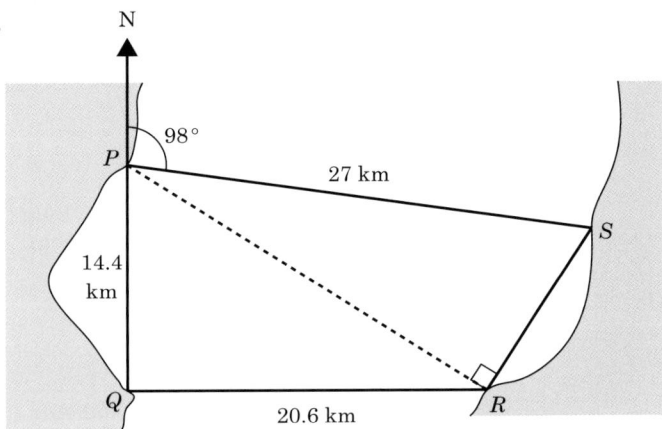

Not to scale

P, Q, R and S are ports on a river delta, as shown in the diagram above.

A ship sails from P, stopping at Q, R and S before returning to P.

a) Q is 14.4 kilometres due south of P and R is 20.6 kilometres due east of Q.

    i) Show by calculation that angle $QPR = 55°$. [2]

    ii) Write down the bearing of R from P. [1]

b) The bearing of S from P is 098° and $SP = 27$ km.

    i) Explain why angle $RPS = 27°$. [1]

    ii) Angle $PRS = 90°$. Calculate the distance RS. [2]

    iii) Find the total distance the ship sails. [1]

c) The total sailing time for the ship is 9 hours.

    Calculate the average sailing speed, in kilometres per hour, for the whole journey. [2]

Answers

Chapter 1 Number 1

Exercise 1.1A
1. 20
2. 400
3. 80
4. 6
5. 6000
6. 20 000
7. 5 000 000
8. 800 000
9. 200
10. 70
11. 10
12. 800
13. 6000
14. 60
15. 400
16. 70 000, 70
17. a) 720
    b) 5206
    c) 16 430
    d) 500 000
    e) 300 090
    f) 8500
    g) 12 000 000 000
18. a) 8753
    b) 3578
19. a) four thousand, six hundred and twenty
    b) six hundred and seven
    c) twenty-five thousand, four hundred
    d) six million, eight hundred thousand
    e) twenty-one thousand, four hundred and twenty-five
20. a) 75 423
    b) 23 574
21. a) 257
    b) 3221
    c) 704
22. a) 1392
    b) 26 611
    c) 257 900
23. a) $\boxed{5}\ \boxed{}\ \boxed{0}$
    b) 52 000
24. a) 2058, 2136, 2142, 2290
    b) 5029, 5299, 5329, 5330
    c) 25 000, 25 117, 25 171, 25 200, 25 500
25. a) 100
    b) 10
26. a) 100, 7
    b) 1000, 10

Exercise 1.2A
1. 3497
2. 785
3. 68 521
4. 212
5. 859
6. 208
7. 270
8. 5000
9. 2528
10. 85
11. 324
12. 52
13. 52
14. 4569
15. 7
16. 1080
17. 524
18. 188
19. 1641
20. 254
21. 2435
22. 41 g
23. 365
24. 856
25. 64 568 mm
26. $639
27. 325 g
28. $2018
29. 1492 g
30. 5800
31. 365
32. 21 200

Exercise 1.2B
1. $684
2. $618
3. $1652
4. $2690
5. $816
6. $5580

Exercise 1.3A
1. a) $\begin{array}{r} 285 \\ +\,514 \\ \hline 799 \end{array}$
   b) $\begin{array}{r} 637 \\ +\,252 \\ \hline 889 \end{array}$
   c) $\begin{array}{r} 635 \\ +\,344 \\ \hline 979 \end{array}$

2. a) $\begin{array}{r} 356 \\ +\,526 \\ \hline 882 \end{array}$
   b) $\begin{array}{r} 224 \\ +\,537 \\ \hline 761 \end{array}$
   c) $\begin{array}{r} 388 \\ +\,425 \\ \hline 813 \end{array}$

3. a) $\begin{array}{r} 48 \\ \times\ \ 3 \\ \hline 144 \end{array}$
   b) $\begin{array}{r} 33 \\ \times\ \ 7 \\ \hline 231 \end{array}$
   c) $\begin{array}{r} 321 \\ \times\ \ \ 5 \\ \hline 1605 \end{array}$

4. a) 150
   b) 15
   c) 9
   d) 552

5. a) $\begin{array}{r} 445 \\ +\,285 \\ \hline 730 \end{array}$
   b) $\begin{array}{r} 427 \\ +\,177 \\ \hline 604 \end{array}$
   c) $\begin{array}{r} 535 \\ +\,264 \\ \hline 799 \end{array}$

6. a) 35
   b) 58
   c) 4
   d) 950

7. a) 72
   b) 108
   c) $\begin{array}{r} 889 \\ -\,346 \\ \hline 543 \end{array}$
   d) $\begin{array}{r} 335 \\ -\,218 \\ \hline 117 \end{array}$

8. a) $4 \times 4 - 4$
   b) $8 \div 8 + 8$
   c) $8 \times 8 + 8$
9. a) −
   b) ÷
   c) ×
   d) ÷
   e) +
10. a) +
    b) −, −
    c) +

Exercise 1.4A

1. 805	**2.** 459	**3.** 650	**4.** 1333	**5.** 2745
6. 1248	**7.** 4522	**8.** 30 368	**9.** 28 224	**10.** 8568
11. 46 800	**12.** 66 281	**13.** 57 602	**14.** 89 516	**15.** 97 525

Exercise 1.4B

1. 32	**2.** 25	**3.** 18	**4.** 13	**5.** 35
6. 22 r 2	**7.** 23 r 24	**8.** 18 r 10	**9.** 27 r 18	**10.** 13 r 31
11. 35 r 6	**12.** 64 r 37	**13.** 151 r 17	**14.** 2961 r 15	

Exercise 1.4C

1. $47.04	**2.** 46	**3.** 7592	**4.** 21, 17c change	**5.** 8
6. $80.64	**7.** $14 million	**8.** $85	**9.** $21 600	

Exercise 1.5A

1. 5 °C	**2.** −1 °C	**3.** −4 °C	**4.** −7 °C	**5.** −4 °C
6. −4 °C	**7.** 4 °C	**8.** 12 °C	**9.** −5 °C	**10.** 0 °C
11. −11 °C	**12.** −15 °C	**13.** fallen 3 °C	**14.** fallen 6 °C	**15.** fallen 10 °C
16. fallen 5 °C	**17.** risen 7 °C	**18.** risen 11 °C	**19.** risen 15 °C	**20.** risen 5 °C
21. fallen 80 °C	**22.** fallen 15 °C	**23. a)** 0, −3	**b)** −5	**c)** −20, −15
24. −17 m	**25. a)** C	**b)** B	**26.** $60	

Exercise 1.5B

1. −4	**2.** −12	**3.** −11	**4.** −3	**5.** −5	**6.** 4
7. −5	**8.** −8	**9.** 19	**10.** −17	**11.** −4	**12.** −5
13. −11	**14.** 6	**15.** −4	**16.** 6	**17.** 0	**18.** −18
19. −3	**20.** −11	**21.** −8	**22.** −7	**23.** 1	**24.** 1
25. 9	**26.** 11	**27.** −8	**28.** 42	**29.** 4	**30.** 15
31. −7	**32.** −9	**33.** −1	**34.** −7	**35.** 0	**36.** 11
37. −14	**38.** 0	**39.** 17	**40.** 3		

Exercise 1.5C

1. −6	**2.** −4	**3.** −15	**4.** 9	**5.** −8	**6.** −15
7. −24	**8.** 6	**9.** 12	**10.** −18	**11.** −21	**12.** 25
13. −60	**14.** 21	**15.** 48	**16.** −16	**17.** −42	**18.** 20
19. −42	**20.** −66	**21.** −4	**22.** −3	**23.** 3	**24.** −5
25. 4	**26.** −4	**27.** −4	**28.** −1	**29.** −2	**30.** 4
31. −16	**32.** −2	**33.** −4	**34.** 5	**35.** −10	**36.** 11
37. 16	**38.** −2	**39.** −4	**40.** −5	**41.** 64	**42.** −27
43. −600	**44.** 40	**45.** 2	**46.** 36	**47.** −2	**48.** −8
49. 160	**50.** −2				

Exercise 1.5D

1. −16.5	**2.** 15	**3.** −17.5	**4.** −2.5	**5.** 15.5	**6.** 18
7. 2	**8.** −6	**9.** 3	**10.** −96	**11.** −14	**12.** 15
13. 3	**14.** −26	**15.** −4	**16.** −4	**17.** −26	**18.** −6
19. −2.5	**20.** 8				

Exercise 1.6A

1. T; 0.3 < 0.31	**2.** F	**3.** T; 0.7 = 0.70	**4.** T; 0.17 < 0.71
5. T; 0.02 > 0.002	**6.** F	**7.** T; $0.1 = \frac{1}{10}$	**8.** T; 5 = 5.00
9. $50 + 7 + \frac{2}{10}$	**10. a)** 235.1 **b)** 67.23		**c)** 98.32 **d)** 3.167

11. 0.2, 0.31, 0.41   **12.** 0.58, 0.702, 0.75   **13.** 0.41, 0.43, 0.432
14. 0.6, 0.609, 0.61   **15.** 0.04, 0.15, 0.2, 0.35   **16.** 0.18, 0.81, 1.18, 1.8
17. 0.061, 0.07, 0.1, 0.7   **18.** 0.009, 0.025, 0.03, 0.2   **19.** CARWASH
20. a) 32.51   **b)** 0.853   **c)** 1.16
21. a) 5.69   **b)** 0.552   **c)** 1.30
22. a) $3.50   **b)** $0.15   **c)** $0.03   **d)** $0.10   **e)** $12.60   **f)** $0.08
23. a) T   **b)** F   **c)** T   **d)** T

Exercise 1.6B

1. 4.3	**2.** 0.7	**3.** 9.4	**4.** 1.2	**5.** 16	**6.** 10.7
7. 17.4	**8.** 128	**9.** 375	**10.** 0.24	**11.** 1.92	**12.** 5.2
13. 0.06	**14.** 1.76	**15.** 3.16	**16.** 105	**17.** 50	**18.** 125

Exercise 1.6C

1. 6.34	**2.** 8.38	**3.** 81.5	**4.** 7.4	**5.** 7245	**6.** 32
7. −6.3	**8.** −142	**9.** −4.1	**10.** 30	**11.** −710	**12.** 39.5
13. 0.624	**14.** 0.897	**15.** 0.175	**16.** −0.236	**17.** 0.127	**18.** −0.705
19. −1.3	**20.** 0.08	**21.** −0.007	**22.** 21.8	**23.** 0.035	**24.** 0.0086
25. −95	**26.** 111.1	**27.** −0.32	**28.** 70	**29.** −5.76	**30.** 9.99
31. 660	**32.** 1	**33.** 0.042	**34.** −6200	**35.** −0.009	**36.** 0.0555

37. a) 0 **b) i)** 5, 2 **ii)** 5, 2, 0 **iii)** 0, 5, 2 and ·

Exercise 1.6D

1. 10.14	**2.** 20.94	**3.** 26.71	**4.** 216.95	**5.** 9.6	**6.** 23.1
7. 9.14	**8.** 17.32	**9.** 0.062	**10.** 1.11	**11.** 4.36	**12.** 2.41
13. 1.36	**14.** 6.23	**15.** 2.46	**16.** 8.4	**17.** 2.8	**18.** 10.3
19. 0.18	**20.** 4.01	**21.** 6.66	**22.** 41.11	**23.** 3.6	**24.** 6.44

Exercise 1.6E

1. 0.06	**2.** 0.15	**3.** 0.12	**4.** 0.006	**5.** 1.8	**6.** 3.5
7. 1.8	**8.** 0.8	**9.** 0.36	**10.** 0.014	**11.** 1.26	**12.** 2.35
13. 8.52	**14.** 3.12	**15.** 0.126	**16.** 127.2	**17.** 0.17	**18.** 0.327
19. 0.126	**20.** 0.34	**21.** 0.055	**22.** 0.52	**23.** 1.3	**24.** 0.001

Exercise 1.6F

1. 2.1	**2.** 3.1	**3.** 4.36	**4.** 4	**5.** 4	**6.** 2.5
7. 16	**8.** 200	**9.** 70	**10.** 0.92	**11.** 30.5	**12.** 6.2
13. 12.5	**14.** 122	**15.** 212	**16.** 56	**17.** 60	**18.** 1500
19. 0.3	**20.** 0.7	**21.** 0.5	**22.** 3.04	**23.** 5.62	**24.** 0.78
25. 0.14	**26.** 3.75	**27.** 0.075	**28.** 0.15	**29.** 1.22	**30.** 163.8
31. 1.75	**32.** 18.8	**33.** 12	**34.** 88	**35.** 580	

Exercise 1.7A

1. 19	**2.** 3	**3.** 36	**4.** 34	**5.** 10	**6.** 58
7. 26	**8.** 51	**9.** 16	**10.** 3	**11.** 10	**12.** 12.8
13. 18.8	**14.** 99	**15.** 0.3	**16.** 35.8	**17.** 52	**18.** 3

Exercise 1.7B

1. 49	**2.** 9	**3.** 34	**4.** 12	**5.** 5	**6.** 4	**7.** 10	**8.** 2	**9.** 30

Exercise 1.7C

1. a) 1, 2, 3, 6 **b)** 1, 3, 5, 15 **c)** 1, 2, 3, 6, 9, 18 **d)** 1, 3, 7, 21 **e)** 1, 2, 4, 5, 8, 10, 20, 40

2. 2, 3, 5, 7, 11, 13, 17, 19 **3.** $2 + 5 = 7$ $2 + 11 = 13$ etc.

4. 101, 151, 293 are prime

5. a) $36 = 2 \times 2 \times 3 \times 3$ **b)** $60 = 2 \times 2 \times 3 \times 5$ **c)** $216 = 2 \times 2 \times 2 \times 3 \times 3 \times 3$

d) $200 = 2 \times 2 \times 2 \times 5 \times 5$ **e)** $1500 = 2 \times 2 \times 3 \times 5 \times 5 \times 5$

6. $1200 = 2$ times $2 \times 2 \times 2 \times 3 \times 5 \times 5$

7. a) 3, 6, 9, 12 **b)** 4, 8, 12, 16 **c)** 10, 20, 30, 40 **d)** 11, 22, 33, 44 **e)** 20, 40, 60, 80

8. a) 32 **b)** 56 **9.** 12, 24, etc. **10. a)** even **b)** odd **c)** even

11.

	Prime number	Multiple of 3	Factor of 16
Number > 5	7	9	8
Odd number	5	3	1
Even number	2	6	4

12. a) 7 **b)** 50 **c)** 1 **d)** 5

13. a) 2, 4, 6, 8, 10, 12 **b)** 5, 10, 15, 20, 25, 30 **c)** 10

14. a) 4, 8, 12, 16 **b)** 12, 24, 36, 48 **c)** 12

15. a) 18   **b)** 24   **c)** 70   **d) 12**   **e) 30**   **f)** 252
16. 12   **17.** 6   **18. a)** 6   **b)** 11   **c)** 9   **d)** 6
e) 12   **f)** 10   **19. a)** 6   **b)** 40   **c)** 11, 22 (or others)   **d)** 2, 5 (or others)
20. a) $30 = 2 \times 3 \times 5$; $165 = 3 \times 5 \times 11$ **b)** 15   **c)** 330
21. a) $315 = 3 \times 3 \times 5 \times 7$; $273 = 3 \times 7 \times 13$ **b)** 21   **c)** 4095

Exercise 1.7D

1. Rational: $\left(\sqrt{17}\right)^2$; 3.14; $\frac{1}{3} + \frac{1}{9}$; $\frac{22}{7}$; $\sqrt{49}$; $\sqrt{100}$

2. a) Many possible answers, e.g. 5, $\frac{9}{2}$, $4\frac{17}{18}$ **b)** Many possible answers, e.g. $\sqrt{23}$

3. a) Both irrational   **b)** Both rational

Exercise 1.8A

1. a) 3   **b)** 12   **c)** 15   **d)** 80   **e)** 2   **f)** 9
2. a) $\frac{3}{4}$ **b)** $\frac{1}{2}$ **c)** $\frac{5}{6}$ **d)** $\frac{2}{3}$ **e)** $\frac{11}{12}$ **f)** $\frac{3}{4}$
3. a) 27   **b)** 25   **c)** 55   **d)** 15   **e)** 64   **f)** 12
4. $\frac{5}{6}$ **5.** $\frac{1}{6}$ **6.** $\frac{2}{3}$ **7.** $\frac{5}{12}$ **8.** $\frac{1}{4}$ **9.** $2\frac{1}{4}$
10. $\frac{9}{10}$ **11.** $\frac{1}{5}$ **12.** $\frac{4}{5}$ **13.** $\frac{13}{14}$ **14.** $\frac{3}{14}$ **15.** $\frac{6}{7}$
16. $\frac{3}{8}$ **17.** $\frac{5}{32}$ **18.** $2\frac{1}{2}$ **19.** $\frac{29}{30}$ **20.** $\frac{2}{15}$ **21.** $\frac{5}{24}$
22. $\frac{16}{21}$ **23.** $\frac{1}{7}$ **24.** $1\frac{2}{7}$ **25.** $\frac{11}{20}$ **26.** $\frac{1}{5}$ **27.** $3\frac{1}{5}$
28. $\frac{13}{24}$ **29.** $\frac{1}{12}$ **30.** $5\frac{1}{3}$ **31.** $\frac{29}{36}$ **32.** $\frac{5}{36}$ **33.** $2\frac{2}{9}$
34. $2\frac{1}{4}$ **35.** $\frac{5}{8}$ **36.** 10   **37.** $3\frac{1}{12}$ **38.** $2\frac{1}{2}$ **39.** $5\frac{5}{8}$
40. $2\frac{1}{3}$ **41.** $\frac{5}{13}$ **42.** 18   **43.** 6   **44.** 3   **45.** $2\frac{2}{3}$
46. 3   **47.** $4\frac{1}{4}$ **48.** $1\frac{5}{27}$ **49.** 6   **50.** $1\frac{5}{28}$ **51.** $2\frac{2}{3}$
52. $\frac{11}{24}$ **53.** $4\frac{5}{6}$

Exercise 1.8B

1. a) $\frac{1}{2}, \frac{7}{12}, \frac{2}{3}$ **b)** $\frac{2}{3}, \frac{3}{4}, \frac{5}{6}$ **c)** $\frac{1}{3}, \frac{5}{8}, \frac{17}{24}, \frac{3}{4}$ **d)** $\frac{5}{6}, \frac{8}{9}, \frac{11}{12}$
2. a) $\frac{1}{2}$ **b)** $\frac{3}{4}$ **c)** $\frac{17}{24}$ **d)** $\frac{7}{18}$ **e)** $\frac{3}{10}$ **f)** $\frac{5}{12}$
3. a) <   **b)** >   **c)** =   **d)** <   **e)** <   **f)** =
4. $48   **5. a)** $\frac{9}{4}$ **b)** $2\frac{1}{7}$ **c)** $\frac{31}{10}$ **6.** $\frac{1}{4}$
7. a) 240 cm   **b)** 123 cm   **8.** $\frac{1}{5}$ **9. a)** 30.5   **b)** $305
10. a) 9   **b)** $\frac{5}{16}$ **11.** $\frac{16}{24}$ **12.** 5   **13.** $45   **14.** $\frac{5}{24}$

Exercise 1.9A

1. 3^4 **2.** 5^2 **3.** 6^3 **4.** 10^5 **5.** 1^7 **6.** 8^4
7. 7^6 **8.** $2^3 \times 5^2$ **9.** $3^2 \times 7^4$ **10.** $3^2 \times 10^3$ **11.** $5^4 \times 11^2$ **12.** $2^2 \times 3^3$
13. $3^2 \times 5^3$ **14.** $2^2 \times 3^3 \times 11^2$ **15. a)** 16   **b)** 36   **c)** 100   **d)** 27
e) 1000   **f)** 256   **g)** 125
16. a) 81   **b)** 441   **c)** 1.44   **d)** 0.04   **e)** 9.61   **f)** 10 000
g) 625   **h)** 75.69   **i)** 0.81   **j)** 6625.96
17. a) 4.41 cm^2 **b)** 0.36 cm^2 **c)** 196 m^2
18. a) 216   **b)** 256   **c)** 243   **d)** 100 000   **e)** 64   **f)** 0.001
g) 8.3521   **h)** 567   **i)** 1250   **19.** 10^{10} **20.** 2^7 **21.** Yes

Exercise 1.9B

1. a) 4   **b)** 6   **c)** 1   **d)** 10   **e)** 15
2. a) 9 cm   **b)** 7 cm   **c)** 12 cm
3. a) 3.16227766   **b)** 5.385164807   **c)** 10.34408043   **d)** 4.438468204   **e)** 49.05099387
f) 7.655063684   **g)** 0.3872983346   **h)** 0.8526429499   **4.** 12.2 cm   **5.** 447 m
6. 7.8 cm   **7. a)** 4   **b)** 5   **c)** 10   **8.** 5.8 cm

Exercise 1.9C

1. $\frac{1}{3}$ 2. $\frac{1}{4}$ 3. $\frac{1}{10}$ 4. 1   5. $\frac{1}{9}$ 6. $\frac{1}{16}$

7. $\frac{1}{100}$ 8. 1   9. $\frac{1}{49}$ 10. 1   11. $\frac{1}{81}$ 12. 1

13. T   14. F   15. T   16. T   17. F   18. F

19. F   20. T   21. T   22. T   23. F   24. F

25. F   26. T   27. T   28. T   29. T   30. T

31. T   32. F   33. a) $\frac{1}{7}$ b) $\frac{1}{23}$ c) 5   d) 80

e) $\frac{3}{2}$ f) $\frac{2}{5}$ g) $\frac{71}{45}$ h) 5

Exercise 1.9D

1. 5^6 2. 6^5 3. 10^9 4. 7^8 5. 3^{10} 6. 8^6

7. 2^{13} 8. 3^4 9. 5^3 10. 7^4 11. 5^2 12. 3^{-4}

13. 6^5 14. 5^{-10} 15. 7^6 16. 7^2 17. 6^5 18. 8^1

19. 5^8 20. 10^2 21. 9^{-2} 22. 3^{-2} 23. 2^4 24. 3^{-2}

25. 7^{-6} 26. 3^{-4} 27. 5^{-5} 28. 8^{-5} 29. 1   30. 2

31. 4   32. 5   33. 0   34. 4   35. 4   36. 10

Exercise 1.9E

1. 3^6 2. 5^{12} 3. 7^{10} 4. 8^{20} 5. 2^{-2} 6. 3^{-4}

7. 7^2 8. 9^{-48} 9. 7^9 10. 3^{12} 11. 5^{12} 12. 6^6

13. a) 5   b) 3   c) 4   d) 2

Exercise 1.10A

1. 4×10^3 2. 5×10^2 3. 7×10^4 4. 6×10 5. 2.4×10^3

6. 3.8×10^2 7. 7×10^{-3} 8. 4×10^{-4} 9. 3.5×10^{-3} 10. 4.21×10^{-1}

11. 5.5×10^{-5} 12. 1×10^{-2} 13. a) 5.64×10^5 b) 1.9×10^8 c) 5×10^{-2}

14. 46 000   15. 370 000 000   16. 92 100 000 000   17. 0.256   18. 0.0037

19. 0.000 000 48   20. 1.4×10^9 21. 1.67×10^{-24} 22. 5.1×10^8 25. 3×10^{10}

23. 0.000 000 000 25   24. 602 300 000 000 000 000 000 000

Exercise 1.10B

1. 1.5×10^7 2. 3×10^8 3. 2.8×10^{-2} 4. 7×10^{-9} 5. 2×10^6

6. 4×10^{-6} 7. 9×10^{-2} 8. 1.5×10^{-7} 9. 3.5×10^{-7} 10. 10^{-16}

11. 8×10^9 12. 2×10^{-6} 13. 4.9×10^{11} 14. 4.4×10^{12} 15. 1.5×10^3

16. 2×10^{17} 17. 1.68×10^{13} 18. 4.25×10^{11} 19. 9.9×10^7 20. 6.25×10^{-16}

21. 2.88×10^{12} 22. 6.82×10^{-7} 23. c, a, b 24. 13   25. 16

26. a) 6×10^2 b) 6.67×10^7 27. 50 min   28. 6×10^2

29. a) 9.46×10^{12} km   b) 144 million km   30. 25 000

Revision exercise 1

1. a) i) 13, 49, 109   ii) 4, 49   iii) 13, 109

   b) i) 27   ii) 33   c) 148, 193   d) 94, 127

2. a) $720   b) 144   c) $1630 profit   3. $80   4. $184

5. 0.8 cm   6. $25.60; $6.70; 4; $55.30   7. a) -11 b) 23   c) -10

d) -20 e) 6   f) -14

8. a) $\frac{14}{15}$ b) $\frac{1}{4}$ c) $\frac{1}{10}$ d) $2\frac{2}{3}$ e) $1\frac{1}{10}$ f) $1\frac{11}{16}$

9. a) 4^5 b) 1^7 c) $2^3 \times 5^2$

10. a) 6^5 b) 7^8 c) 3^7 d) 10^3 e) 5^4 f) 2^{-1}

11. a) 5×10^4 b) 6.1×10^5 c) 3×10^{-4} d) 1.5×10^{-3} e) 1×10^7

12. a) 3×10^{10} b) 4×10^4 c) 8×10^6 d) 4.5×10^7 13. 8.64×10^5

Examination-style exercise 1

1. $-29\,°C$ 2. 12   3. a) 12   b) 5   4. 20

5. a) 36   b) 27   c) 31   d) 34

6. a) 10   b) 3   c) -2 7. a) 1.689×10^6 b) 4.26×10^{-2} mm

8. 3.54×10^{-3} 9. a) 12   b) $\frac{13}{8} - \frac{6}{8} = \frac{7}{8}$ 10. 14   11. 25

Chapter 2 Algebra 1

Exercise 2.1A
1. a) 2   **b) i)** 3   **ii)** 1   **2. a)** 4   **b) i)** 9   **ii)** -5 **c)** 7
3. a) 5   **b) i)** 7   **ii)** -3 **c)** -6
4. 3 is also a term so there are three terms. The coefficient of x is 1, not zero, and the coefficient
    of y is -7, not 7.
5. a) i) $x + 3$ **ii)** $4x$ **b)** $x - 8$
6. a) $a + b$ **b)** $a + 5b$ **c)** $6b$ **7. a)** p^2 **b)** $4p$
8. a) xy **b)** $2x + 2y$

Exercise 2.1B
1. $7x + 2y$ **2.** $5x + 3y$ **3.** $2a - 3b$ **4.** $2q - 5p$ **5.** $6x^2 - 5x$
6. $3a^2 + 6a$ **7.** $3x + 2y + 5xy$ **8.** $p^2 - 3q^2 + 4pq$ **9.** $4e - 3c - 3d$
10. $11a - 18b - 6ab + 2a^2 + 4b^2$

Exercise 2.1C
1. -5 **2.** 8   **3.** -17 **4.** 8   **5.** -2
6. -27 **7.** 1   **8.** -22 **9.** -22 **10.** -22
11. -10 **12.** -2 **13.** 23   **14.** -44 **15.** 26
16. 25   **17.** -4 **18.** 0   **19.** -16 **20.** 22
21. -5 **22.** 30   **23.** 13   **24.** 25   **25.** 40
26. 3   **27.** -5 **28.** -12 **29.** -34 **30.** 2
31. 12   **32.** 39   **33.** 40   **34.** 7   **35.** 3
36. 10   **37.** 51   **38.** -2 **39.** 1   **40.** 11

Exercise 2.2A
1. a) 17; add 4   **b)** 27; subtract 3   **c)** 48; $\times 2$
   d) 30; add 5, 6, 7, etc.   **e)** 12.5; $\div 2$ **f)** 121; add 11
2. a) 11; add 5   **b)** 16; $\times 2$ **c)** -1; add 3
   d) 2.4; $\div 10$ **e)** 11; add 1, 2, 3, etc.   **f)** 0; subtract 4
3. a) 7, 11, 18, 29, 47, 76, 123   **b)** 12, 19, 31, 50   **c)** 1, 1, 2, 3, 5, 8, 13, 21, 34, 55
4. a) $6 \times 7 = 6 + 6^2$, $7 \times 8 = 7 + 7^2$ **b)** $10 \times 11 = 10 + 10^2$, $30 \times 31 = 30 + 30^2$
5. $5 + 9 \times 1234 = 11\,111$
    $6 + 9 \times 12\,345 = 111\,111$
    $7 + 9 \times 123\,456 = 1\,111\,111$
6. 63, 3968   **7.** 3, 5, 5   **8.** Yes
9. a) $1^3 + 2^3 + 3^3 + 4^3 = (1 + 2 + 3 + 4)^2 = 10^2 = 100$;
      $1^3 + 2^3 + 3^3 + 4^3 + 5^3 = (1 + 2 + 3 + 4 + 5)^3 = 15^2 = 225$;
      $1^3 + 2^3 + 3^3 + 4^3 + 5^3 + 6^3 = (1 + 2 + 3 + 4 + 5 + 6)^2 = 21^2 = 441$
   b) $(1 + 2 + 3 + \ldots + 10)^2 = 55^2 = 3025$
10. a) 16   **b)** 15   **c)** 26   **d)** 25
   e) The nth odd number is one less than the nth even number   **f)** 113
   g) i) 90   **ii)** 105   **iii)** 199   **iv)** 437
11. a) i) 24   **ii)** 36   **iii)** 75   **b) i)** 23   **ii)** 35   **iii)** 59
   c) i) 28   **ii)** 39   **iii)** 50   **iv)** 88   **d) i)** 40   **ii)** 21   **iii)** 31   **iv)** 50
12. a) 2, 3, 4, 5, etc.   **b)** The differences increase by 1 each time
   c) i) 36   **ii)** 78   **iii)** 120

Exercise 2.3A
1. $3n + 2$ **2. a)** $4n + 1$ **b)** $6n - 4$ **3.** 51; $5n + 1$ **4. a)** $4n + 2$ **b)** 82
5. a) $2n + 3$; 103   **b)** $4n - 1$; 199   **c)** $6n - 4$; 296   **6.** $5n - 2$ **7.** $-3n + 18$ **8.** $6n + 1$
9. $-2n + 16$ **10.** $n + 5$ **11.** $-3n + 16$ **12.** $8n - 3$ **13.** $-5n + 32$ **14.** $-3n + 44$
15. $10n + 3$ **16.** $3n + 3$ **17.** $-5n + 136$
18. a) $-8n + 34$; -126 **b)** $-3n + 19$; -41 **c)** $-9n + 57$; -123
19. 72   **20.** All of the terms in the sequence are even
21. a) $2n$ **b)** $2n - 1$ **c)** $2n + 2n - 1 = 4n - 1$ which is odd

Exercise 2.3B
1. $w = b + 4$ 2. $w = 2b + 6$ 3. $w = 2b - 12$ 4. $m = 2t + 1$ 5. $m = 3t + 2$
6. $s = t + 2$ 7. a) $p = 5n - 2$ b) $k = 7n + 3$ c) $w = 2n + 11$
8. a) $y = 3n + 1$ b) $h = 4n - 3$ c) $k = 3n + 5$ 9. $m = 8c + 4$

Exercise 2.3C
1. 5, 8, 13, 20, 29   2. 2, 6, 12, 20, 30   3. 3, 10, 29, 66, 127   4. 0, 6, 24, 60, 120
5. $n^2 + 3$ 6. $2n^2$ 7. $n^2 - 1$ 8. $\frac{1}{2}n^2$ 9. $n^2 - 7$
10. $-n^2$ 11. $-n^2 + 1$ 12. $n^3 + 1$ 13. $2n^3$ 14. $n^3 - 2$

Exercise 2.4A
1. $3x + 9$ 2. $4x - 8$ 3. $10x + 5$ 4. $4a + 28$ 5. $12x + 6$
6. $50 - 10x$ 7. $12x + 15$ 8. $27 + 9x$ 9. $5y - 10$ 10. $7a - 14$
11. $22x - 11y$ 12. $24x + 16y$ 13. $4x + 2x^2$ 14. $3x^2 - 2x$ 15. $4y^2 + 7y$
16. $2a^2 - 6a$ 17. $12x - 18x^2$ 18. $12y^2 + 20y$ 19. $15a^2 - 20a$ 20. $42p - 49p^2$
21. $12x^2 + 3x^3$ 22. $4x^3 - 12x^2$ 23. $20x - 1$ 24. $17x - 17$ 25. $6x^2 + 4x$
26. $8x + 27$ 27. $10x + 16$ 28. $18x - x^2$

Exercise 2.4B
1. $x^2 + 3x + 2$ 2. $x^2 + 10x + 24$ 3. $x^2 + x - 6$ 4. $x^2 + 4x - 12$ 5. $x^2 - 2x - 35$
6. $x^2 - 5x - 24$ 7. $x^2 - 3x + 2$ 8. $x^2 - 8x + 15$ 9. $x^2 - 17x + 72$ 10. $2x^2 + 5x + 2$
11. $3x^2 - x - 2$ 12. $2x^2 + 5x - 12$ 13. $4x^2 + 8x + 3$ 14. $x^2 + 10x + 25$

Exercise 2.4C
1. $2(3x + 2y)$ 2. $3(3x + 4y)$ 3. $2(5a + 2b)$ 4. $4(x + 3y)$
5. $5(2a + 3b)$ 6. $6(3x - 4y)$ 7. $4(2u - 7v)$ 8. $5(3s + 5t)$
9. $8(3m + 5n)$ 10. $9(3c - 8d)$ 11. $4(5a + 2b)$ 12. $6(5x - 4y)$
13. $3(9c - 11d)$ 14. $7(5u + 7v)$ 15. $4(3s - 8t)$ 16. $8(5x - 2t)$
17. $12(2x + 7y)$ 18. $4(3x + 2y + 4z)$ 19. $3(4a - 2b + 3c)$ 20. $5(2x - 4y + 5z)$
21. $4(5a - 3b - 7c)$ 22. $8(6m + n - 3x)$ 23. $7(6x + 7y - 3z)$ 24. $3(2x^2 + 5y^2)$
25. $5(4x^2 - 3y^2)$ 26. $7(a^2 + 4b^2)$ 27. $9(3a + 7b - 4c)$ 28. $6(2x^2 + 4xy + 3y^2)$
29. $8(8p - 9q - 5r)$ 30. $12(3x - 5y + 8z)$ 31. $a(a + 4)$ 32. $x(3 + 4x)$
33. $x(4x - 1)$ 34. $x(7 - 3x)$ 35. $2x(x + 2)$ 36. $3x(2 - x)$
37. $4x(3 + 4x^2)$ 38. $5x^2(5 - 3x)$ 39. $10x^2(3x + 1)$ 40. $10y^2(8y - 3)$

Exercise 2.5A
1. 12   2. 9   3. 18   4. 4   5. 17
6. -5 7. 6   8. -7 9. 4   10. 8
11. 17   12. -5 13. 5   14. 6   15. 2
16. $\frac{4}{5}$ 17. $2\frac{1}{3}$ 18. $7\frac{1}{2}$ 19. $1\frac{5}{6}$ 20. 0
21. $\frac{5}{9}$ 22. 1   23. $\frac{1}{5}$ 24. $\frac{2}{7}$ 25. $\frac{3}{4}$
26. $\frac{2}{3}$ 27. $1\frac{1}{4}$ 28. $1\frac{1}{5}$ 29. $1\frac{5}{9}$ 30. $\frac{1}{3}$
31. $\frac{1}{2}$ 32. $\frac{1}{10}$ 33. $-\frac{3}{8}$ 34. $\frac{9}{50}$ 35. $\frac{1}{2}$
36. $\frac{3}{5}$ 37. $-\frac{4}{9}$ 38. 0   39. $4\frac{5}{8}$ 40. $-1\frac{3}{7}$
41. $2\frac{1}{3}$ 42. $\frac{3}{4}$ 43. 1   44. $3\frac{3}{5}$ 45. $\frac{1}{3}$
46. $2\frac{1}{14}$ 47. -1 48. $-\frac{5}{6}$ 49. $8\frac{1}{4}$ 50. -55

Exercise 2.5B
1. $2\frac{3}{4}$ 2. $1\frac{2}{3}$ 3. 2   4. $\frac{1}{5}$ 5. $\frac{1}{2}$
6. 2   7. $5\frac{1}{3}$ 8. $1\frac{1}{5}$ 9. 0   10. $\frac{2}{9}$
11. $1\frac{1}{2}$ 12. $\frac{1}{6}$ 13. $1\frac{1}{3}$ 14. $\frac{6}{7}$ 15. $\frac{4}{7}$

16. 7 **17.** $\frac{5}{8}$ **18.** 5 **19.** $\frac{2}{5}$ **20.** $\frac{1}{3}$

21. 4 **22.** -1 **23.** 1 **24.** $\frac{6}{7}$ **25.** $1\frac{1}{4}$

26. 1 **27.** $\frac{7}{9}$ **28.** $-1\frac{1}{2}$ **29.** $\frac{2}{9}$ **30.** $-1\frac{1}{2}$

Exercise 2.6A

1. 3 **2.** 5 **3.** $10\frac{1}{2}$ **4.** -8 **5.** $\frac{1}{3}$

6. $-4\frac{1}{2}$ **7.** $3\frac{1}{3}$ **8.** $3\frac{1}{2}$ **9.** $3\frac{2}{3}$ **10.** -2

11. $-5\frac{1}{2}$ **12.** $4\frac{1}{5}$ **13.** $\frac{3}{7}$ **14.** $\frac{7}{11}$ **15.** $4\frac{4}{5}$

16. 5 **17.** 9 **18.** $-2\frac{1}{3}$ **19.** $\frac{2}{5}$ **20.** $\frac{3}{5}$

Exercise 2.6B

1. 24 **2.** 15 **3.** -10 **4.** 21 **5.** 21

6. $2\frac{2}{3}$ **7.** $4\frac{3}{8}$ **8.** $1\frac{1}{2}$ **9.** $3\frac{3}{4}$ **10.** $1\frac{1}{3}$

11. $3\frac{3}{5}$ **12.** 2 **13.** -24 **14.** -70 **15.** 200

16. $\frac{3}{5}$ **17.** $\frac{4}{7}$ **18.** $\frac{11}{12}$ **19.** $\frac{6}{11}$ **20.** $\frac{2}{3}$

21. $\frac{5}{9}$ **22.** $\frac{7}{9}$ **23.** $1\frac{1}{3}$ **24.** $\frac{1}{2}$ **25.** $\frac{2}{3}$

26. 3 **27.** $1\frac{1}{2}$ **28.** $\frac{5}{8}$ **29.** $\frac{7}{19}$ **30.** $-\frac{3}{5}$

31. $8\frac{1}{4}$ **32.** -500 **33.** $-\frac{98}{99}$ **34.** 6 **35.** 30

Exercise 2.6C

1. 2 **2.** 3 **3.** 2 **4.** 2 **5.** 2 **6.** 3 **7.** 6

Exercise 2.7A

1. 3 **2.** 17 **3.** 14 **4.** 82 **5.** $-\frac{1}{2}$

6. $17\frac{2}{3}$ **7.** $\frac{1}{6}$ **8.** 5 **9.** 12 **10.** $3\frac{1}{3}$

11. $4\frac{2}{3}$ **12.** 6

Exercise 2.8A

1. a) a^3 **b)** n^4 **c)** s^5 **d)** $p^2 \times q^3$ **e)** b^7
2. a) x^7 **b)** y^8 **c)** a^5 **d)** x^4 **e)** x^5
f) p^6 **g)** y^{-3} **h)** x^4 **i)** x^3 **j)** x^6
k) a^{15} **l)** n^{14} **m)** y^9 **n)** x^{-3}

Exercise 2.8B

1. a) $6a^5$ **b)** $20n^4$ **c)** $14x^5$ **d)** $24y^7$ **e)** $5n^7$ **f)** $12y^2$
g) $9p^5$ **h)** $10p^6$ **i)** $8x^6$ **j)** $27a^6$ **k)** $16y^6$ **l)** $25x^8$
2. 3 **3.** 1 **4.** 3 **5.** 0 **6.** 3 **7.** 1
8. 2 **9.** 3 **10.** -1 **11.** -1 **12.** 0 **13.** 2
14. 4 **15.** 0 **16.** -1 **17.** 0

Exercise 2.9A

1. a) $<$ **b)** $>$ **c)** $>$ **d)** $>$ **e)** $>$ **f)** $>$
2. a) $x > 2$ **b)** $x \leqslant 5$ **c)** $x < 100$ **d)** $-2 \leqslant x \leqslant 2$ **e)** $x \geqslant -6$ **f)** $3 < x \leqslant 8$

3. a)

b)

c)

d)

e)

4. a) $A \geqslant 16$ **b)** $3 < A \leqslant 70$ **c)** $150 < T < 175$ **d)** $h \geqslant 1.5$

5. a) True   **b)** True   **c)** True

6. a) =   **b)** <   **c)** =   **d)** <   **e)** =   **f)** <

7. −3, −2, −1, 0, 1   **8.** 8, 9, 10, 11   **9.** 3, 4, 5, 6, 7, 8

10. a) 16   **b)** −16   **c)** 20   **d)** −5

Revision exercise 2

1. a) 30, 37; add 7   **b)** 12, 10; subtract 2   **c)** 7, 10; add 3   **d)** 8, 4; divide by 2
    e) 26, 33; add 3, 4, 5, etc.

2. a) 9   **b)** 50   **c)** $7 \times 11 - 6 = 72 - 1$

3. a) 4   **b)** 19

4. a) $1 + 2 + 3 + 4 + 5 + 4 + 3 + 2 + 1 = 5^2$ **b)** $... + 4 + ... + 9 + ... + 1 = 9^2$
    $1 + ... + 6 + ... + 1 = 6^2$

5. a) $l = 2d - 4$ **b)** 149   **c)** No: all the terms are even

6. a)

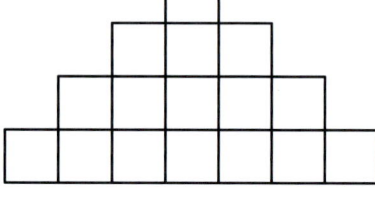

    b) 1, 4, 9, 16   **c)** Square numbers   **d)** 49

7. a)

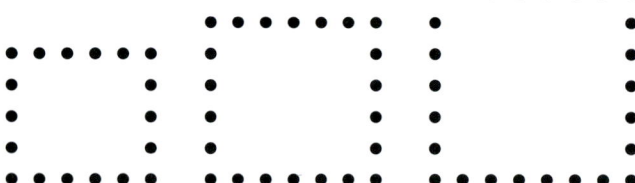

    b) $3 \to 14, 4 \to 18, 5 \to 22, 6 \to 26$ **c) i)** 42   **ii)** 62   **d)** $n = 4x + 2$

8. a) 9   **b)** 11   **c)** 3   **d)** 7

9. a) 7   **b)** $\frac{1}{4}$ **c)** $\frac{4}{5}$ **10. a)** $6x + 15 < 200$ **b)** 29

11. a) 2   **b)** 2   **c)** 4   **12. a)** x^8 **b)** n^9 **c)** $12a^3$

Examination-style exercise 2

1. a) 25   **b)** 45   **c)** $4n + 5$ **d)** No: all the terms are odd

2. a) i) −9, −14   **ii)** Subtract 5   **iii)** $16 - 5n$ **b)** $5n - 9$ **c)** 7

3. a) 35   **b)** 120   **c)** $n^2 + 2n$ **4.** $2x(2y - 1)$ **5.** $9x^2 - 6xy$

6. a) $6c - 16d$ **b)** $a(b - a)$ **7. a)** 6   **b)** 11   **c)** $-\frac{5}{3}$

8. a) $x - 5$ **b)** $2x + 7$ **c)** $2x + 7 = 4(x - 5)$ **d)** \$13.50

9. 20   **10.** $-\frac{3}{7}$

11. a) 2   **b)** −4   **12. a)** 1   **b)** x^{12}

13. a) 1   **b)** p^9 **c)** x^{-12} **14.** A, B and D   **15.** −3, −2, −1, 0, 1, 2, 3, 4

Chapter 3 Shape and space 1

Exercise 3.1A
1. 7.3 cm
2. 7.9 cm
3. 8.0 cm
4. 10.3 cm
5. 6.4 cm
6. 6.8 cm
7. 9.0 cm
8. 9.6 cm
9. 7.6 cm
10. 8.7 cm
11. 8.2 cm
12. 5.3 cm

Exercise 3.1B
1. $63°$
2. $35°$
3. $62°$
4. $30°$
5. $37°$
6. $94°$
7. 4.9 cm

Exercise 3.1C
1. (a), (b), (d)
2. Student nets
3. Student nets
4. (d) Square-based pyramid
5. Student constructions
6. Student nets
7. Student nets

Exercise 3.2A
1. $70°$
2. $100°$
3. $70°$
4. $100°$
5. $44°$
6. $80°$
7. $40°$
8. $48°$
9. $40°$
10. $35°$
11. $a = 40°, b = 140°$
12. $x = 108°, y = 72°$

Exercise 3.2B
1. $50°$
2. $70°$
3. $29°$
4. $30°$
5. $70°$
6. $42°$
7. $40°$
8. $a = 55°, b = 70°$
9. $60°$
10. $x = 122°, y = 116°$
11. $135°$
12. $30°$
13. $154°$
14. $75°$
15. $x = 30°$
16. $28°$

Exercise 3.2C
1. $72°$ Alternate angles
2. $98°$ Co-interior angles
3. $80°$ Vertically opposite angles and co-interior angles
4. $74°$ Corresponding angles
5. $86°$ Alternate angles
6. $88°$ Co-interior angles and vertically opposite angles
7. $x = 95°$ Alternate angles, $y = 50°$ Alternate angles
8. $a = 87°$ Co-interior angles, $b = 74°$ Alternate angles
9. $a = 65°$ Vertically opposite and co-interior angles, $c = 103°$ Co-interior and vertically opposite angles
10. $a = 68°$ Alternate angles, $b = 42°$ Alternate angles
11. $y = 65°$ Alternate angles, $z = 50°$ Co-interior angles
12. $a = 55°$ Co-interior angles, $b = 75°$ Angles in triangle, $c = 50°$ Alternate angles

Exercise 3.2D
1. $108°$ Angles in quadrilateral add to $360°$
2. $50°$ Angles in quadrilateral add to $360°$
3. $76°$ Angles in quadrilateral add to $360°$
4. $270°$ Angles in quadrilateral add to $360°$
5. $a = 119°$ Angles in quadrilateral add to $360°$,
   $b = 25°$ Angles on straight line add to $180°$, Angles in triangle add to $180°$
6. $c = 70°$ Angles in quadrilateral add to $360°$, Angles on straight line add to $180°$,
   $d = 60°$ Angles on straight line add to $180°$, Angles in triangle add to $180°$
7. $a = 45°, b = 67\frac{1}{2}°$
8.

Polygons		
Name	Number of sides	Angle at centre
Quadrilateral	4	$90°$
Pentagon	5	$72°$
Hexagon	6	$60°$
Heptagon	7	$51.4°$ (1 d.p.)
Octagon	8	$45°$
Nonagon	9	$40°$
Decagon	10	$36°$

Exercise 3.2E

1. $42°$ Angles in triangle add to $180°$
2. $68°$ Angles on straight line add to $180°$, Angles in triangle add to $180°$
3. $100°$ Angles in triangle add to $180°$, Angles on straight line add to $180°$
4. $73°$ Equal angles in isosceles triangle
5. $120°$ Angles around a point add to $360°$
6. $52°$ Equal angles in isosceles triangle
7. $100°$ Angles in quadrilateral add to $360°$
8. $a = 70°$, $b = 60°$ Corresponding angles
9. $x = 58°$ Corresponding angles, $y = 109°$ Co-interior angles
10. $66°$ Angles on straight line add to $180°$, Angles in triangle add to $180°$, Equal angles in isosceles triangle
11. $65°$ Alternate angles, Angles in triangle add to $180°$, Equal angles in isosceles triangle
12. $e = 70°$, $f = 30°$ Angles in triangle add to $180°$, Equal angles in isosceles triangle
13. $x = 72°$, $y = 36°$, Angles on straight line add to $180°$, Equal angles in isosceles triangle
14. $a = 68°$ Angles in triangle add to $180°$, $b = 72°$ Corresponding angles, $c = 68°$ Corresponding angles
15. $28.5°$ Equal angles in isosceles triangle, Angles on straight line add to $180°$, Angles in triangle add to $180°$
16. $20°$ Equal angles in isosceles triangle, Angles on straight line add to $180°$, Angles in triangle add to $180°$
17. $x = 62°$ Corresponding angles, $y = 28°$ Vertically opposite angles, Angles on straight line add to $180°$
18. $34°$ Equal angles in isosceles triangle, Alternate angles, Angles on straight line add to $180°$
19. $58°$ Angles in triangle add to $180°$, Alternate angles, Equal angles in isosceles triangle
20. $x = 60°$ Co-interior angles, $y = 48°$ Co-interior angles, Angles on straight line add to $180°$, Angles in triangle add to $180°$
21. $a = 65°$, $b = 40°$ Equal angles in isosceles triangle
22. $x = 49°$ Alternate angles, $y = 61°$ Angles on straight line add to $180°$, Angles in triangle add to $180°$
23. $a = 60°$ Angles in equilateral triangle, $b = 40°$ Equal angles in isosceles triangle
24. $136°$ Equal angles in isosceles triangle, Angles around a point add to $360°$
25. $80°$ Alternate angles, Vertically opposite angles, Angles around a point add to $360°$
26. $x = 65°$ Angles in triangle add to $180°$, $y = 35°$ Alternate angles, $z = 55°$ Angles in right-angled triangle
27. $26°$ Alternate angles, Angles in triangle add to $180°$, Angles on straight line add to $180°$, Equal angles in isosceles triangle

Exercise 3.3A

1. a) Trapezium   b) Square   c) Parallelogram
2. a) (6, 6)   b) (6, 4)   c) (0, 6), (1, 6), (3, 6)
3.

	How many lines of symmetry?	How many pairs of opposite sides are parallel?	Diagonals always equal?	Diagonals are perpendicular?
Square	4	2	Y	Y
Rectangle	2	2	Y	N
Kite	1	0	N	Y
Rhombus	2	2	N	Y
Parallelogram	0	2	N	N
Arrowhead	1	0	N	Y (outside)

4. a) $50°$ b) $100°$ c) $90°$ d) $130°$ e) $80°$ f) $95°$
5. Many possible answers   6. Several possible answers
7. Kite   8. Kite, parallelogram
9. a) Yes, yes, no   b) Can make *any* kind of triangle

Exercise 3.3B

1. C(−3, −3), D(−4, 2)   2. a) $34°$ b) $56°$
3. a) $35°$ b) $35°$ 4. a) $72°$ b) $108°$ c) $80°$
5. a) $40°$ b) $30°$ c) $110°$ 6. a) $116°$ b) $32°$ c) $58°$
7. a) $55°$ b) $55°$ 8. a) $26°$ b) $26°$ c) $77°$

9. a) 52° **b)** 64° **c)** 116° **10.** 110°
11. a) 54° **b)** 72° **c)** 36° **12. a)** 60° **b)** 15° **c)** 75°

Exercise 3.3C
1. 5 **2.** Student drawings **3.** 5 **4.** 6
5. a) Square **b)** Equilateral triangle **6.** Student drawings **7.** No

Exercise 3.4A
1. a) $a = 80°$, $b = 70°$, $c = 65°$, $d = 86°$, $e = 59°$ **2. a)** $a = 36°$ **b)** 144°
3. a) i) 40° **ii)** 20° **iii)** 8° **iv)** 6° **b) i)** 140° **ii)** 160° **iii)** 172° **iv)** 174°
4. $p = 101°$, $q = 79°$, $x = 70°$, $m = 70°$, $n = 130°$ **5.** 24 **6.** 9 **7.** 20
8. 20 **9. a)** 720 **b)** 1440 **10.** 14 **11.** 132

Exercise 3.4B
1. a) Student diagrams **b)** All of the angles are 90°
2. 90° Angle in a semi-circle **3.** 65° Angle in a semi-circle, Angles in triangle add to 180°
4. 45°, Angle in a semi-circle, Equal angles in isosceles triangle
5. 90° Angle between tangent and radius
6. $e = 40°$ Angle in a semi-circle, Angles in triangle add to 180°,
   $f = 50°$ Angle between tangent and radius
7. $g = 30°$ Angle in a semi-circle, Angles in triangle add to 180°
8. $h = 90°$ Angle in a semi-circle, $i = 60°$ Angles in triangle add to 180°
9. $j = 49°$ Angle between tangent and radius, Angles in triangle add to 180°
10. 45° Angle in a semi-circle, Equal angles in isosceles triangle
11. $l = 60°$, $m = 50°$ Angle in a semi-circle, Angles in triangle add to 180°
12. $n = 40°$ Angle in a semi-circle, Angles in triangle add to 180°
13. 50° Angle between tangent and radius, Angles in quadrilateral add to 360°

Exercise 3.5A
1. a) 1 **b)** 1 **2. a)** 1 **b)** 1 **3. a)** 4 **b)** 4
4. a) 2 **b)** 2 **5. a)** 0 **b)** 6 **6. a)** 0 **b)** 2
7. a) 0 **b)** 2 **8. a)** 4 **b)** 4 **9. a)** 0 **b)** 4
10. a) 4 **b)** 4 **11. a)** 6 **b)** 6
12. a) Infinite **b)** Infinite **13.** Student drawings

Exercise 3.5B
1. 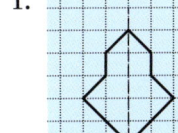 **2.** 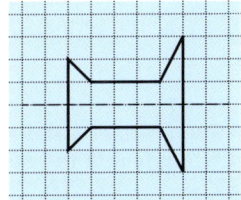 **3.** 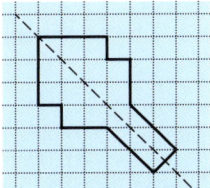 **4.**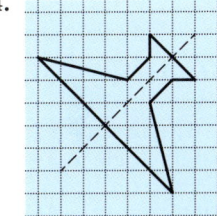

5. 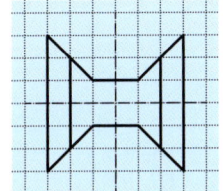 **6.** 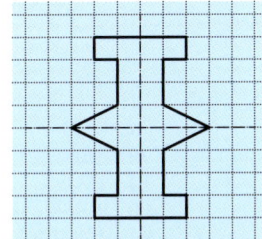 **7.** 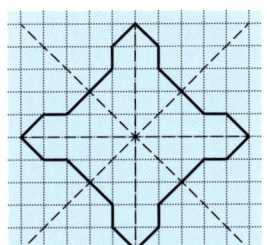 **8.**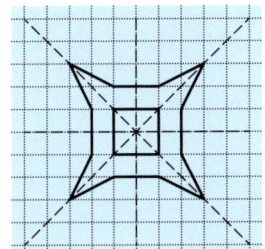

Exercise 3.6A
1. $AB = 5$ cm **5. b)** 12 km **6.** Yes (they are 3.6 km apart) **8.** 6.5 m
10. a) i) 14 m **ii)** 6 m **iii)** 4 m **b)** 8 m **c)** 14 m **d)** 2 cm **e)** 12 m **f)** 42 m²

Exercise 3.7A
1. a) 090°   b) 270°   c) 180°   d) 000°   e) 225°   f) 045°
2. A 035°, B 070°, C 155°, D 220°, E 290°, L 340°
3. A 040°, B 065°, C 130°, D 160°, E 230°, F 330°   4. 045°

Exercise 3.7B
1. a) $147\frac{1}{2}$°   b) 122°   c) 090°   2. a) 286°   b) 225°   c) 153°
3. a) 061°   b) $327\frac{1}{2}$°   4. a) 302°   b) 344°   c) 045°
5. a) 259°   b) 332°   c) 068°

Exercise 3.7C
Student drawings

Exercise 3.7D
Student drawings

Exercise 3.7E
1. 11.5 km   2. 14.1 km   3. a) 12.5 km   b) 032°
4. 6.9 km   5. a) 8.5 km   b) 074°
6. a) 8.4 km   b) 029°   7. b) 5.2 h   8. No

Revision exercise 3
1. (a) and (c)   2. a) 40°   b) 100°   3. b) 85.5 km (±1.5 km)
4. a) 220°   b) 295°   5. 93°, 39° and 48° (±2°)
6.   7. 15 sides

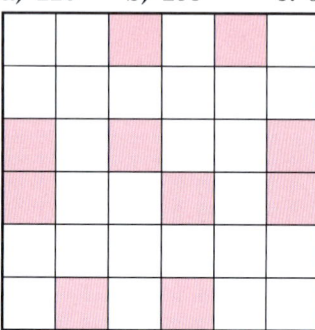

8. $c = 96°$, $d = 84°$, $e = 53°$, $f = 53°$. There are many possible ways of finding each answer.

Examination-style exercise 3
1. 36°
2. a) 67° Alternate angles   b) 23° Angle between tangent and radius
   c) 90° Angle in a semi-circle   d) 67° Angles in triangle add to 180°   3. Order 2
4. a) 140°   b) $p = 48$, $q = 84$, $r = 48$, $s = 48$, $t = 84$ c) 27°
5. 322°   6. a) $85 + (180 - 120) = 145°$ b) 265°

Chapter 4 Number 2

Exercise 4.1A
1. 2.35   2. 0.814   3. 26.2   4. 35.6   5. 113
6. 211   7. 0.825   8. 0.0312   9. 5.9   10. 1.2
11. 0.55   12. 0.72   13. 0.14   14. 1.8   15. 25
16. 31   17. 486.7   18. 500.4   19. 2.889   20. 3.113
21. 0.071 54   22. 3.041   23. 2464   24. 488 900   25. 0.513
26. 5.8   27. 66   28. 587.6   29. 0.6   30. 0.07
31. 5.84   32. 88   33. 2500   34. 52 700   35. 0.006
36. 4.40

Exercise 4.1B

1. 5.38	**2.** 11.05	**3.** 0.41	**4.** 0.37	**5.** 8.02
6. 87.04	**7.** 9.01	**8.** 0.07	**9.** 8.4	**10.** 0.7
11. 0.4	**12.** 0.1	**13.** 6.1	**14.** 19.5	**15.** 8.1
16. 7.1	**17.** 8.16	**18.** 3.0	**19.** 0.545	**20.** 0.0056
21. 0.71	**22.** 6.83	**23.** 0.8	**24.** 19.65	**25.** 0.0714
26. 60.1	**27.** −7.3	**28.** 3.80		

29. a) i) 5.9 cm by 3.3 cm   **ii)** 5.1 cm by 2.9 cm   **b) i)** 19.5 cm^2 **ii)** 14.8 cm^2

Exercise 4.2A

1. B	**2.** C	**3.** B	**4.** A	**5.** B
6. C	**7.** B	**8.** B	**9.** A	**10.** B
11. A	**12.** A	**13.** C (or B)	**14.** About 20	

15. a) True   **b)** True   **c)** True   **d)** False   **e)** False
    f) True   **g)** True   **h)** False   **i)** True

Exercise 4.2B

1. B	**2.** C	**3.** B	**4.** B **5.** A
6. B	**7.** B	**8.** A	**9.** No – it will be approximately $450

10. He got it wrong. Correct answer is $10.45 each
11. a) (Say) 200 g per paper: 3250 papers per tree   **b)** About 5×10^{11}
    c) For discussion
12. Yes: $20 \times 200 = 4000$ **13.** $10 \times 1 + 15 \times 1 + 20 \times 2.5 = 10 + 15 + 50 = 75$

Exercise 4.3A

1. b) $7.5 \leqslant V < 8.5$ **c)** $71.5 \leqslant m < 72.5$ **d)** $3.15 \leqslant t < 3.25$ **e)** $5.75 \leqslant r < 5.85$
2. 83.5 cm   **3.** 5.25 kg   **4. a)** 8.45 cm   **b)** 4.25 cm
5. 173.5 cm   **6.** 3.65 kg   **7.** 92.5 million miles (= 92 500 000 miles)
8. 20.625 seconds   **9.** 17.85 g
10.

	Lower bound	Upper bound
a)	5.55 cm	5.65 cm
b)	36.5 m	37.5 m
c)	0.265 mg	0.275 mg
d)	225 °C	235 °C
e)	314.5 km	315.5 km

Exercise 4.4A

1. 8	**2.** 32	**3.** 30	**4.** 4	**5.** 0
6. 6	**7.** 5	**8.** 1	**9.** 47	**10.** 6
11. 3	**12.** 1851	**13.** 6.889	**14.** 1.214	**15.** 6.619
16. 3.306	**17.** 2.303	**18.** 41.73	**19.** 8.163	**20.** 0.1090
21. 0.5001	**22.** 20.63	**23.** 10.09	**24.** 45.66	**25.** 52.86
26. 22.51	**27.** 5.479	**28.** 5.272	**29.** 0.2116	**30.** 57.19
31. 19.90	**32.** 6.578	**33.** 10.16	**34.** 0.082 80	**35.** 1855
36. 2.367	**37.** 1.416	**38.** 7.261	**39.** 0.3934	**40.** −0.7526
41. 2.454	**42.** 40 000	**43.** 3.003		

Exercise 4.4B

1. 0.57	**2.** 3.45	**3.** 431	**4.** 19.3	**5.** 0.22
6. 3942.7	**7.** 53	**8.** 18.4	**9.** 0.059	**10.** 1.1
11. 6140	**12.** 127.89	**13.** 20.3	**14.** 47.6	**15.** 599.1
16. 0.16				

Exercise 4.4C

1. a) 2.218 hours   **b)** 5.711 hours   **c)** 1.114 hours
2. a) 2 hours 38 minutes 24 seconds   **b)** 3 hours 52 minutes 48 seconds
    c) 8 hours 17 minutes 24 seconds

3. a) 4 hours 22 minutes 12 seconds   **b)** 14 hours 32 minutes 24 seconds
4. 530   **5.** $73.20   **6.** $26 875   **7.** $8.60   **8.** 2 hours 15 minutes

Revision exercise 4
1. a) 2.088   **b)** 3.043
2. a) 0.340   **b)** 4.08×10^{-6} **c)** 64.9   **d)** 0.119
3. a) 600   **b)** 9000 or 10 000   **c)** 3   **d)** 60
4. 2.1×10^{24} tonnes
5. a) 0.5601   **b)** 3.215   **c)** 0.6161   **d)** 0.4743
6. About 92 g   **7.** About 3 g
8. a) 18.72   **b)** 89.18   **c)** 63.99   **d)** 144.78   **e)** 31.16
    f) 48.248   **g)** 9.24   **h)** 1.92   **i)** 4.08
9. a) 44.5 cm   **b)** 45.5 cm
10. a) 0.2865 seconds   **b)** 0.2875 seconds

Examination-style exercise 4
1. 1 390 000 km
2. No: Min height of bridge is 4.35 metres and max height of lorry is 4.385 metres
3. $9450 \leqslant d < 9550$
4. a) 0.1234759102   **b)** 0.12   **5.** 40   **6.** $9000
7. a) 0.1977988324   **b)** 0.198   **8.** $7.25 \leqslant m < 7.35$

Chapter 5 Handling data

Exercise 5.1A
1. a) Mean = 6, median = 5, mode = 4   **b)** Mean = 9, median = 7, mode = 7
    c) Mean = 6.5, median = 8, mode = 9   **d)** Mean = 3.5, median = 3.5, mode = 4
2. 2 °C   **3. a)** 3   **b)** 3
4. Many possible answers   **5.** 70.4, 73.25, No   **6.** 6
7. a) 1.6 m   **b)** 1.634 m   **8. a)** 51 kg   **b)** 50 kg
9. a) 7.2   **b)** 5   **c)** 6
10. a) Mean = $47 920, median = $22 500, mode = $22 500   **b)** The mean is skewed by one large number.
11. a) Mean = 157.1 kg, median = 91 kg
    b) Mean. No: over three quarters of the sheep are below the mean weight.
12. a) Mean = 74.5 cm, median = 91 cm   **b)** Yes
13. a) 3.38   **b)** 8   **c) i)** The median number is much larger for children than adults
    ii) The range for children was bigger than the range for adults
14. a) 3.475   **b)** 7   **c)** On average, people have more coins than notes. The number
                        of notes that people have is more consistent/less varied.
15. a) 3.025; 3; 3   **b)** 17.75; 17; 17   **c)** 3.38; 4; 4
16. a) 5.17   **b)** 5   **17.** 78 kg   **18.** 35.2 cm
19. a) 2   **b)** 9
20. a) 20.4 m   **b)** 12.8 m   **c)** 1.66 m
21. 55 kg   **22.** 12   **23.** Mean = 17, median = 3; The median is more representative
24. The median   **25.** 15, 20, 31

Exercise 5.2A
1. a) Sharon   **b)** $11   **c)** Half of a $ symbol
2. a) 5   **b)**

Make	Number of cars	
Ford	20	
Renault	30	
Toyota	25	
Audi	15	

3.

House	Number of letters
1	
2	
3	
4	
5	
6	

Key: represents 4 letters

Exercise 5.2B

1. a)

Stem	Leaf						
2	1	6	9				
3	3	6	7				
4	2	5	5	8	8		
5	2	3	3	6			
6	0	2	2	2	4	6	8
7	1	4	6				

Key
2 | 6 means a score of 26

b) 53   **c)** 55

2. a) 12   **b)** 14.35 seconds   **c)** 2.8 seconds   **d)** 15.2 seconds

3. a)

Girls				Boys	
7	5	1	16		
		4	17	5	7
	8	5	18	2	8
	4	3	19	3	
		8	20		
		0	21	0	9
			22	2	7
			23	1	

Key
4 | 17 | 7 means 17.4 cm for the girls and 17.7 cm for the boys
b) Girls 18.65 cm, Boys 20.15 cm
c) Girls 4.9 cm, Boys 5.6 cm

4. a)

		Horror			Action		
	9	8	5	8	8		
9	5	4	2	0	9	0	3	9
			5	0	10	0	6
				11	0	0	9
				12	1		

Key
2 | 9 | 0 means 92 minutes for Horror and 90 minutes for Action
b) Horror 93 minutes, Action 103 minutes
c) On average, the action films are longer.

Exercise 5.3A

1. **a)** $425 **b)** $150 **c)** $250 **d)** $75
2. **a)** $13 333.30 **b)** $15 000 **c)** $6 666.70 **d)** $10 000 **e)** $12 000
3. **a) i)** $21 600 000 **ii)** $8 000 000 **b)** 10° **c)** $1 000 000
4. **a) i)** 8 min **ii)** 34 min **iii)** 10 min **b)** 18°
5. **a)** Lucy **b)** 20 min **c)** 22 min

Exercise 5.3B

1. **a) i)** 45° **ii)** 200° **iii)** 110° **iv)** 5° **b)** Student diagrams
2. **a)** $\frac{3}{10}, \frac{4}{10}, \frac{1}{5}, \frac{1}{10}$ **b)** Student diagrams
3. $x = 60°, y = 210°$ **4.** Barley 60°, Oats 90°, Rye 165°, Wheat 45°
5. **a)** 180° **b)** 36° **c)** 90° **d)** 54°

Exercise 5.4A

1. **a)** 5 **b)** 19 **c)** 23 **d)** 55 **e)** $\frac{6}{23}$
   f) 2 **g)** $6 - 1 = 5$
2. **a)** **b)** Student diagrams

Score	Tally	Frequency
2	||	2
3	||||	4
4	⊮	5
5	⊮||	7
6	⊮||||	9
7	⊮⊮|||	13
8	⊮|	6
9		0
10	⊮	5
11	⊮|	6
12	|||	3

3. **a)** £3000 approx. **b)** Profits increase in months before Christmas. Very few sales after Christmas.

Exercise 5.4B

1. **a)** Dual bar chart to show the favourite type of music of some boys and girls **b)** Composite bar chart to show the favourite type of music of some boys and girls

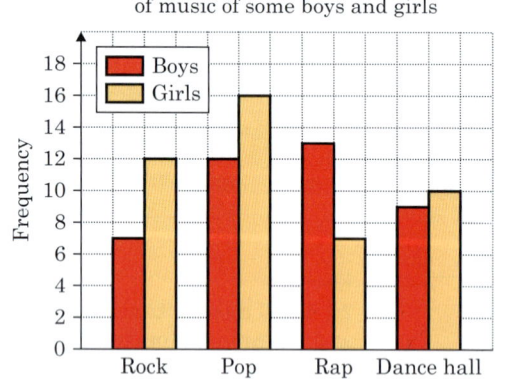
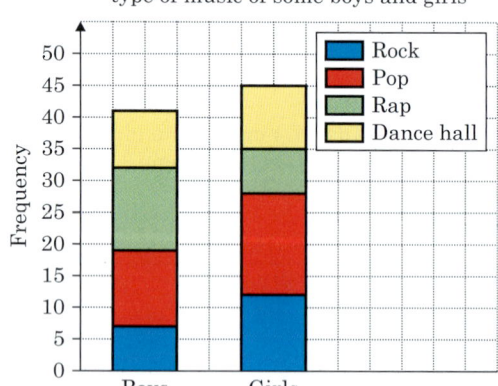

c) The composite bar chart (can compare the total height of the bar)

2. a)

	Curry	Kebab	Dhal
Adults	12	15	8
Children	9	12	16

b)

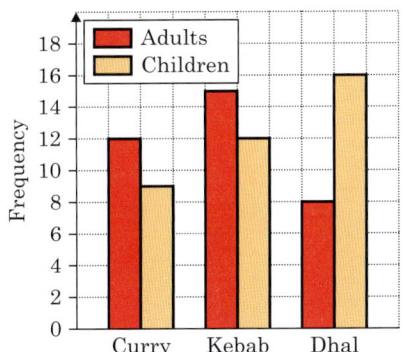

Dual bar chart showing the favourite foods of some adults and children

c)

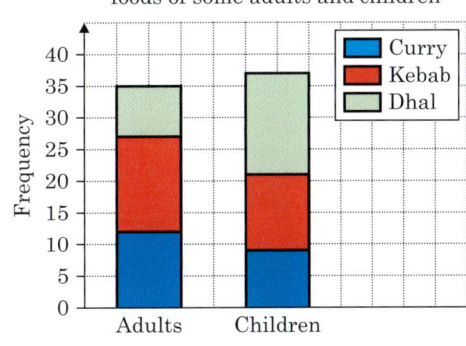

Composite bar chart showing the favourite foods of some adults and children

d) The dual bar chart (can directly compare the height of the kebab bars)

3. a)

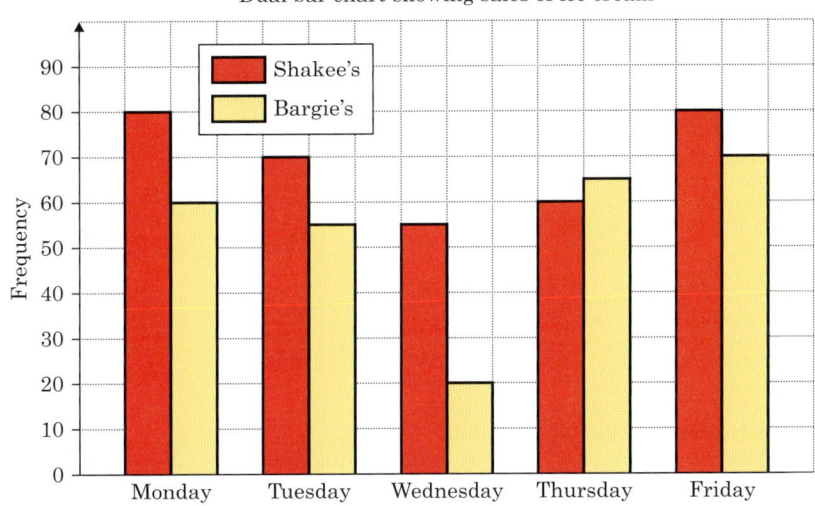

Dual bar chart showing sales of ice cream

b) Wednesday – the bar is much lower than the Shakee's bar and also much lower than the other Bargie's' bars

c)

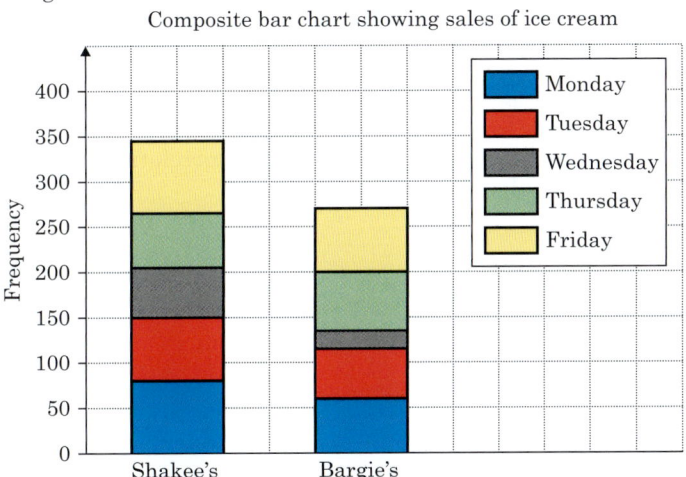

Composite bar chart showing sales of ice cream

d) Shakee's – the bar is much taller than for Bargie's

4. a)

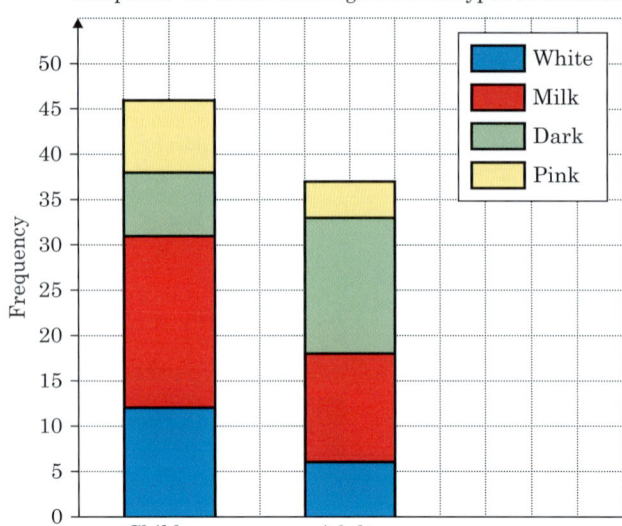

Composite bar chart showing favourite types of chocolate

b) There were 9 more children than adults in the survey.

Exercise 5.4C

1. a) Group B   **b)** 5   **c)** 35   **d)** 56   **e)** $\frac{1}{8}$

2. a) D   **b)** A   **c)** A   **d)** C   **e)** B

3.

Mass	Tally	Frequency
$60 \leqslant m < 70$	||||	4
$70 \leqslant m < 80$	ЖІ	6
$80 \leqslant m < 90$	Ж	5
$90 \leqslant m < 100$	ЖЖ	10
$100 \leqslant m < 110$	ЖЖІІІІ	14
$110 \leqslant m < 120$	ЖЖІ	11

Student diagrams

Field A

4. a) Student tables   **b)** Student diagrams   **c)** No significant change.

Exercise 5.5A

1. a) Student bar charts   **b)** Student frequency polygons

2. a) 25   **b)** 90   **3.** Student diagrams

4. a) 62   **b)** Sport B has more heavy people. Sport A has a much smaller range of weights compared with sport B.   **c)** For discussion

5. a) Plants with fertilizer are significantly taller   **b)** No significant effect

Exercise 5.6A

1. a) Strong positive correlation   **b)** No correlation   **c)** Weak negative correlation

2. a) No correlation   **b)** Strong positive correlation

    c) No correlation   **d)** Strong negative correlation

3. There are only four data points shown. More data needs to be gathered to further support the inference.

4. While there is correlation, this does not imply causation. It is more likely that the correlation is coincidental, or that both variables are linked to a third factor, e.g. age.

5. Student diagrams

Exercise 5.6B

1. Student diagrams; $p \approx 11$ **2.** Student diagrams; $p \approx 9$

3. Student diagrams; not possible

4. a) b) Student diagrams   **c)** About 26   **d)** About 43

5. a) Student diagrams **b)** About 34 m.p.g. **c)** About 64 m.p.h.
6. a) Student diagrams **b)** About 20

Revision exercise 5

1. a) 5.89 **b)** 6 **c)** 7 **2.** 1.55 m
3. a) i) 560 kg **ii)** 57 kg **b)** 50 kg **4. a)** 84 **b)** 19.2
5. a) 25 **b)** 75 **c)** 20
6. Dubai 108°, Egypt 45°, Greece 90°, Turkey 36°, India 81°
7. a) 29 **b)** 23
c) Wrong; the diagram does not show the days on which each amount of rain fell.
8. a) F **b)** Possible **c)** Possible
9. a)

Stem	Leaf
0	7 9 9
1	2 2 2 3 4 4 7
2	0 3

Key
1 | 2 means 12 °C

b) i) 12.5 °C **ii)** 12 °C **iii)** 16 °C

10. a) c)

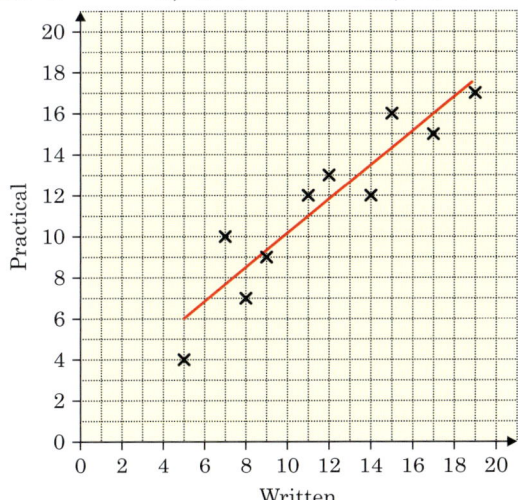

11.

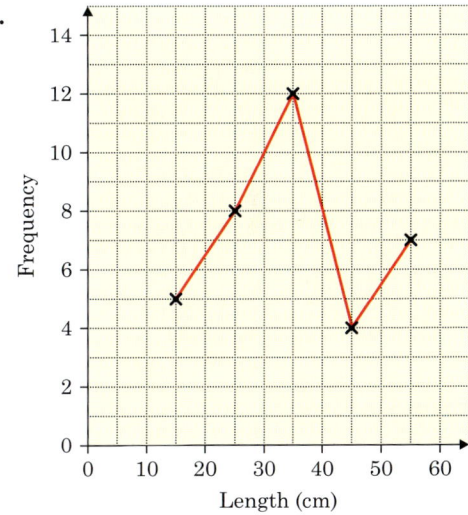

b) Positive correlation
d) 10

Examination-style exercise 5

1. a) 45°, 5, 75° **b)** Student diagrams
2. a) i) 35 **ii)** 7 **iii)** 8 **iv)** 7.71
b) i) 72 **ii)** Student diagrams **3.** Negative **4.** Positive
5. a) 756 **b)** 759 **c)** 29
6. a) i) 163.4 or 163 **ii)** 24
b) i) to ii) Student diagrams **iii)** About 163 cm
iv) Positive **v)** The larger a person's foot length, the greater their height
7. a) i) 3, 6, 8, 7, 6, 1, 1, 2 **ii)** 5.71 **iii)** 7 **iv)** 5 **v)** 5.5
b) i) 12, 25, 19, 2 **ii)** 5 and 6
8. a) i) 2016 Median = 44.8 seconds, 2020 Median = 44.15 seconds
ii) 2016 Range = 1.2 seconds, 2020 Range = 1.6 seconds
b) The median was lower in 2020 so, on average, the 2020 athletes were faster. The range, however, was lower in 2016, so the 2020 athletes were not more consistent.

Chapter 6 Number 3

Exercise 6.1A
1. **a)** 0.25   **b)** 0.4   **c)** 0.375   **d)** 0.35   **e)** 2.625   **f)** 9.7
2. **a)** $\frac{1}{5}$ **b)** $\frac{9}{20}$ **c)** $\frac{9}{25}$ **d)** $\frac{1}{8}$ **e)** $1\frac{1}{20}$ **f)** $\frac{7}{1000}$
3. **a)** 25%   **b)** 10%   **c)** 72%   **d)** 7.5%   **e)** 2%   **f)** 180%
4. '70% of ...'   **5.** $\frac{1}{5}$ of $5000

6. **a)** $\frac{1}{4}$, 0.25, 25%   **b)** $\frac{1}{5}$, 0.2, 20%   **c)** $\frac{4}{5}$, 0.8, 80%   **d)** $\frac{1}{100}$, 0.01, 1%
   e) $\frac{3}{10}$, 0.3, 30%   **f)** $4\frac{1}{20}$, 4.05, 405%

7. **a)** 45%, $\frac{1}{2}$, 0.6   **b)** 4%, $\frac{6}{16}$, 0.38   **c)** $\frac{1}{10}$, 11%, 0.111
8. 0.58   **9.** 1.42   **10.** 0.65   **11.** 1.61   **12.** 0.07
13. 0.16   **14.** 3.64   **15.** 0.60   **16.** 62.5%

Exercise 6.2A
1. $12   **2.** $8   **3.** $10   **4.** $3   **5.** $2.40   **6.** $104
7. $45   **8.** $72   **9.** $244   **10.** $89.60   **11.** $42   **12.** $88
13. 8 kg   **14.** 72 kg   **15.** 272 g   **16.** 45 m   **17.** 2040 km   **18.** $710
19. 4.94 kg   **20.** 12 060 g   **21.** $204   **22. a)** 0.144   **b)** 0.0165   **23.** $157.50
24. 3600   **25.** 286   **26.** 5 months   **27.** $696

Exercise 6.3A
1. $63   **2.** $59.80   **3.** $792   **4.** $132   **5.** $42   **6.** $45.75
7. $736   **8.** $77.55   **9.** $1960   **10.** $110.30   **11.** $12.03   **12.** $9.49
13. $7.35   **14.** $7.01   **15.** $12.34   **16.** $16.92

Exercise 6.3B
1. $35.20   **2.** $5724   **3.** $171.50   **4.** $8.84   **5.** 2.828 kg   **6.** $58.50
7. 59 400   **8.** $9.52   **9.** 3.348 kg   **10.** $2762.50
11. **a)** $28   **b)** $21   **12. a)** $480   **b) i)** $162   **ii)** $10 530

Exercise 6.4A
1. **a)** $216   **b)** $115.50   **c)** 2 years   **d)** 5 years
2. **a)** $2295 interest   **b)** $9045
3. 7.5% annual rate   **4.** 1200   **5.** 3.5   **6.** 4.5

Exercise 6.4B
1. **a)** $2180   **b)** $2376.20   **c)** $2590.06   **2. a)** $5550   **b)** $6838.16   **c)** $8425.29
3. $13 107.96   **4. a)** $36 465.19   **b)** $38 288.45
5. **a)** $9540   **b)** $10 719   **c)** $16 118   **6. a)** $14 033   **b)** $734   **c)** $107 946
7. $9211.88

Exercise 6.4C
1. $7250   **2.** $8800   **3.** $18 800   **4.** $3640

Exercise 6.4D
1. 4c   **2.** 11c   **3.** 9c   **4.** $0.10   **5.** 4c   **6.** $0.18   **7.** 6c
8. 13c   **9.** $1.75   **10.** 36c   **11.** 2.5c   **12.** $7.25   **13.** 15c   **14.** $5
15. 16c   **16.** $0.89   **17.** 14c   **18.** 2c   **19.** $3.60   **20.** 15c

Exercise 6.5A
1. 8% increase   **2.** 10% increase   **3.** 25% increase   **4.** 2% increase   **5.** 4% increase
6. 25% decrease   **7.** 20% decrease   **8.** $12\frac{1}{2}$% decrease   **9.** $33\frac{1}{3}$% decrease   **10.** 80% decrease

Exercise 6.5B
1. 36.4% 2. 19.0% 3. 19.4% 4. 22.0% 5. 12.2%
6. 7.7% 7. 35.3% 8. 30.8% 9. 5.2% 10. 14.1%

Exercise 6.5C
1. 12% 2. 29% 3. 16% 4. 0.25% 5. 15%
6. 61.1% 7. 15% 8. 13.7% 9. 1.5% 10. 23.8%

Exercise 6.5D
1. a) 25c b) $12.80 c) $2.80 d) 28%
2. a) $10 b) i) $4.20 ii) 42%
3. a) i) 120 cm ii) 75 cm iii) 10 000 cm² iv) 9 000 cm² b) 10%
4. a) $57 500 b) i) $61 900 ii) 7.7%
5. 250 m²

Exercise 6.6A
1. 200 m 2. 500 m 3. b) 200 m c) 1 km d) 0.6 km 4. 63 m
5. 24 km 6. 120 m 7. 150 cm 8. 125 cm 9. 28 cm 10. 5.9 cm
11. a) 60 cm b) 84 cm c) 56 cm d) 140 cm e) 100 cm 12. 2.5 cm
13. 4.5 minutes

Exercise 6.6B
1. a) $5:6$ b) $1:3$ c) $2:5$ d) $3:2$ e) $1:2:4$
f) $1:2:10$ g) $1:2:8$ h) $5:2$ i) $9:14$ j) $2:1$ k) $25:1$
2. $10, $20 3. $45, $15 4. 330 g, 550 g 5. $480, $600 6. 36, 90
7. $10, $20, $30 8. $70 9. $50 10. 3250

Exercise 6.6C
1. 8 2. 5 3. 9 4. 30 g zinc, 40 g tin
5. 24 6. 2.4 kg 7. 6 8. 18 cm
9. 22.5 cm 10. 7.2 cm 11. 300 g 12. $5:3$
13. $200 14. 42c 15. $175 000 16. $\frac{1}{4}$ m³
17. 25% 18. 20%
19. 562.5 g butter; 500 g sugar; 10 eggs; 1.25 lemons; 2.5 teaspoons vanilla; 625 g flour
20. Carton B 21. Box 2 22. 19

Exercise 6.7A
1. $28 2. $6 3. $3 4. 400 5. 10 min 6. $4.80
7. $8.40 8. 12 days 9. 12 hours 10. 1 day 11. 9 hours 12. 24 men
13. 20 hours 14. a) 800 b) 2400 c) $1\frac{1}{2}$ d) 100

Exercise 6.7B
1. $3.66 2. $19.04 3. 12 litres 4. 10 hours 5. 10 days
6. 28 min 7. 6 hours 8. 3655 g 9. 400 bottles 10. 165 min
11. 11 grapefruit 12. £24.35 13. 5 men 14. 55 litres
15. a) 180 hens b) 9 days c) 10 hens d) 6 days
16. 7500 batteries 17. 1 h 15 min 18. $2n$ hours

Exercise 6.7C
1. a) €20.40 b) £61.60 c) 30 760 pesos d) 124.02 rupees e) 342.70 yen
f) 0.28 dinars
2. a) $490.20 b) $2840.91 c) $0.09 d) $2903.23 e) $0.81
f) $474.68
3. £0.96 4. Cheaper in UK by $5.74 5. €405.38
6. Kuwait £425.81, France £474.51, Japan £543.36 7. €451.90

Exercise 6.7D
1. a) i) 2.5 kg ii) 3.6 kg iii) 0.9 kg b) i) 4.4 lb ii) 6.6 lb iii) 3.3 lb c) 2.2 lb d) 3.2 kg
2. DM 0.6 less 3. a) 10 °C b) 68 °F c) −18 °C

Revision exercise 6
1. 95c for 1 lb
2. 20.75 litres
3. **a)** 1 : 50 000   **b)** 1 : 4 000 000
4. **a)** $143 000   **b)** 198   **c)** $715   **d)** $141 570   **e)** $1430 less
5. **a)** 2 cm   **b)** 8 m   6. **a)** $13   **b)** $148   **c)** $1700
7. **a)** $1.98   **b)** 760 km   8. **a)** $\frac{3}{5}$, 0.6, 60%   **b)** $\frac{3}{4}$, 0.75, 75%   **c)** $\frac{1}{20}$, 0.05, 5%   **d)** $\frac{1}{8}$, 0.125, $12\frac{1}{2}$%

Examination-style exercise 6
1. **a)** 36%   **b)** 35   2. $131   3. **a)** £36.90   **b)** $72.45
4. **a)** $1480   **b)** $1610.21   5. $1.80   6. 16.75 km
7. **a)** $18.80   **b)** 20.89%   8. 5 : 1   9. 0.38, $\frac{2}{5}$, 42%
10. **a)** $\frac{6}{100}$ **b)** 62%   **c)** 0.062 and 6.2%
11. 0.08, 8%, $\frac{8}{100}$ 12. €192.31
13. Joseph $25, Maria $21, Rebecca $14
14. **a) ii)** Aida $8000, Bernado $6000, Cristiano $2000
   b) i) $18 000   **ii)** $\frac{3}{16}$ **iii)** $14 500   **c)** 20.83%

Chapter 7 Shape and space 2

Exercise 7.1A
1. 0.85 m   2. 2400 m   3. 63 cm   4. 0.25 m   5. 0.7 cm
6. 20 mm   7. 1200 m   8. 700 cm   9. 580 m   10. 0.815 m
11. 0.65 km   12. 2.5 cm   13. 5000 g   14. 4200 g   15. 6400 g
16. 3000 g   17. 800 g   18. 0.4 kg   19. 2000 kg   20. 0.25 kg
21. 500 kg   22. 620 kg   23. 0.007 t   24. 1.5 kg   25. 0.8 litres
26. 2000 ml   27. 1 litre   28. 4500 ml   29. 6000 ml   30. 3000 cm^3
31. 2000 litres   32. 5500 litres   33. 900 cm^3 34. 0.6 litres   35. 15 000 litres
36. 0.24 litres   37. 0.28 m   38. 550 cm   39. 0.305 kg   40. 46 m
41. 0.016 litres   42. 0.208 m   43. 2.8 cm   44. 0.27 m   45. 0.788 km
46. 14 000 kg   47. 1300 g   48. 0.09 m^3 49. 2900 kg   50. 0.019 litres
51. (For discussion)

Exercise 7.1B
1. 200 mm^2 2. 4500 mm^2 3. 16 cm^2 4. 0.48 cm^2 5. 30 000 cm^2
6. 260 000 cm^2 7. 0.86 m^2 8. 0.076 m^2 9. 5 000 000 m^2 10. 4.5 km^2
11. 8000 mm^3 12. 21 000 mm^3 13. 48 cm^3 14. 6 000 000 cm^3 15. 28 m^3

Exercise 7.1C
(For discussion)
1. **a)** Nelson's Column 56 m   **b)** Empire State Building 450 m   **c)** Mount Everest 8700 m
2. **a)** No   **b)** No   3. **a)** No   **b)** Probably
4. **a)** Yes   **b)** About 6.7 times

Exercise 7.2A
1. 08 00   2. 21 30   3. 18 00   4. 05 30   5. 19 40
6. 22 00   7. 19 15   8. 22 45   9. 08 30   10. 04 15
11. 02 25   12. 13 30   13. 19 20   14. 06 50   15. 07 10
16. 23 58   17. 21 30   18. 11 55   19. 15 30   20. 01 00
21. 10 30   22. 00 20   23. 19 00   24. 12 06   25. 00 50
26. 7.00 a.m.   27. 7.30 p.m.   28. 11.20 a.m.   29. 4.45 a.m.   30. 8.30 p.m.
31. 9.15 p.m.   32. 9.10 a.m.   33. 11.45 a.m.   34. 11.10 p.m.   35. 8.00 p.m.
36. Noon   37. 1.40 a.m.   38. 4.00 a.m.   39. 7.07 a.m.   40. 1.13 p.m.
41. 12.15 p.m.   42. 12.30 p.m.   43. 3.45 p.m.   44. 4.20 p.m.   45. 5.16 a.m.

Exercise 7.2B
1. 1 h 10 min	**2.** 2 h 10 min	**3.** 55 min	**4.** 35 min	**5.** 3 h 20 min
6. 1 h 45 min	**7.** 2 h 20 min	**8.** 1 h 53 min	**9.** 53 min	**10.** 5 h 25 min
11. 1 h 50 min	**12.** 3 h 14 min	**13.** 1 h 15 min	**14.** 4 h 35 min	**15.** 2 h 20 min
16. 12 h	**17.** 2 h	**18.** 7 h 20 min	**19.** 3 h 15 min	**20.** 7 h 05 min
21. 5 h	**22.** 4 h 40 min	**23.** 8 h 30 min	**24.** 2 h 25 min	**25.** 8 h 10 min
26. 19 h	**27.** 19 h	**28.** 17 h 30 min	**29.** 9 h 25 min	**30.** 34 h

Exercise 7.2C
1. a) 25 min	**b)** 45 min	**c)** 1 h 6 min	**2.** 7	**3.** 0954
4. 1135	**5.** 15 min	**6.** 1252		
7. a) 1 h 42 min	**b)** 2 h 32 min	**c)** 3 h 43 min		
8. 2	**9.** 1407	**10.** 0921	**11.** 27 min	**12.** 1201
13. 1800	**14.** 1302	**15.** 2140	**16.** 52	**17.** 90
18. 26	**19.** 304	**20.** 10 June		

Exercise 7.3A
1. a) $2\frac{1}{2}$ hours   **b)** $3 \text{ h } 7\frac{1}{2}$ min   **c)** 75 seconds   **d)** 4 hours

2. a) 20 m/s   **b)** 30 m/s   **c)** $83\frac{1}{2}$ m/s   **d)** 108 km/h   **e)** 79.2 km/h
  f) 1.2 cm/s   **g)** 90 m/s   **h)** 25 mph
3. a) 75 km/h   **b)** 4.52 km/h   **c)** 7.6 m/s   **d)** 400 m/s   **e)** 1×10^5 m/s
  f) 200 km/h   **g)** 3 km/h
4. a) 110 000 m   **b)** 10 000 m   **c)** 56 400 m   **d)** 4500 m   **e)** 50 400 m
  f) 1.6×10^6 m   **g)** 96 000 m
5. 12 km/h   **6.** 132 km   **7.** 1 hour 36 minutes
8. a) $3 \text{ h } 7\frac{1}{2}$ min   **b)** 76.8 km/h   **9. a)** 4 hours 27 minutes   **b)** 23.6 km/h
10. 46 km/h

Exercise 7.3B
1. a) 0.8   **b)** 16   **c)** 390   **d)** 4800   **e)** 1460
2. a) 1.5   **b)** 2.5   **c)** 90   **d)** 36   **3.** 13.3 litres per minute
4. 12.6 kWh/day   **5.** 8.33 minutes   **6. a)** 2500   **b)** 66.7
7. 250 N/m²   **8.** 1.5 m²   **9.** 14 000 000   **10.** 41.2 people/km²
11. 1.33 g/cm³   **12.** 3.25 grams   **13.** 833 cm³
14. $50 per hour   **15.** $176.25   **16.** 37.5 hours

Exercise 7.4A
1. $A = 24$ cm²; $P = 20$ cm   **2.** $A = 14$ cm²; $P = 17.8$ cm   **3.** 36 cm²
4. 77 cm²   **5.** 54 cm²   **6.** 25 cm²
7. 36 cm²   **8.** 48 cm²   **9.** $A = 51$ cm²; $P = 40$ cm
10. $A = 36$ cm²; $P = 32$ cm   **11.** $A = 24$ cm²; $P = 28$ cm   **12.** $A = 24$ cm²; $P = 28$ cm
13. $A = 57$ cm²; $P = 36$ cm   **14.** $A = 48$ cm²; $P = 52$ cm   **15.** $A = 36$ cm²; $P = 34$ cm
16. $A = 38$ cm²; $P = 36$ cm

Exercise 7.4B
Questions **1** to **7** answers are in square units.
1. a) 10, 6, 3   **b)** 36   **c)** 17
2. a) 5, 14, 6   **b)** 42   **c)** 17
3. $13\frac{1}{2}$ **4.** $14\frac{1}{2}$ **5.** 24   **6.** 22   **7.** 21

Exercise 7.4C
1. 42 cm²   **2.** 22 cm²   **3.** 103 cm²   **4.** 60.5 cm²
5. 143 cm²   **6.** 4 000 000 m²; 400 hectares   **7.** $252

Exercise 7.5A

1. 11π cm	**2.** 8π cm	**3.** 12π cm	**4.** 5π cm	**5.** 28.3 cm
6. 53.4 m	**7.** 44.6 m	**8.** 72.3 m	**9.** 8.48 m	**10.** 212
11. 400 m	**12.** 22.6 cm	**13.** 823 m	**14.** 643 cm	

Exercise 7.5B

1. 30.25π cm^2	**2.** 25π cm^2	**3.** 9π m^2	**4.** 12.25π m^2	**5.** 113 cm^2	**6.** 201 cm^2
7. 19.6 m^2	**8.** 380 cm^2	**9.** 29.5 cm^2	**10.** 125 m^2	**11.** 21.5 cm^2	**12.** 4580 g
13. a) 30	**b)** 1508 cm^2	**c)** 508 cm^2	**14.** 40.8 m^2	**15.** 118 m^2	

16. No; she has squared the π too. The correct answer is 36π cm^2.

Exercise 7.5C

1. 23.1 cm	**2.** 38.6 cm	**3.** 20.6 m	**4.** 8.23 cm	**5.** 28.6 cm	**6.** 39.4 m
7. 17.9 cm	**8.** 28.1 m	**9.** 24.8 cm	**10.** 46.3 m	**11.** 28.8 cm	

Exercise 7.5D

1. 35.9 cm^2	**2.** 84.1 cm^2	**3.** 37.7 cm^2	**4.** 74.6 cm^2	**5.** 13.7 cm^2	**6.** 25.1 cm^2
7. a) 12.5 cm^2	**b)** 50 cm^2	**c)** 78.5 cm^2	**d)** 28.5 cm^2		

Exercise 7.5E

1. 2.39 cm	**2.** 4.46 m	**3.** 1.11 m	**4.** 4.15 cm	**5.** 3.48 cm	**6.** 3.95 m
7. 2.55 m	**8.** 4.37 cm	**9.** 4.62 cm	**10.** 5.75 cm	**11.** 15.9 cm	**12.** 5.09 cm
13. 9.2 m	**14.** 58.6 cm	**15.** 5.39 cm	**16.** 17.8 cm	**17.** 195 km	**18.** 3.95 cm
19. 215 m^2	**20.** 3.88 m	**21.** 575 m^2	**22.** 5.41 cm	**23.** 4.5 m	

Exercise 7.6A

1. a) $\frac{5}{6}\pi$ cm **b)** $\frac{5}{6}\pi + 10$ cm **c)** $\frac{25}{12}\pi$ cm^2

2. a) 4π cm **b)** $4\pi + 16$ cm **c)** 16π cm^2

3. a) 9π cm **b)** $9\pi + 12$ cm **c)** 27π cm^2 **4. a)** $\frac{55}{4}\pi$ cm **b)** $\frac{55}{4}\pi + 15$ cm **c)** $\frac{825}{16}\pi$ cm^2

5. a) 7.07 cm^2 **b)** 19.5 cm^2 **6. a)** 8.50 cm **b)** 8.90 cm

7. a) 12 cm **b)** 30° **8. a)** 30° **b)** 10.5 cm

Exercise 7.7A

1. 150 cm^3 **2.** 480 cm^3 **3.** Volume = 300 cm^3 Surface area = 320 cm^2

4. Volume = 56 m^3 Surface area = 142 m^2 **5.** Volume = 280 cm^3 Surface area = 336 cm^2

6. 145 cm^3 **7.** 448 cm^3 **8.** 108 m^3 **9. a)** 248 cm^2 **b)** 120

Exercise 7.7B

1. $V = 20\pi$ cm^3; $A = 28\pi$ cm^2 **2.** $V = 36\pi$ cm^3; $A = 42\pi$ cm^2 **3.** $V = 63\pi$ cm^3; $A = 60\pi$ cm^2

4. $V = 763$ cm^3; $A = 467$ cm^2 **5.** $V = 157$ cm^3; $A = 220$ cm^2 **6.** $V = 385$ cm^3; $A = 297$ cm^2

7. $V = 770$ cm^3; $A = 528$ cm^2 **8.** $V = 176$ m^3; $A = 188$ m^2 **9.** $V = 228$ m^3; $A = 273$ m^2

10. $V = 486$ cm^3; $A = 368$ cm^2 **11.** 113 litres **12.** 141 cm^3, 25.1 cm^3

Exercise 7.7C

1. $V = 96\pi$ cm^3; $A = 96\pi$ cm^2 **2.** $V = \frac{500}{3}\pi$ cm^3; $A = 100\pi$ cm^2 **3.** $V = 288\pi$ cm^3; $A = 144\pi$ cm^2

4. $V = 100\pi$ cm^3; $A = 90\pi$ cm^2 **5.** $V = 268$ cm^3; $A = 201$ cm^2 **6.** $V = 4220$ cm^3; $A = 3690$ cm^2

7. $V = 0.00419$ m^3; $A = 0.126$ m^2 **8.** $V = 56.5$ cm^3; $A = 84.8$ cm^2 **9.** 93.3 cm^3

10. 48 m^3 **11.** 92.4 cm^3 **12.** 5 m

13. 2.43 cm **14.** 23.9 cm **15. a)** 0.36 cm **b)** 0.427 cm

16. a) 15 cm^2 **b)** 96 cm^2 **17.** 95 cm^2

Exercise 7.7D

1. 2400 cm^3 **2. a)** 200 m^2 **b)** 2400 m^3 **3.** 770 cm^3

4. a) 2.25 cm^2 **b)** 0.451 cm^3 **c)** 4510 cm^3 **5. a)** 62 800 cm^3 **b)** 500 cm^3 **c)** 125

6. 8 cm^3 **7. a)** 76 cm^2 **b)** 30 400 cm^3 **c)** 237 kg **d)** 33

8. No. This is the approximate number of bricks that would fill a space of size 6 m by 4 m by 2.5 m. But Rahim only needs to build the walls; he doesn't need to fill the space.

9. 53 times **10.** 98 min

Revision exercise 7
1. **a)** 0.005 m/s   **b)** 1.6 s   **c)** 173 km
2. 12 km   **3. a)** 1810 s   **b)** 72.4 s
4. 17 kg   **5.** $33\frac{1}{3}$ km/h
6. **a)** 91.5 cm^2 **b)** 119 cm^2 **7.** 17.7 cm
8. **a)** 198 cm^3 **b)** 1357 mm^3 **c)** 145   **9.** Both arrive at the same time.
10. 3.43 cm^2, 4.57 cm^2 **11.** 9.95 cm   **12.** 25   **13.** 5.14 cm^2 **14.** 30 cm
15. 84.48 grams   **16.** 50 N/m^2 **17.** 3090 people/km^2
18. SA $= 64\pi$ cm^2; V $= \frac{256}{3}\pi$ cm^3 **19. a)** 3.98 cm   **b)** 75.6 cm^3

Examination-style exercise 7
1. 3 h 29 min   **2. a)** 40   **b)** 9
3. **a) i)** 2 000 000   **ii)** 4   **iii)** 100   **b)** 16.8 minutes or 16 minutes 48 seconds
4. **a)** 128.6796351 cm^2 or $\frac{1024}{25}\pi$ cm^2 **b)** 129 cm^2 **5.** 13.7 cm^2 **6.** 16.8 cm
7. **a)** 180 m   **b)** 57.3 m   **8.** 3.11 m or 311 cm   **9. a)** 204 800π cm^3 **b)** 0.643
10. **a)** 140 cm   **b)** 736 cm^2 **11.** 13.5 m
12. **a)** 154 cm^3 **b)** 2 h 51 minutes   **c)** 33.6   **d) i)** \$10.60   **ii)** 30.3%
13. **a)** 141.37 cm^2 **b)** 13.42 cm   **14.** $\frac{2}{3}$ **15. a)** 1.7 g/cm^3 **b)** 1334 N/m^2
16. **a)** 5.97 m^2 **b)** 26 hours and 11 minutes

Chapter 8 Algebra 2

Exercise 8.1A
1. **a–d**

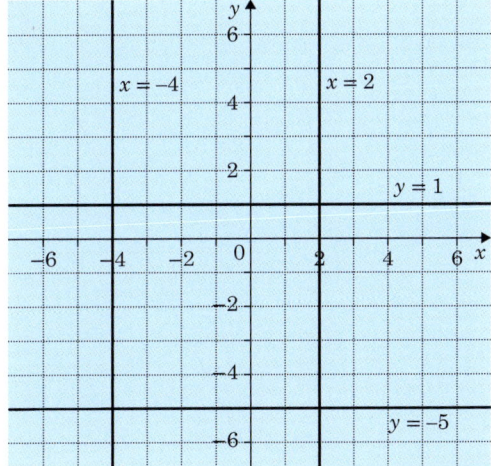

2.

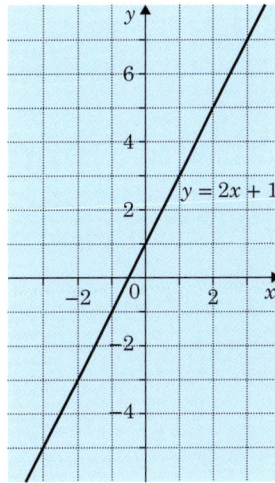

3.

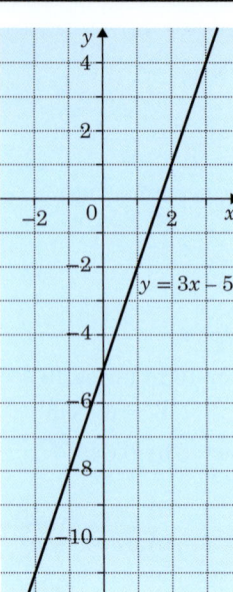

$y = 3x - 5$

4.

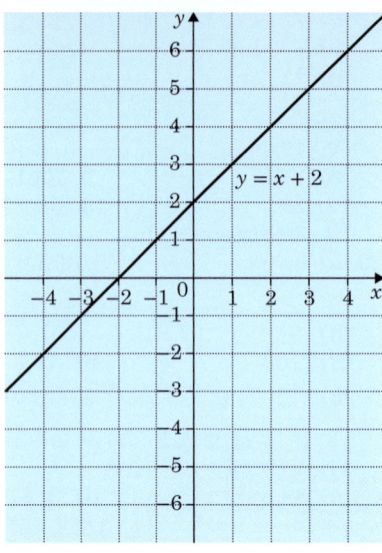

$y = x + 2$

5.

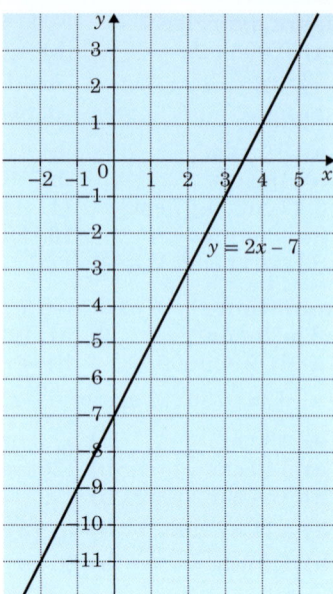

$y = 2x - 7$

6.

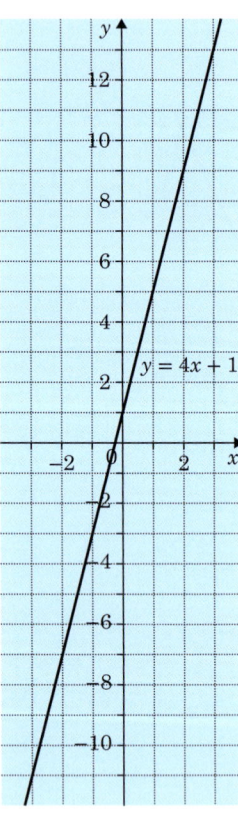

$y = 4x + 1$

7.

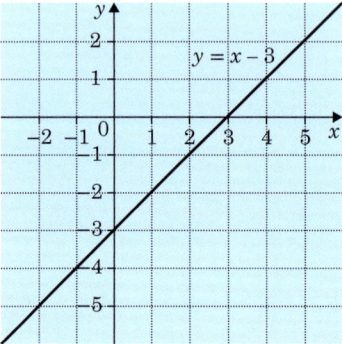

$y = x - 3$

8.

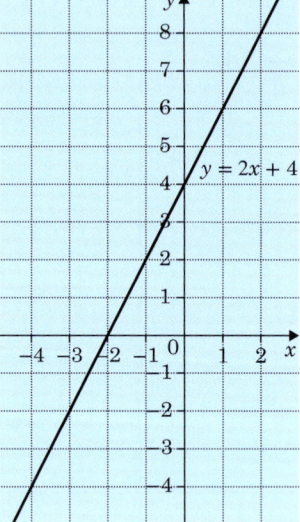

$y = 2x + 4$

9.

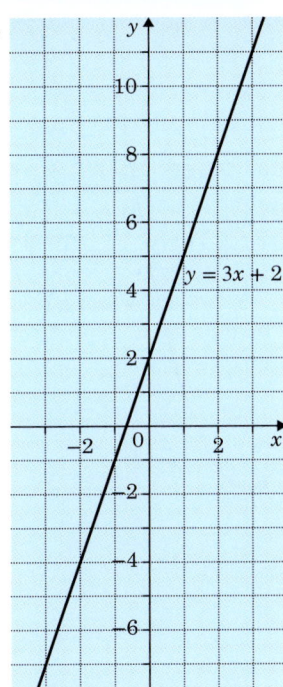

$y = 3x + 2$

10.

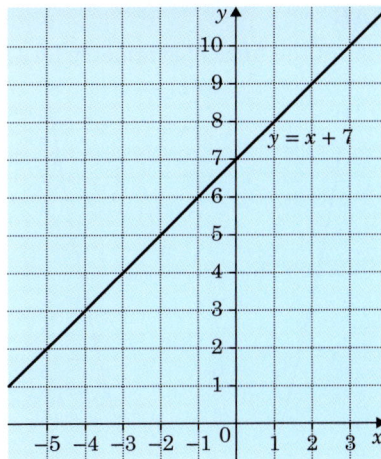

$y = x + 7$

11.

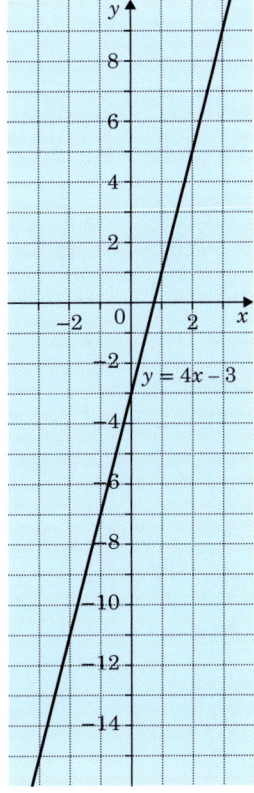

$y = 4x - 3$

12.

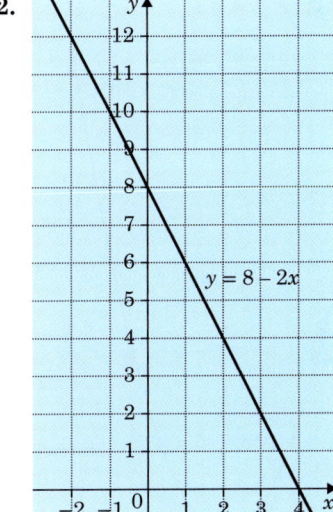

$y = 8 - 2x$

Exercise 8.1B

The equations of the lines of symmetry for Questions **1** to **12** are given.

1.

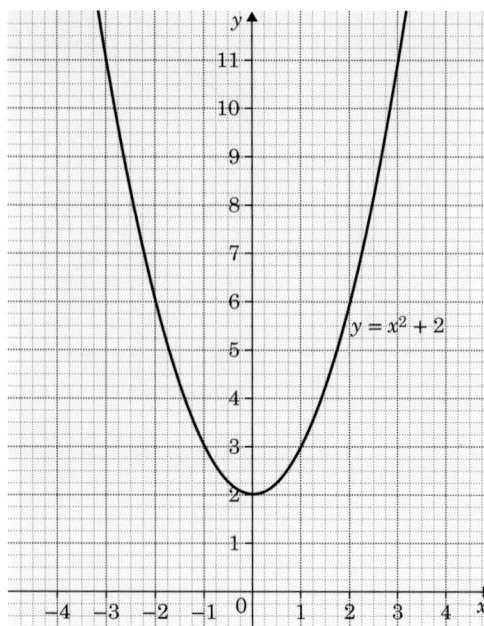

$y = x^2 + 2$

Line of symmetry: $x = 0$

2.

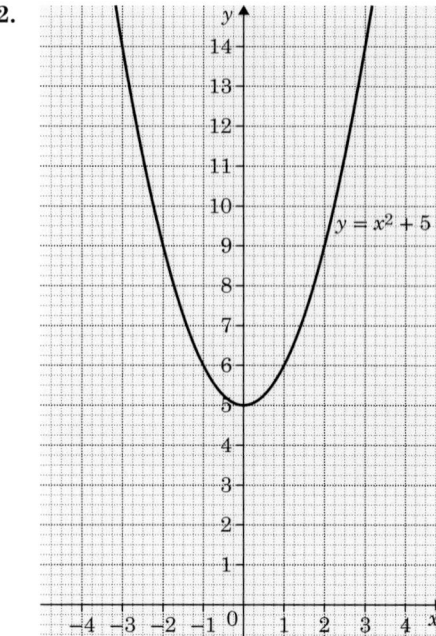

$y = x^2 + 5$

Line of symmetry: $x = 0$

3.

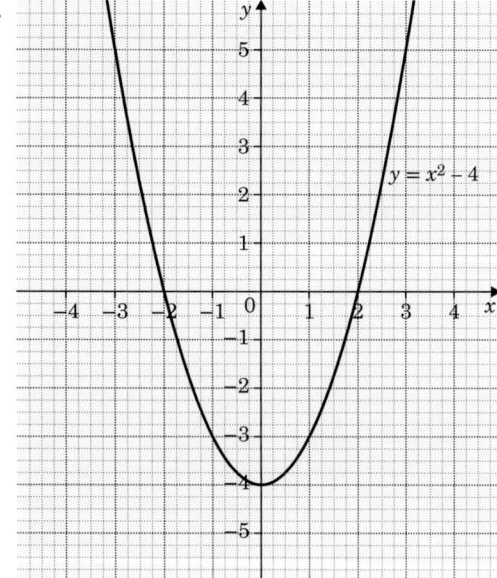

$y = x^2 - 4$

Line of symmetry: $x = 0$

4.

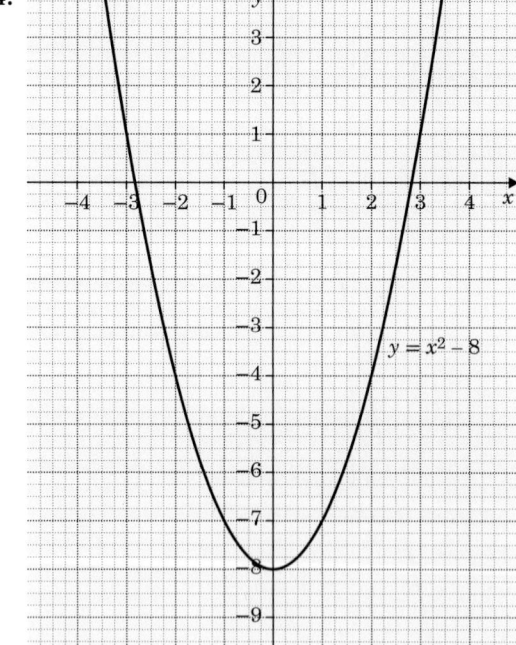

$y = x^2 - 8$

Line of symmetry: $x = 0$

5.

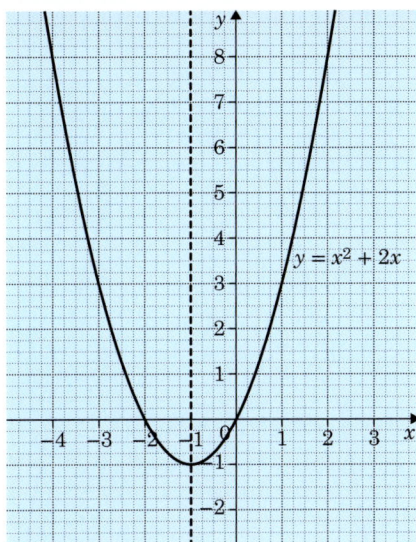

$y = x^2 + 2x$

Line of symmetry: $x = -1$

6.

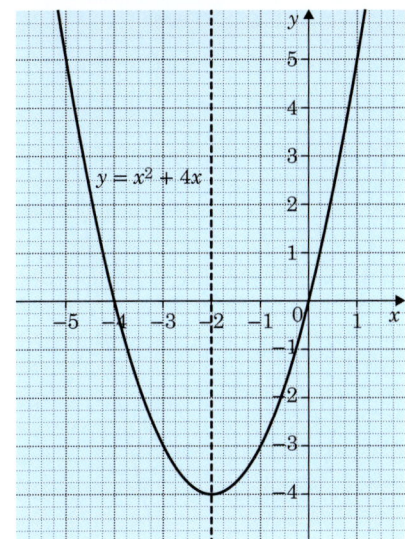

$y = x^2 + 4x$

Line of symmetry: $x = -2$

7.

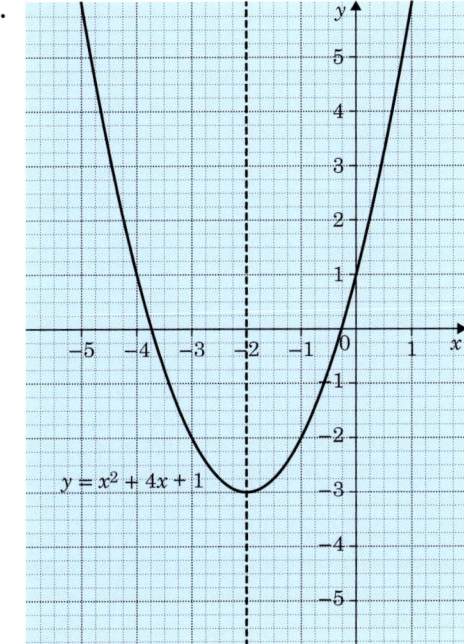

$y = x^2 + 4x + 1$

Line of symmetry: $x = -2$

8.

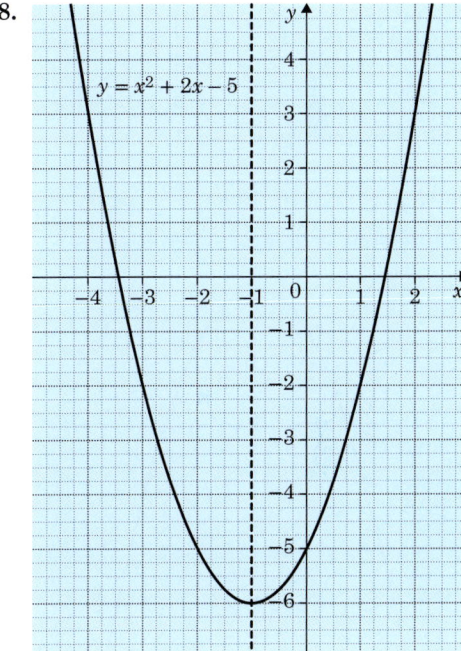

$y = x^2 + 2x - 5$

Line of symmetry: $x = -1$

9.

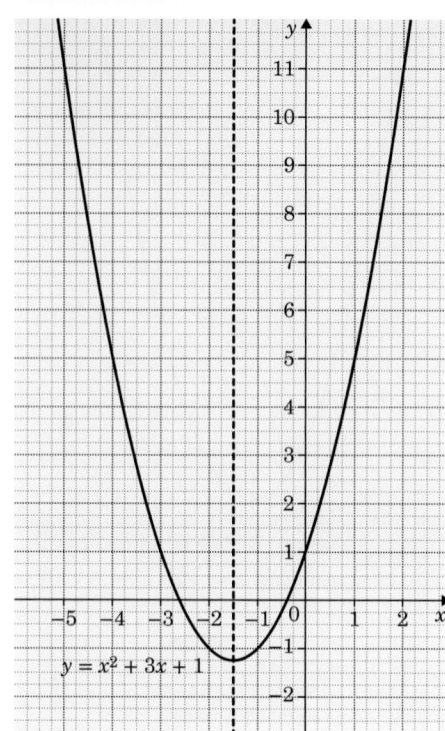

Line of symmetry: $x = -1.5$

10.

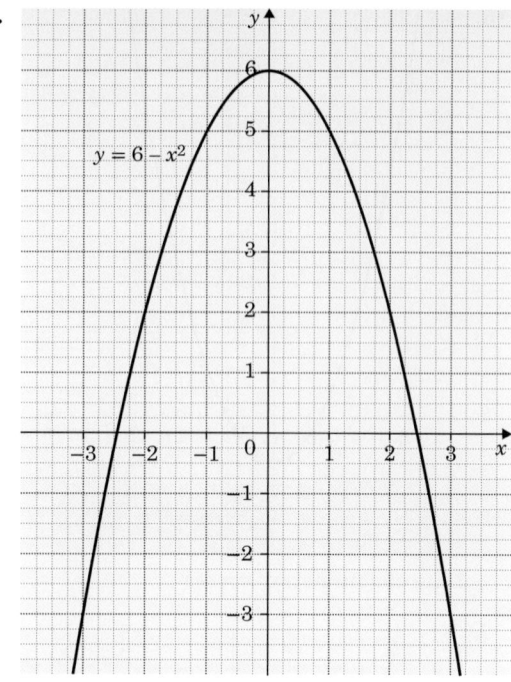

Line of symmetry: $x = 0$

11.

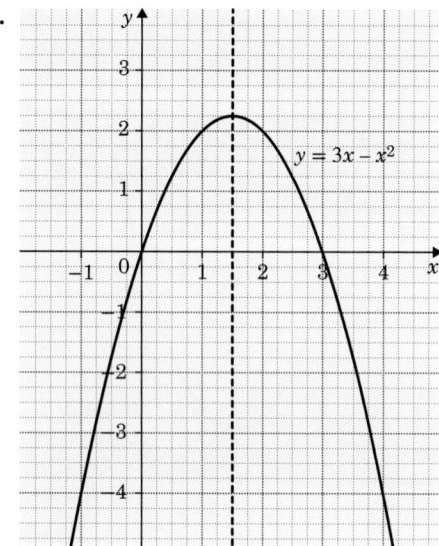

Line of symmetry: $x = 1.5$

12.

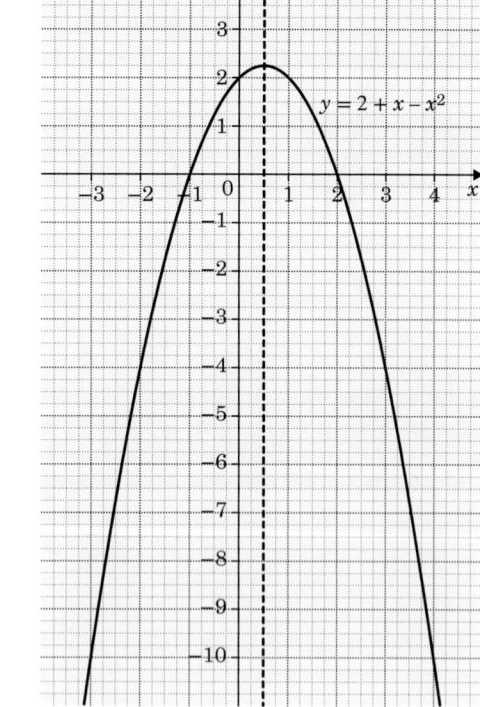

Line of symmetry: $x = 0.5$

13.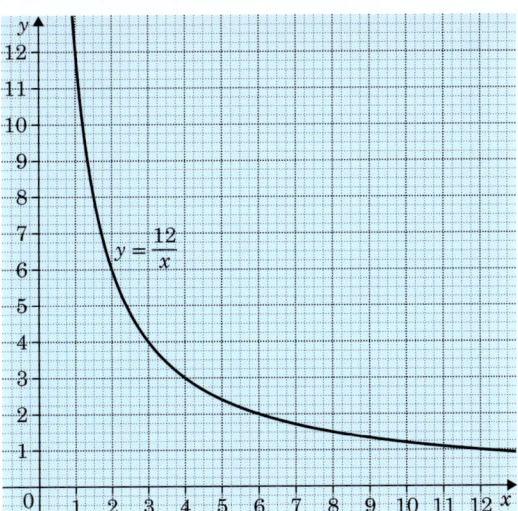

14.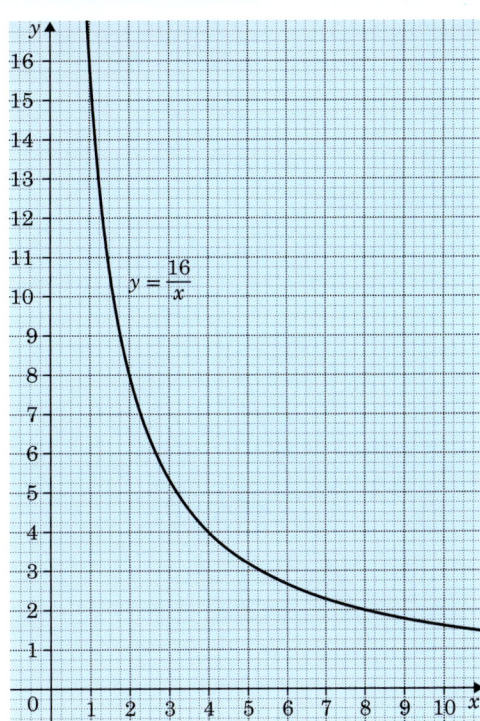

15. a) Half of the perimeter = 7 cm, so the width = $(7 - x)$ cm
Area = length × width = $x × (7 - x)$, so A = $x(7 - x)$

b)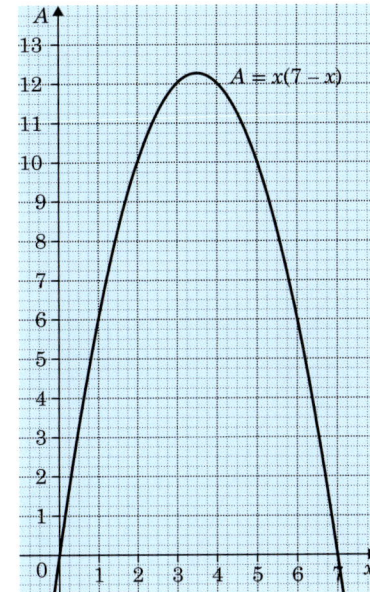

c) 10.7 cm² **d)** 5.3 × 1.7 **e)** 12.25 cm² **f)** 3.5 × 3.5

16. Positive x^2: B and C; negative x^2: A and D

17. No: the graph is not symmetrical

Exercise 8.2A

1. a) 3   **b)** 2   **c)** $\frac{1}{2}$ **d)** $\frac{1}{3}$

2. a) -1 **b)** -2 **c)** -2 **d)** $-\frac{2}{3}$ **e)** 6   **f)** $-\frac{9}{4}$

3. a) $\frac{5}{2}$ **b)** $-\frac{3}{5}$ **c)** $\frac{5}{2}$ **d)** $-\frac{3}{5}$

Exercise 8.2B

1.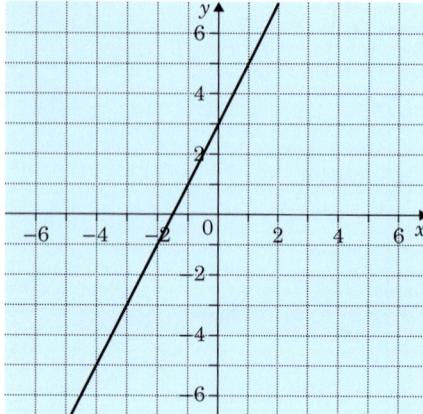

2.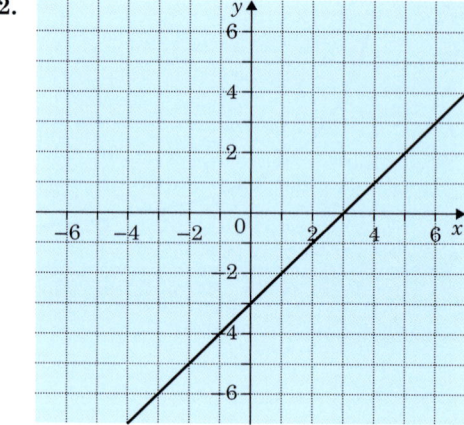

3.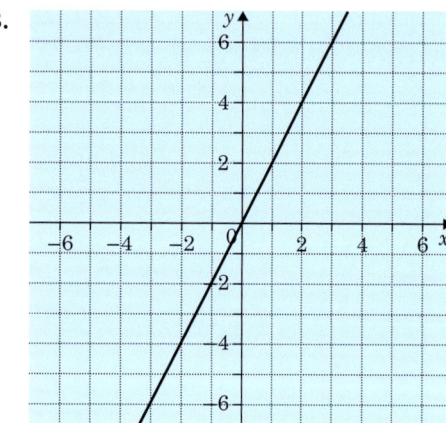

4.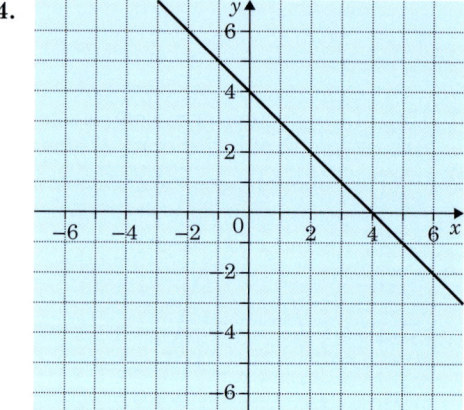

5.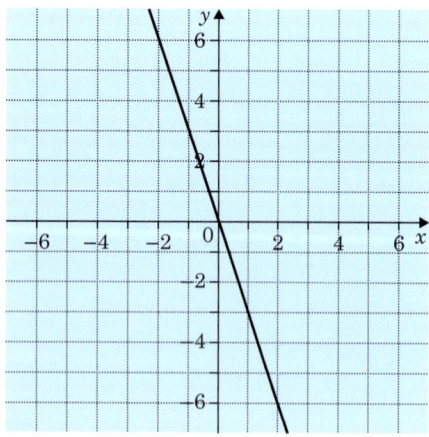

6.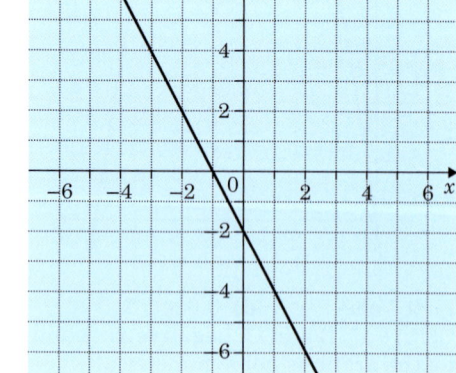

7. (The gradient is given first) A: 2, -2;   B: $\frac{1}{3}$, 1;   C: -1, -4;   D: $-\frac{5}{2}$, 5;   E: 0, -5

Exercise 8.2C
In Questions **1** to **10** gradient is given first and the y-intercept is second.

1. 2, −3   **2.** 3, 2   **3.** −1, −4   **4.** 1, 3   **5.** −2, −4

6. −3, 2   **7.** −7, 4   **8.** 2, −1   **9.** −1, 3   **10.** −2, 7

11. −3 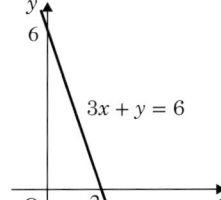   **12.** 3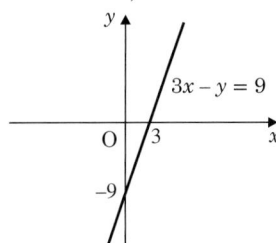

13. 2 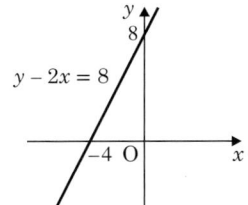   **14.** −2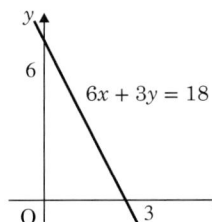

Exercise 8.2D
1. $y = 3x$ **2.** $y = 5$ **3.** $y = 3 - x$ **4.** $y = 2x + 3$ **5.** $y = 2x - 1$

6. $y = -x - 4$ **7.** A: $y = 3x - 4$ B: $y = x + 2$ **8.** C: $y = \frac{2}{3}x - 2$ D: $y = -2x + 4$

9. a) $y = 2x + 6$ **b)** $y = -x + 3$ **10. a)** $y = 3x + 1$ **b)** $y = x - 2$

Exercise 8.3A
1. a) 3.6/3.7 and −1.6/−1.7   **b)** 2.4, −0.4   **c)** −1, 3

2. a) 2.4, −0.4   **b)** 0, 2   **3. a)** 0, 3   **b)** 3.8, −0.8

4. a) −1.6, 3.6   **b)** 2.4, −0.4   **5. a)** 6.5, 0.5   **b)** No intersection

Exercise 8.4A
1. a) 40 km   **b)** 60 km   **c)** Gap, Sisteron   **d)** 15 min

     e) i) 11 00   **ii)** 13 45   **f) i)** 40 km/h   **ii)** 60 km/h   **iii)** 100 km/h

2. a) 25 km   **b)** 15 km   **c)** 10 00   **d)** 1 h

     e) i) 20 km/h   **ii)** 6.67 km/h   **iii)** 30 km/h   **iv)** 40 km/h

3. a) i) 14 00   **ii)** 13 45   **b) i)** 15 45   **ii)** Towards Aston

     c) i) 15 mph   **ii)** 40 mph   **iii)** 40 mph   **iv)** 20 mph   **d)** 16 07

4. a) 45 min   **b)** 09 15   **c)** 60 km/h   **d)** 100 km/h   **e)** 57.1 km/h

5. a) 09 00   **b)** 64 km/h   **c)** 40 km/h   **d)** 70 km   **e)** 80 km/h

6. 11 05   **7.** 12 42   **8.** 12 35

Exercise 8.4B
1. a) 740c   **b)** $280   **c)** $14 000   **3. a) i)** 08 00   **ii)** 21 00

4. a) 2   **b) i)** 40 km   **ii)** 24 km   **iii)** 72 km   **iv)** 8 km

     c) i) 40 miles   **ii)** 35 miles   **iii)** 10 miles   **iv)** 20 miles

5. a) i) €28   **ii)** €112   **iii)** €70

     b) i) $40   **ii)** $60   **iii)** $100   **c)** $110

6. a) 30 litres
   b) i) 8 km/l   **ii)** 6 km/l
   c) 6.7 km/l
   d) 30 litres

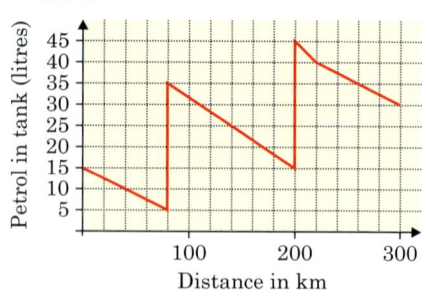

7. 180 km

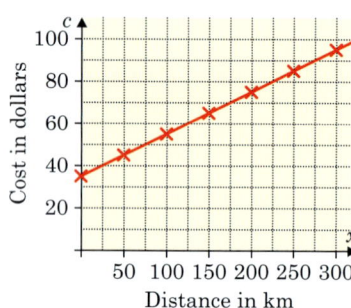

8. C: 18, 33, 48, 63; 2.5 h

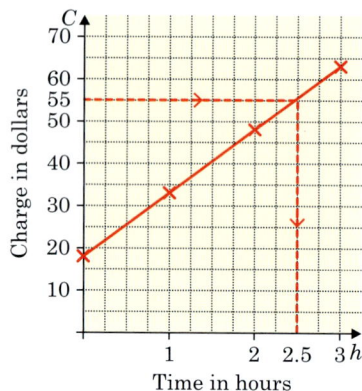

Exercise 8.4C

1. B   **2.** D   **3. a)** C   **b)** A   **c)** D   **d)** B
4. a) i) B   **ii)** A   **b)** 8 s to 18 s   **c)** About 15 s   **d)** About 9 s   **e)** B   **f)** A
6. a) Runners slow down for handover   **b)** Baton dropped at third handover

Revision exercise 8

1. A, $y = 6$; B, $y = x - 3$; C, $y = 10 - x$; D, $y = 3x$
2. Student's graph
3. A, walker; B, cyclist to Abu Dhabi; C, car; D, train; E, market stall; F, cyclist from Abu Dhabi
4. a) i) Consett   **ii)** Durham   **iii)** Consett   **b) i)** 55 km   **ii)** 40 km
   c) i) 80 km/h   **ii)** 55 km/h   **iii)** 70 km/h   **iv)** 80 km/h   **d)** $1\frac{3}{4}$ h
5. a), d)   **b)** −3 and 1   **c)** $x = -1$ **e)** (2, 5) and (−1, −4)

Examination-style exercise 8

1. a) c　　　　　　**b)** $y = 2x + 4$　　　　**c)** Missing numbers 6, 10, 1

d)

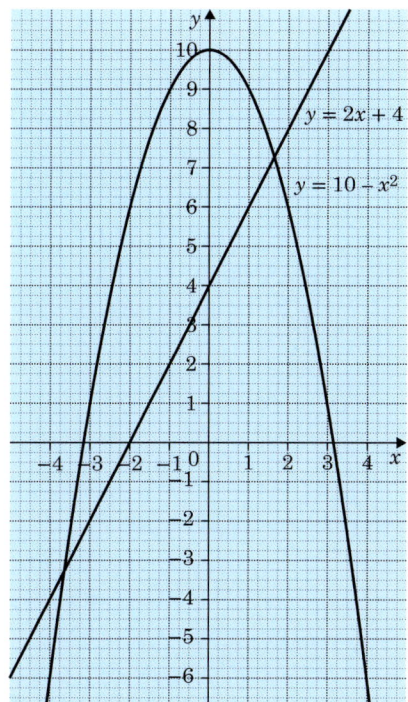

e) $(-3.6, -3.3)$ $(1.6, 7.3)$

2. a) i) 6, −2, 13

ii)

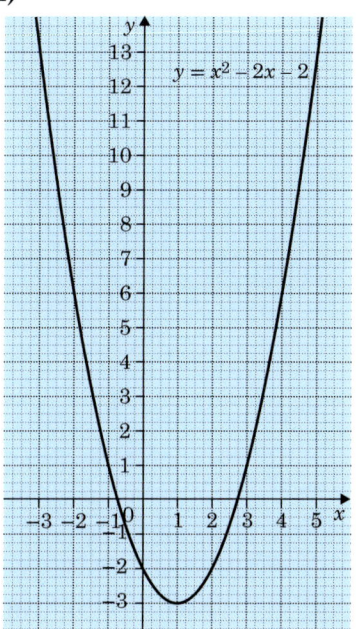

b) i) 8, 2

ii)

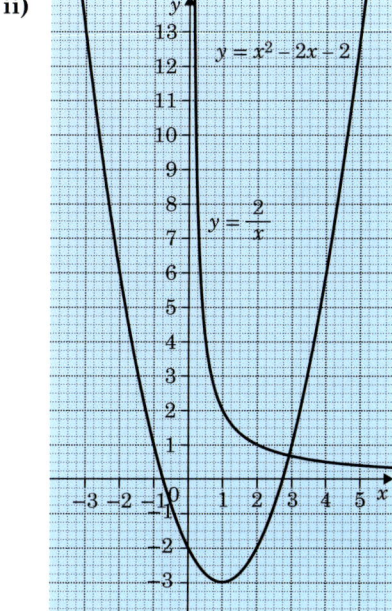

iii) −0.7, 2.7　　**iv)** −0.4, 2.4　　　　　　　　**iii)** 2.9 to 3.0

3. a) $y = 2x - 3$　　**b)** $y = 2 - 3x$　　**4. a)** -2　　　　　　**b)** $y = -2x + 8$

5. a) i) $x + 3$　　　　**ii)** $x(x + 3)$　　　**c) i)** $-3, -9, -3$

c) ii), e) i)

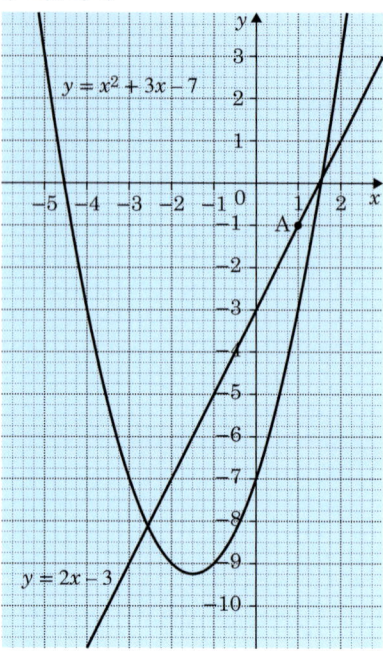

d) i) 1.5 to 1.6, -4.5 to -4.6　　　　**ii)** 4.5 to 4.6

e) ii) $y = 2x - 3$

6. a) 12　　　　　　**b)** 20 km/h　　　**c)** 28 km

7. a) i) 30

a) ii), b)

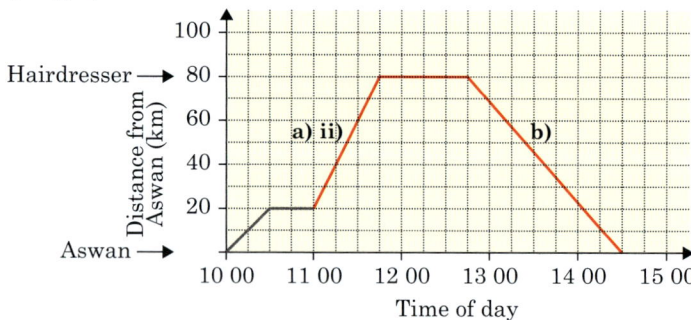

9 Probability

Exercise 9.1A
Student's answers

Exercise 9.1B
Student's answers

Exercise 9.1C

1. a) $\frac{3}{8}$　　**b)** $\frac{5}{8}$　　　**c)** 0　　　**2. a)** $\frac{1}{9}$　　　**b)** $\frac{1}{3}$　　**c)** $\frac{4}{9}$　　**d)** $\frac{2}{9}$　　**e)** 0

3. a) $\frac{5}{11}$　　**b)** $\frac{2}{11}$　　**c)** $\frac{4}{11}$　　**d)** 0　　**4. a)** $\frac{4}{17}$　　**b)** $\frac{3}{17}$　　**c)** 0

5. a) $\frac{4}{17}$ **b)** $\frac{8}{17}$ **c)** $\frac{5}{17}$ **6. a)** $\frac{1}{10}$ **b)** $\frac{3}{10}$ **c)** $\frac{3}{10}$

7. a) $\frac{3}{13}$ **b)** $\frac{5}{13}$ **c)** $\frac{8}{13}$ **8. a) i)** $\frac{5}{13}$ **ii)** $\frac{6}{13}$ **b) i)** $\frac{5}{12}$ **ii)** $\frac{1}{12}$

9. $\frac{9}{20}$ **10.** $\frac{1}{7}$ **11. a) i)** $\frac{1}{4}$ **ii)** $\frac{1}{4}$ **iii)** $\frac{1}{4}$ **b)** $\frac{1}{4}$ **c)** $\frac{6}{27} = \frac{2}{9}$

12. a) $\frac{9}{53}$ **b)** $\frac{10}{53}$ **13.** 24

Exercise 9.1D

1. 50   **2.** 25   **3.** 50

4. 40   **5. a)** $\frac{3}{8}$ **b)** 25   **6.** $\frac{1}{2}$

Exercise 9.2A

1. $\frac{4}{5}$ **2. a)** $\frac{7}{20}$ **b)** $\frac{13}{20}$ **3.** 76%

4. 0.494   **5. a)** $\frac{1}{4}$ **b)** $\frac{3}{4}$ **c)** $\frac{1}{4}$ **d)** $\frac{3}{4}$ **e)** 0   **f)** 1

6. a) 0.3   **b)** 0.9   **7. a) i)** 0.24   **ii)** 0.89   **b)** 575   **8.** 2

9. No: the events overlap since 2 is both even and prime

Exercise 9.3A

1. Student's own experiment results

2. Mike. with a large number of spins he should get zero with a probability of about $\frac{1}{10}$.

3. a) 0.12   **b)** 0.28   **c)** 0.6   **4.** 0.35 and 54

Exercise 9.4A

1. a) i) 8   **ii)** 3   **iii)** 4   **b)** 18   **c)** 7

2. a) i) 9   **ii)** 5   **iii)** 4   **iv)** 11   **b)** 31

3. a)

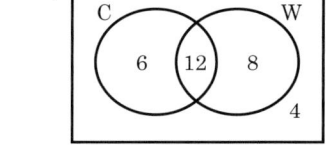

b) i) 18   **ii)** 12   **iii)** 26   **iv)** 4

4. a)

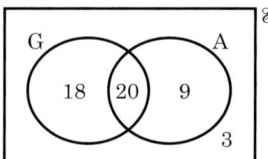

b) i) 27   **ii)** 10   **iii)** 17   **iv)** 40

Exercise 9.4B

1. a) $\frac{16}{25}$ **b)** $\frac{7}{25}$ **c)** $\frac{3}{25}$ **d)** $\frac{6}{25}$

2. a) i) $\frac{14}{30}$ **ii)** $\frac{7}{30}$ **iii)** $\frac{8}{30}$

b) Children who like carrots but do not eat them

3. a)

G   A $\mathscr{E}$
18 20 9
3

b) i) $\frac{9}{50}$ **ii)** $\frac{38}{50}$ **iii)** $\frac{29}{50}$

4. a)

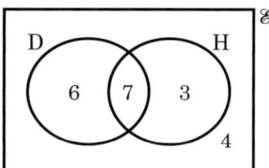

b) i) $\frac{7}{20}$ **ii)** $\frac{3}{20}$ **iii)** $\frac{7}{20}$

5. a)

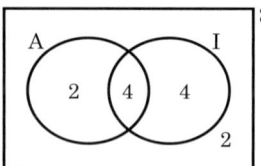

b) i) $\frac{4}{12}$ **ii)** $\frac{8}{12}$ **iii)** $\frac{4}{12}$

Exercise 9.5A

1. HH, HT, TH, TT **a)** $\frac{1}{4}$ **b)** $\frac{1}{2}$

2. Student's sample space diagram showing totals

   a) $\frac{1}{12}$ **b)** $\frac{1}{36}$ **c)** $\frac{5}{18}$ **d)** $\frac{1}{6}$ **e)** $\frac{1}{6}$

   The most likely total is 7.

3. a) $\frac{5}{25}$ **b)** $\frac{5}{25}$ **c)** $\frac{15}{25}$ **d)** $\frac{8}{25}$

Exercise 9.5B

1. a) $\frac{49}{100}$ **b)** $\frac{9}{100}$ **2. a)** $\frac{9}{64}$ **b)** $\frac{15}{64}$ **3. a)** $\frac{49}{100}$ **b)** $\frac{9}{100}$

4. a) $\frac{1}{216}$ **b)** $\frac{125}{216}$ **c)** $\frac{25}{72}$ **d)** $\frac{91}{216}$

5. a) $\frac{216}{1000}$ **b)** $\frac{64}{1000}$ **c)** $\frac{64}{1000}$ **d)** $\frac{936}{1000}$ **6. a)** $\frac{27}{64}$ **b)** $\frac{1}{64}$

Revision exercise 9

1. a) $\frac{3}{8}$ **b)** $\frac{5}{8}$ **c)** 0   **2. a)** $\frac{2}{11}$ **b)** $\frac{5}{11}$ **c)** $\frac{9}{11}$ **d)** $\frac{6}{11}$ **3.** $\frac{1}{6}$

4. $\frac{5}{16}$

5. a)

		\multicolumn{6}{c}{**Second dice**}

		1	**2**	**3**	**4**	**5**	**6**
First dice	**1**	2	2	3	4	5	6
	2	2	4	6	8	10	12
	3	3	6	9	12	15	18
	4	4	8	12	16	20	25
	5	5	10	15	20	25	30
	6	6	12	18	24	30	36

b) i) $\frac{1}{9}$ **ii)** $\frac{1}{12}$ **iii)** 0   **6.** 50

7. a) $\frac{1}{2}$ **b)** Spinner has 6 sectors shaded   **c)** Spinner has six 1s, four 2s, and two 3s

8. a) i) $\frac{1}{9}$ **ii)** $\frac{6}{9}$ **b) i)** $\frac{1}{8}$ **ii)** $\frac{4}{8}$ **iii)** 1

9. a) $\frac{1}{64}$ **b)** $\frac{27}{64}$ **c)** $\frac{9}{64}$ **d)** $\frac{27}{64}$ The sum is 1.

10. a) Yes: you cannot spin both an odd number and an even number at the same time
   b) No: you can spin a 2 which is both even and prime

Examination-style exercise 9

1. $\frac{9}{30} = \frac{3}{10}$ **2. a) i)** $\frac{4}{10}$ **ii)** 0 **b)** $\frac{7}{12}$ **3. a)** 15% **b) i)** $\frac{4}{15}$ **ii)** $\frac{2}{3}$ **iii)** 0

4. a) $\frac{3}{7}$ **b)** $\frac{5}{18}$ **5. a) i)** $\frac{31}{36}$ **ii)** 0 **iii)** 1 **b)** $\frac{17}{99}$ **c)** Paulo's

6. a) $\frac{10}{12}$ **b)** $\frac{4}{12}$ **c)** 1

7. a)

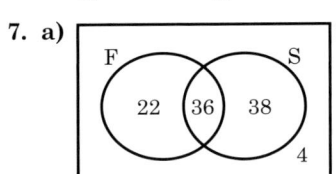

b) i) $\frac{36}{100}$ **ii)** $\frac{38}{100}$ **iii)** $\frac{26}{100}$

Chapter 10 Algebra 3

Exercise 10.1A
1. $\frac{3}{4}$ **2.** $\frac{1}{4}$ **3.** $1\frac{3}{8}$ **4.** $1\frac{1}{4}$ **5.** 7

6. a) $3\frac{3}{5}$ **b)** $\frac{3}{4}$ **7. a)** 41 **b)** 31 **8.** 29

9. a) 53 **b)** 65 **10.** 55, 56, 57 **11.** 41, 42, 43, 44

12. a) i) $x - 3$ **ii)** $2(x - 3)$ **b)** $x = 12\frac{1}{2}$ **13.** $x = 3$

Exercise 10.2A
1. 36 **2.** 29 **3.** 8 **4.** 18 **5.** 84

6. 9×10^{12} **7.** 165 **8.** $\sqrt{181}$ **9.** 1.62 **10.** 650

Exercise 10.3A
1. $e - b$ **2.** $m + t$ **3.** $a + b + f$ **4.** $A + B - h$ **5.** y

6. $b - a$ **7.** $m - k$ **8.** $w + y - v$ **9.** $\frac{b}{a}$ **10.** $\frac{m}{h}$

11. $\frac{a + b}{m}$ **12.** $\frac{c - d}{k}$ **13.** $\frac{e + n}{v}$ **14.** $\frac{y + z}{3}$ **15.** $\frac{r}{p}$

16. $\frac{h - m}{m}$ **17.** $\frac{a - t}{a}$ **18.** $\frac{k + e}{m}$ **19.** $\frac{m + h}{u}$ **20.** $\frac{t - q}{e}$

21. $\frac{v^2 + u^2}{k}$ **22.** $\frac{s^2 - t^2}{g}$ **23.** $\frac{m^2 - k}{a}$ **24.** $\frac{m + v}{m}$ **25.** $\frac{c - a}{b}$

26. $\frac{y - t}{s}$ **27.** $\frac{z - y}{c}$ **28.** $\frac{a}{h}$ **29.** $\frac{2b}{m}$ **30.** $\frac{cd - ab}{k}$

31. $\frac{c + ab}{a}$ **32.** $\frac{e + cd}{c}$ **33.** $\frac{n^2 - m^2}{m}$ **34.** $\frac{t + ka}{k}$ **35.** $\frac{k + h^2}{h}$

36. $\frac{n - mb}{m}$ **37.** $2a$ **38.** $\frac{d - ac}{c}$ **39.** $\frac{e - mb}{m}$

Exercise 10.3B

1. mt **2.** en **3.** ap **4.** amt **5.** abc

6. ey^2 **7.** $a(b + c)$ **8.** $t(c - d)$ **9.** $m(s + t)$ **10.** $k(h + i)$

11. $\dfrac{ab}{c}$ **12.** $\dfrac{mz}{y}$ **13.** $\dfrac{ch}{d}$ **14.** $\dfrac{em}{n}$ **15.** $\dfrac{hb}{e}$

16. $c(a + b)$ **17.** $m(h + k)$ **18.** $\dfrac{mu}{y}$ **19.** $t(h - k)$ **20.** $(z + t)(a + b)$

21. $\dfrac{e}{7}$ **22.** $\dfrac{e}{a}$ **23.** $\dfrac{h}{m}$ **24.** $\dfrac{bc}{a}$ **25.** $\dfrac{ud}{c}$

26. $\dfrac{m}{t^2}$ **27.** $\dfrac{b^2c^2}{a^2}$

Exercise 10.4A

1. a) $x = 3, y = 7$ **b)** $x = 1, y = 3$ **c)** $x = 11, y = -1$ **2.** $x = 2, y = 4$ **3.** $x = 2, t = 3$
4. $x = 3, y = 1$ **5.** $x = 1, y = 5$ **6.** $x = 5, y = 3$
7. a) $x = 4, y = 0$ **b)** $x = 1, y = 6$ **c)** $x = -2, y = -3$ **d)** $x = 8, y = -1$ **e)** $x = -0.6, y = 1.2$

Exercise 10.4B

1. $x = 2, y = 1$ **2.** $x = 4, y = 2$ **3.** $x = 3, y = 1$ **4.** $x = -2, y = 1$
5. $x = 3, y = 2$ **6.** $x = 5, y = -2$ **7.** $x = 2, y = 1$ **8.** $x = 5, y = 3$
9. $x = 3, y = -1$ **10.** $a = 2, b = -3$ **11.** $a = 5, b = \frac{1}{4}$ **12.** $a = 1, b = 3$
13. $m = \frac{1}{2}, n = 4$ **14.** $w = 2, x = 3$ **15.** $x = 6, y = 3$ **16.** $x = \frac{1}{2}, z = -3$
17. $m = 1\frac{15}{17}, n = \frac{11}{17}$ **18.** $c = 1\frac{16}{23}, d = -2\frac{12}{23}$

Exercise 10.4C

1. $x = 2, y = 4$ **2.** $x = 1, y = 4$ **3.** $x = 2, y = 5$ **4.** $x = 3, y = 7$
5. $x = 5, y = 2$ **6.** $a = 3, b = 1$ **7.** $x = 1, y = 3$ **8.** $x = 1, y = 3$
9. $x = -2, y = 3$ **10.** $x = 4, y = 1$ **11.** $x = 1, y = 5$ **12.** $x = 0, y = 2$
13. $x = \frac{5}{7}, y = 4\frac{3}{7}$ **14.** $x = 1, y = 2$ **15.** $x = 2, y = 3$ **16.** $x = 4, y = -1$
17. $x = 3, y = 1$ **18.** $x = 1, y = 2$ **19.** $x = 2, y = 1$ **20.** $x = -2, y = 1$

Exercise 10.4D

1. $5\frac{1}{2}, 9\frac{1}{2}$ **2.** $6, 3$ or $2\frac{2}{5}, 5\frac{2}{5}$ **3.** $4, 10$ **4.** $10.5, 7.5$
5. $a = 2, c = 7$ **6.** $m = 4, c = -3$ **7.** $a = 30, b = 5$
8. TV \$200, DVD player \$450 **9.** White $2\,$g; Brown $3\frac{1}{2}\,$g
10. $2 \times 15, 5 \times 25$ **11.** $10c \times 14, 50c \times 7$

Revision exercise 10

1. $x = 8$, perimeter $= 60$ cm **2.** 11 **3.** \$6

4. $\dfrac{5}{24}$ **5. a)** $s = t(r + 3)$ **b)** $r = \dfrac{s - 3t}{t}$
6. a) $c = 5, d = -2$ **b)** $x = 2, y = -1$

Examination-style exercise 10

1. a) $4p - 5q$ **b)** 7 **c) i)** $2j + 2k$ **ii)** $2j + 2k = 144$ **iii)** 48
  d) i) $\frac{1}{3}$ **ii)** $r = wp + t$
2. $x = 2, y = 1.5$
3. a) i) 12 **ii)** 288 cm^2 **b) i)** 1.5 **ii)** $5z + 2 = 11z - 1$ **iii)** 0.5
  c) i) $a - b = 3, 4a + b = 17$ **ii)** $a = 4, b = 1$

4. a) i) 363 **ii)** 7.5 **iii)** $m = \dfrac{2E}{v^2}$ **b)** $x = 3, y = 1$

Chapter 11 Shape and space 3

Exercise 11.1A
Student's diagrams

Exercise 11.1B

1.

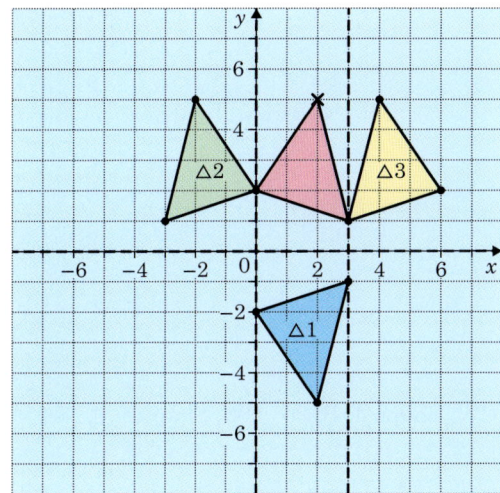

2. a)-c)

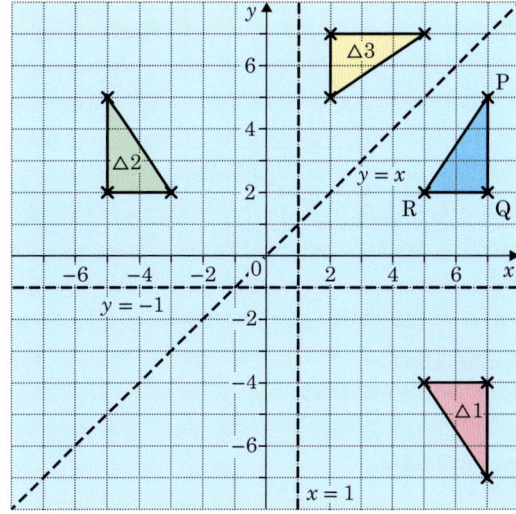

   d) (7, −7), (−5, 5), (5, 7)

3. a)-c)

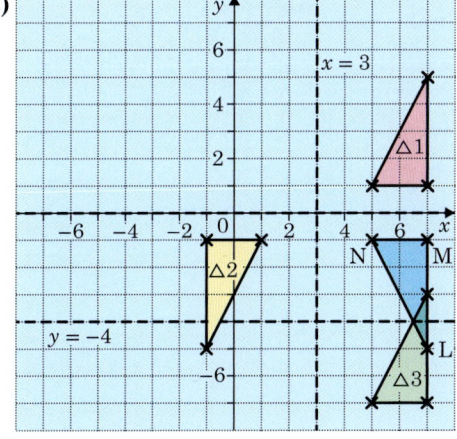

   d) (7, 5), (−1, −5), (7, −3)

4. a)-f)

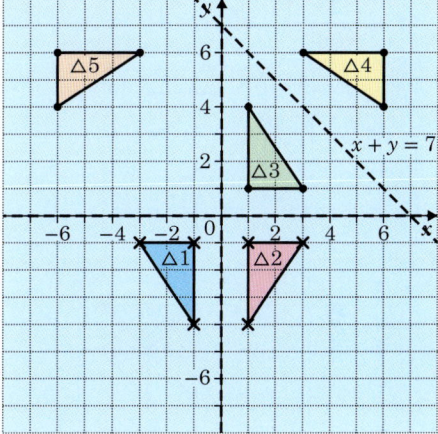

   g) (−3, 6), (−6, 6), (−6, 4)

5. a)-f)

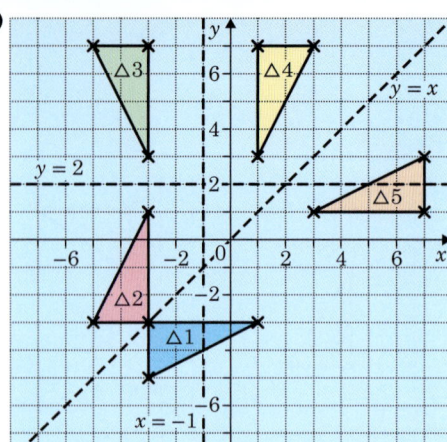

g) $(3, 1), (7, 1), (7, 3)$

6. a) $y = 0$ (x-axis)   **b)** $x = 1$ **c)** $y = 1$

Exercise 11.1C
1–6. Student's rotations
7. Shape 1: C, 90° CW; Shape 2: B, 180°; Shape 3: A, 90° ACW; Shape 4: B, 90° CW; Shape 5: F, 180°

Exercise 11.1D
1. a)-c)

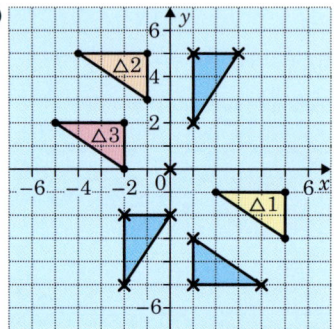

2. a)-d)

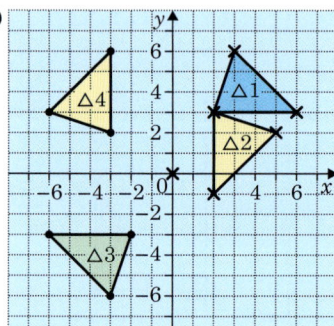

3. a)-d)

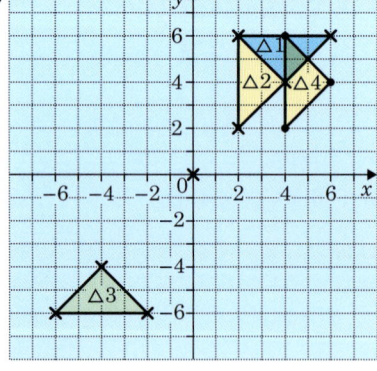

Exercise 11.1E
1. $(3, 2.5)$ **2.** $(0, 0)$ **3.** $(2, 1)$

Exercise 11.2A
1. a) Yes   **b)** No   **c)** Yes   **d)** Yes   **2.** 78 mm
3. $y = 24$ cm, $z = 67.5$ cm

4.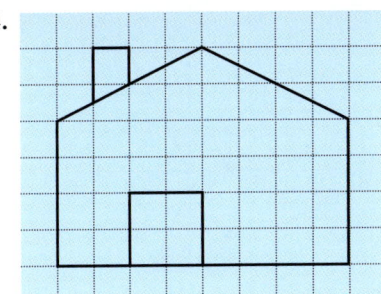

5.

6. a) 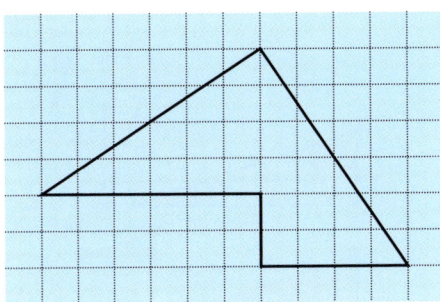 **b)**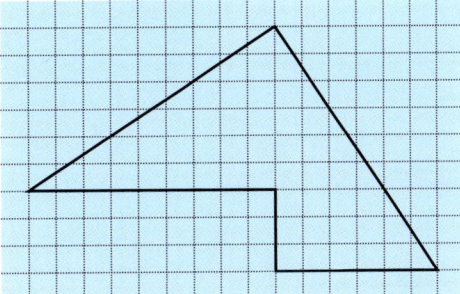

Angles are $a = 34°$ and $b = 90°$; In an enlargement, the angles in a shape are unchanged.

7. OA′ = 2 × OA, OB′ = 2 × OB

8.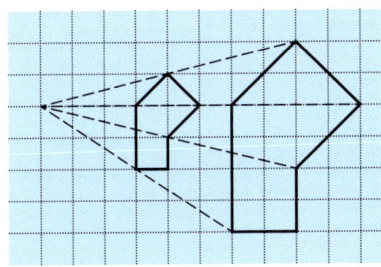

9. a) 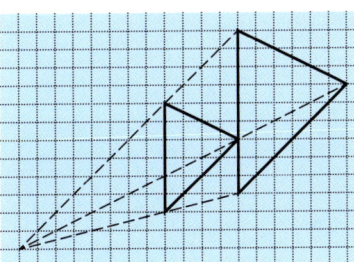 **b)** Scale factor = $1\frac{1}{2}$

Exercise 11.2B

1. 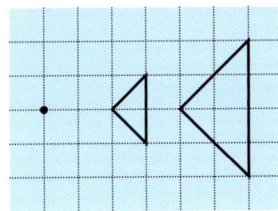 **2.** 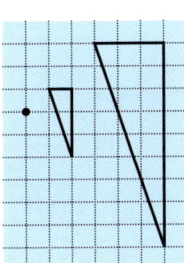 **3.** 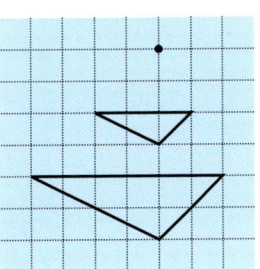 **4.**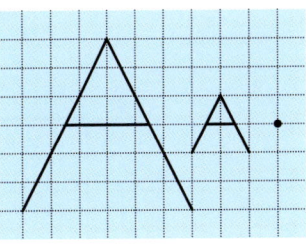

5.

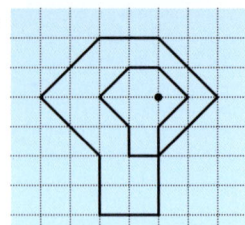

6.

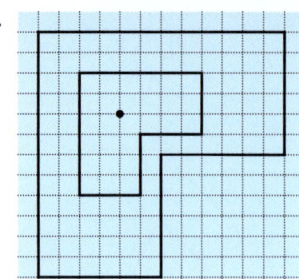

7. a)–d)

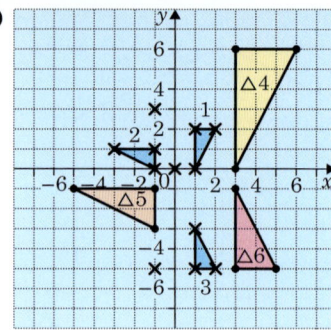

e) $(3, 0), (-5, -1), (3, -1)$

8. a)–d)

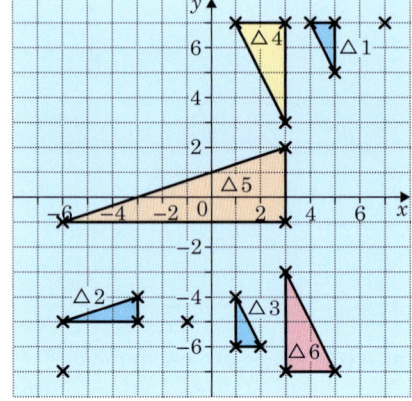

e) $(3, 3), (-6, -1), (3, -3)$

9. a)–d)

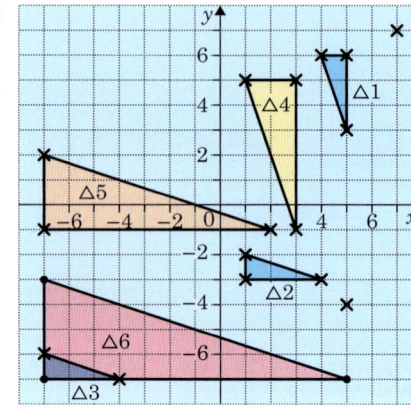

e) $(3, -1), (2, -1), (5, -7)$

Exercise 11.2C

1.

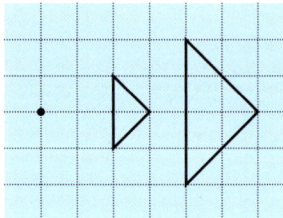

2.

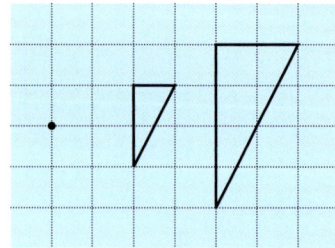

3.

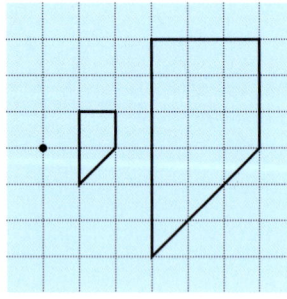

4. a)–e)

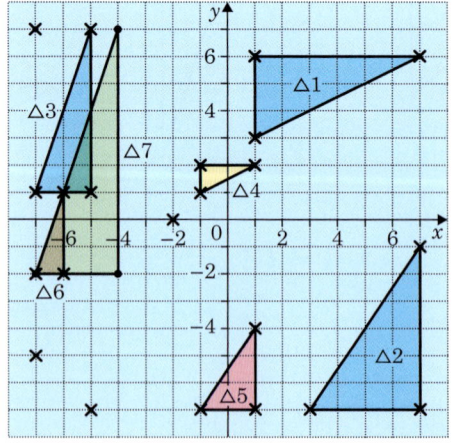

Exercise 11.3A

1. a) $\begin{pmatrix} 4 \\ 6 \end{pmatrix}$ b) $\begin{pmatrix} 6 \\ 4 \end{pmatrix}$ c) $\begin{pmatrix} 6 \\ 0 \end{pmatrix}$ d) $\begin{pmatrix} 6 \\ 0 \end{pmatrix}$ e) $\begin{pmatrix} 5 \\ -2 \end{pmatrix}$

   f) $\begin{pmatrix} 1 \\ 2 \end{pmatrix}$ g) $\begin{pmatrix} -2 \\ 5 \end{pmatrix}$ h) $\begin{pmatrix} 2 \\ -2 \end{pmatrix}$ i) $\begin{pmatrix} -4 \\ -3 \end{pmatrix}$ j) $\begin{pmatrix} 2 \\ -6 \end{pmatrix}$

   k) $\begin{pmatrix} 1 \\ -8 \end{pmatrix}$ l) $\begin{pmatrix} -6 \\ -1 \end{pmatrix}$ m) $\begin{pmatrix} 0 \\ -4 \end{pmatrix}$ n) $\begin{pmatrix} 6 \\ 1 \end{pmatrix}$

2.

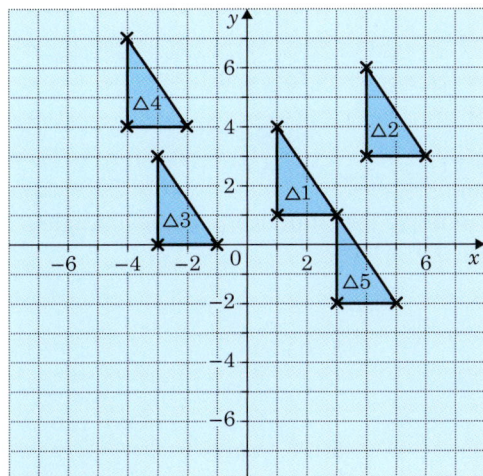

Exercise 11.3B

1. b) Reflection in line 2 c) Rotation 180° about O
2. b) Rotation 90° clockwise about (0, 0) c) Rotation 180° about (0, 0)
3. a)-b)

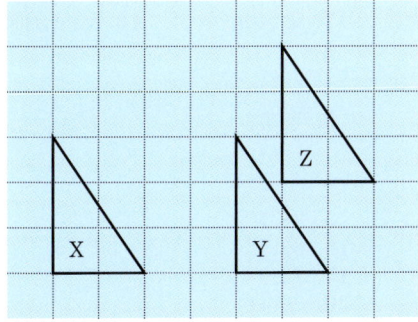

   c) Translation using vector $\begin{pmatrix} 1 \\ 2 \end{pmatrix}$

   d) Translation using vector $\begin{pmatrix} 5 \\ 2 \end{pmatrix}$

4. Yes: both are reflections in the line $y = 3\frac{1}{2}$

5. a) Rotation 90° anticlockwise, centre (0, 0) b) Reflection in $y = 0$

   c) Rotation 180°, centre (0, 0) d) Translation, vector $\begin{pmatrix} 9 \\ 0 \end{pmatrix}$

6. a) Rotation 90° anticlockwise about (0, 0) b) Reflection in the line $x = 3$

   c) Enlargement, scale factor 3, centre (0, 6) d) Translation using vector $\begin{pmatrix} -5 \\ 1 \end{pmatrix}$

Exercise 11.4A
1. 10 **2.** $\sqrt{17}$ **3.** $\sqrt{113}$ **4.** $\sqrt{32}$ **5.** 4.24 **6.** 9.90
7. 4.58 **8.** 5.20 **9.** 9.85 cm **10.** 7.07 cm **11.** 3.46 m **12.** 40.3 km
13. 9.49 cm **14.** 5.39 units **15.** Yes: $PQ = PR = \sqrt{34}$; $QR = 6$. **16.** 6.63 **17.** 5.57
18. 8.72 **19.** 5.66 **20.** 6.63 **21.** 2.24
22. a) i) 13 **ii)** 25 **iii)** 9
  c) a increases by 2; b increases by increasing multiplies of 4 (8, 12, 16); $c = b + 1$
  d) 11, 60, 61; 13, 84, 85; 15, 112, 113
23. Philip, since $10^2 + 11^2 \neq 15^2$

Exercise 11.5A
1. C only **2.** $n = 12$ **3.** $x = 9$ **4.** $a = 2\frac{1}{2}$, $e = 3$ **5.** $x = 6.75$
6. $x = 3.2$ **7.** $t = 5.25$, $y = 5.6$ **8.** 7.7 cm **9.** No
10. a) Yes **b)** No **c)** No **d)** Yes **e)** Yes
  f) No **g)** No **h)** Yes **11. b)** 11.2 **c)** 4.2
12. $y = 6$ **13.** $a = 6$ **14.** $f = 4.5$ **15.** 16 m
16. a) Angle ABC = angle EDC (alternate angles); Angle BAC = angle DEC (alternate angles); Angle
  ACB = angle DCE (opposite angles) **b) i)** 5 cm **ii)** 3 cm
17. 10.8 m **18.** AO = 2 cm, DO = 6 cm

Exercise 11.6A
1. 0.643 **2.** 0.819 **3.** 0.466 **4.** 0.985 **5.** 0.259
6. 2.050 **7.** 0.777 **8.** 0.956 **9.** 0.213

Exercise 11.6B
1. 3.0 cm **2.** 5.4 cm **3.** 3.1 cm **4.** 7.0 cm **5.** 73.1 cm
6. 15.4 cm **7.** 5.3 cm **8.** 8.0 cm **9.** 11.6 cm **10.** 11.4 cm
11. 961.3 cm **12.** 0.9 cm **13.** 46.0 cm **14.** 34.9 cm **15.** 9.4 cm
16. 8.2 cm **17.** 35.6 cm **18.** 80.2 cm **19.** 4.9 cm **20.** 7.0 cm

Exercise 11.6C
1. 18.4 **2.** 9.1 **3.** 10.7 **4.** 17.1 **5.** 13.7 **6.** 125.8
7. 6.9 **8.** 11.8 **9.** 17.6 **10.** 11.4 **11.** 5, 5.6 **12.** 13.1, 27.8
13. 4.3 **14.** 3.5 **15.** 26.2 **16.** 8.8

Exercise 11.6D
1. 53.1° **2.** 45.6° **3.** 26.6° **4.** 23.6° **5.** 78.5°
6. 54.5° **7.** 34.8° **8.** 32.9° **9.** 13.0°

Exercise 11.6E
1. 38.7° **2.** 48.6° **3.** 31.0° **4.** 54.5° **5.** 38.7°
6. 17.5° **7.** 38.9° **8.** 59.0° **9.** 41.3° **10.** 62.7°
11. 54.3° **12.** 66.0° **13.** 48.2° **14.** 12.4° **15.** 72.9°
16. 56.9° **17.** 36.9° **18.** 41.8° **19.** 78.0° **20.** 89.4°

Exercise 11.6F
1. 68.0° **2.** 3.65 m **3.** 14.0 m **4.** 20.6° **5.** 56.7 m
6. 15.3 m **7.** 90.3 cm **8.** 4.32 cm **9.** 7.66 cm **10.** 65.5 km
11. 189 km **12.** 25.7 km **13.** 36.4° **14.** 10.3 cm **15.** $a = 72°$, 8.23 cm
16. 71.1°

Revision exercise 11
1. 4.1 cm **2. a)** 14.1 cm, 48.3 cm square **b)** 1930 cm^2
3. a) $1\frac{2}{3}$ **b)** 20 cm
4. a) Reflection in the x-axis **b)** Reflection in $x = -1$ **c)** Reflection in the y-axis
  d) Rotation, centre (0, 0), 90° clockwise **e)** Reflection in $y = -1$
5. a) Enlargement; scale factor $1\frac{1}{2}$, (1, −4) **b)** Translation $\begin{pmatrix} 11 \\ 10 \end{pmatrix}$
  c) Enlargement; scale factor $\frac{1}{2}$, (−3, 8) **d)** Enlargement; scale factor 3, (−2, 5)

6.

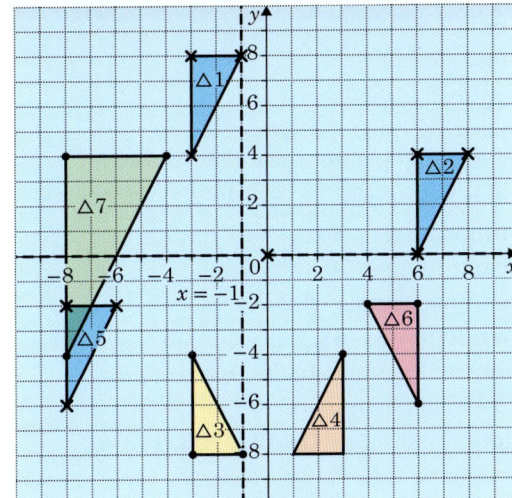

7. a) 7.2 cm   **b)** 10.2 cm   **c)** 7.3 cm   **8.** 0.335 m

9. a) 45.6°   **b)** 58.0°   **c)** 3.89 cm   **d)** 33.8 m

10. 4.8   **11. a)** 1.2   **b)** 1.6

12. No: corresponding sides are not in the same ratio/not enlarged with the same scale factor.

13. 7.6   **14. a)** Corresponding angles are equal   **b)** 4

Examination-style exercise 11

1. a) i) Translation $\begin{pmatrix} -7 \\ -4 \end{pmatrix}$ **ii)** Enlargement scale factor 3, centre (0, 0)

   b)-c)

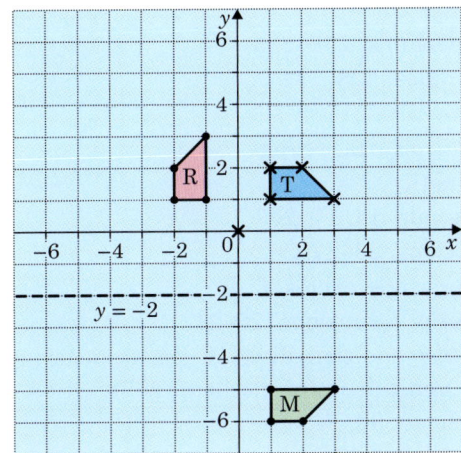

2. a) Translation $\begin{pmatrix} 2 \\ -4 \end{pmatrix}$ **b)** Reflection in y-axis   **c)** Rotation 90° anticlockwise about (0, 0)

3. 1.2 m   **4. a)** Similar   **b)** 6 cm   **c)** 320°

5. a) Similar   **b)** 145°   **6.** 3.31 m   **7.** 430 m

8. a) i) 56.3°   **ii)** 123.7°   **b)** 14.4 m   **c)** 34.4 m, 48 m^2

9. a) 90 m^2 **b)** 14.3 m   **c)** 18.5 to 18.6 m^2 **d)** 20.6 m

10. a) ii) 125°   **b) ii)** 12.26 km   **iii)** 74.2 km   **c)** 8.24 to 8.25 km/h

Index